Genomics in Africa

Africans have the most diverse genomes of all human populations. However, only limited genomic studies from Africans are currently available to inform the development of disease prevention strategies, early detection tools, and treatment options. This trajectory is now changing, following several initiatives in Africa including the H3Africa.

To date, there are many books on genomic studies globally but only a few are representative of the work that is being carried out in Africa and led by African researchers. This book provides insights on African genomics research into infectious diseases, cardiometabolic traits, cancer, mental health, pharmacogenomics, and metagenomics and reflects the broad experiences of researchers across Africa in ethics, biobanking, and capacity building for genomic studies.

Key Features:

- Covers the development of disease prevention strategies, early detection tools, and treatment options.
- Offers an overview of the funding landscape and capacity-building efforts in Africa.
- Provides a ten-year roadmap for research leaders, funders, and governments who are keen to plan for genomics work in Africa.
- Reviews the translation of genomic studies from lab to real-world applications.
- Discusses non-communicable diseases, infectious diseases, and mental illness.

Genomics in Africa

Edited by
Segun Fatumo and Tinashe Chikowore

CRC Press
Taylor & Francis Group
Boca Raton London New York

CRC Press is an imprint of the
Taylor & Francis Group, an **informa** business

Designed cover image: Segun Fatumo and Tinashe Chikowore

First edition published 2027
by CRC Press
2385 NW Executive Center Drive, Suite 320, Boca Raton FL 33431

and by CRC Press
4 Park Square, Milton Park, Abingdon, Oxon, OX14 4RN

CRC Press is an imprint of Taylor & Francis Group, LLC

ISBN: 978-1-032-54735-0 (hbk)
ISBN: 978-1-032-56648-1 (pbk)
ISBN: 978-1-003-43658-4 (ebk)

DOI: 10.1201/9781003436584

Typeset in Sabon
by Apex CoVantage, LLC

Contents

Foreword

HARNESSING OUR COMMON AFRICAN GENOMES TO IMPROVE HEALTH GLOBALLY

Africa, as the cradle of modern humans, holds unparalleled genetic diversity. However, fewer than 2% of sequenced human genomes come from individuals of African ancestry. The authors argue that investing in African genomics is a scientific imperative, driven by three major pillars: *Ancestry*, *ecology*, and *equity*.

Ancestry: All humans share common origins in Africa, with approximately 5–15% of our genetic makeup stemming from ancient hominid lineages (so-called the ghost DNA due to our current incapacity to read its sequences). Moreover, the more than 300,000-year human history in Africa is marked by various waves of migrations within African, out of African, and back to Africa, and the related genetic admixtures have resulted in a rich pool of unique genetic variants. Some populations, especially in Ethiopia, even exhibit Neanderthal genetic introgression due to historical migration patterns back from the Middle Eastern populations to Africa. For example, studies such as those by the H3Africa consortium and identified millions of previously undocumented variants, most of which are population specific. This extreme diversity offers vast opportunities to uncover genetic factors that contribute to health and disease but are currently underexplored due to the limited genomics data from African populations.

Ecology: Africa's vast ecological, linguistic, and cultural diversity has exerted strong selective pressures on its populations, shaping genetic variation in distinct ways. Several well-documented examples include sickle cell trait (*HBB*), which offers protection against malaria; *APOL1* (G1/G2 variants), which guards against trypanosome infections but increases risk of kidney disease (now recommended with therapeutic manipulation with either antisense oligonucleotide or small molecule therapy); and lactase persistence mutations, which allow digestion of milk in adulthood among populations of select African ancestry, such as the –13910*T variant west of Lake Chad (Fulani and Tuareg pastoralists), the –13915*G variant east of Lake Chad (mostly Arabic-speaking pastoralists), and the African-specific variant (rs2814778) in the Duffy antigen receptor for chemokines (*DARC/ ACKR1*) gene, which offers malarial resistance but is potentially associated with highly aggressive triple-negative breast cancer among populations of African ancestry. These genetic adaptations highlight how African genetic variants have functional implications, particularly for infectious diseases and metabolic conditions. Recent findings also indicate dozens of additional variants under natural selection that may be important for genetic medicine globally.

Equity: There is a critical need to address the underrepresentation of African genomes in biomedical research. Doing so can reduce health disparities and accelerate the discovery of treatments for diseases that disproportionately affect African populations. For instance, sickle cell disease, the most common monogenic disorder globally, affects around 300,000 newborns annually, 75% of whom are in Africa. Genomic studies in African populations can help identify modifiers of disease severity (e.g., foetal haemoglobin regulation) and co-inherited mutations (e.g., 3.7 Kb deletion in α-thalassemia genes (*HBA1/2*) that influence chronic kidney disease complications). Additionally, many Mendelian diseases remain unexplained in African contexts due to the lack of genetic data. For example, common mutations in the *GJB2* gene linked to hearing loss in other populations are rare in Africa. Novel genes like *INPP4B*, *CCDC141*, and *UBFD1* associated with non-syndromic hearing impairment have been discovered in Ghana and Mali, illustrating the potential for uncovering new therapeutic targets.

The book, edited by Drs. Segun Fatumo and Tinashe Chikowore, showcases the growing body of knowledge on African genetic variation. Through the work of African and global experts, it underscores the urgent need to invest in African genomics research. Increasing the number of sequenced African genomes will refine tools for predicting disease-related variants, improve diagnoses, and enhance precision medicine. To achieve this, large-scale, inclusive genomics initiatives across the diverse African populations are essential. Ultimately, investing in African genomics promotes both scientific innovation and health equity worldwide.

Ambroise Wonkam, MD, PhD
Johns Hopkins University School of Medicine

About the Book

Africans harbour the most genetically diverse genomes of any human population on earth, a reflection of the continent's deep evolutionary history and the origin of modern *Homo sapiens*. For decades, this extraordinary diversity has been profoundly under-represented in global genomic research, leaving African populations largely excluded from the scientific advances transforming medicine elsewhere. That is changing. Landmark continental initiatives, including the Human Heredity and Health in Africa (H3Africa) Consortium, have catalysed a new era of Africa-led genomic science, building infrastructure, generating data, and training a generation of researchers determined to ensure genomic medicine serves all of humanity.

Genomics in Africa brings together the breadth and depth of this emerging field, written and led by African researchers working at the frontiers of genomic discovery. The book spans the full landscape of African genomic research: infectious diseases, cardiometabolic traits, cancer, sickle cell disease, and mental health. It also addresses the unique contextual dimensions of doing genomics on the continent, including ethics and community engagement, biobanking, pharmacogenomics, metagenomics, and capacity building. These chapters reflect not only the science being produced but also the lived experience of researchers navigating the challenges and opportunities of conducting genomic research in Africa.

This volume is an authoritative and accessible reference for researchers, clinicians, public health practitioners, policymakers, and students with an interest in genomics, global health, and precision medicine. Advancing equity in genomic research is a prerequisite for a future in which the benefits of genomic medicine are shared across all populations. *Genomics in Africa* centres African voices and African data to make that case, directly and without apology

About the Editors

Professor Segun Fatumo, PhD, FAAS, FHEA

Professor Segun Fatumo is a globally recognized genomic epidemiologist and statistical geneticist whose work is fundamentally changing the landscape of human genetics. He is the Professor and the Chair of Genomic Diversity at Queen Mary University of London and the Head of Non-Communicable Disease Genomics at the MRC/UVRI and LSHTM Uganda Research Unit. From these platforms, he leads one of the most ambitious cross-continental research programes in genomic science, placing African populations at the center of a global effort to make precision medicine work for everyone.

His scientific mission is clear and urgent. For too long, the genetic architecture of disease has been studied almost exclusively in European-ancestry populations, producing tools, risk scores, and drug targets that serve a fraction of humanity. Professor Fatumo has spent more than two decades dismantling that imbalance. His research programe spans genome-wide association studies, multi-omics integration, polygenic risk score development, Mendelian randomization, and advanced computational genomics, all directed at uncovering the genetic determinants of non-communicable diseases in African and other under-represented populations. His deepest expertise lies in chronic kidney disease and cardiometabolic traits, areas where African populations carry a disproportionate disease burden and where genomic insight is most urgently needed.

He holds a PhD in bioinformatics and trained as a postdoctoral researcher across three of the world's leading institutions: the University of Georgia in the United States of America, the University of Cambridge, and the Wellcome Sanger Institute in the United Kingdom. That training gave him a command of both computational biology and large-scale population genomic epidemiology, a combination that has proved central to the breadth and rigor of his research programe.

Professor Fatumo is Principal Investigator and director of KidneyGenAfrica-a pan-African partnership that delivers research and training excellence in genomics of kidney disease and decodes the genetic basis of chronic kidney disease across the continent. He leads BCX-Africa, investigating blood cell trait genomics in African populations, and co-leads the Nigerian 100K Genome Project, a transformative effort to bring over 100,000 Nigerians into genomic research and map the genetic roots of the NCDs that disproportionately affect them. Professor Fatumo has been awarded multi-million-dollar research grants from the MRC, the NIH, and the Wellcome Trust and other leading grant agencies. More than that, Professor Fatumo's research programes have built something durable: research infrastructure, training pipelines, and scientific communities across Africa that will generate discovery long after individual grants have closed.

Capacity building is not peripheral to Professor Fatumo's agenda. It is the agenda. He has been instrumental in establishing genomic research infrastructure where little previously existed, creating career pathways for early-career scientists, and embedding a culture of African scientific ownership across the continent's emerging genomics ecosystem and building data and biobank resources. The next generation of African genomic scientists is being shaped, in no small part, by his direct investment in their training using rich African genome resources.

His influence extends far beyond the research environment. A respected global science ambassador, Professor Fatumo has brought African genomics to audiences worldwide through features in the BBC, *The Telegraph*, *WIRED*, *Science Magazine*, and *Forbes*. He delivers keynote lectures at the world's leading scientific conferences and serves as guests and associate editor for leading journals. He sits on the Scientific Advisory Boards of Genomics England, the Berlin Institute of Health at Charité, the GWAS Catalog, and GEN-Impact, where he provides strategic guidance on population diversity, data ethics, and the global implementation of genomic medicine. These are the rooms where the direction of the field is decided. He is in them, and he shapes what comes out.

Professor Fatumo is a Fellow of the Higher Education Academy and teaches across multiple postgraduate programes in the UK and Africa. He supervises PhD researchers on both continents and leads genomic analysis workshops across Africa and the UK, training the scientists who will carry this field into its next chapter.

The recognition is commensurate with the contribution. He is a Fellow of the African Academy of Sciences, one of the continent's most prestigious scientific honors, awarded for sustained and outstanding contributions to research. He is also a recipient of the MRC Impact Prize, recognizing his leadership in advancing African representation in genomic research and his pioneering work on genetic risk prediction in African populations.

Professor Fatumo co-edits this volume and co-authors two chapters. His presence in this book is, in many ways, a reflection of the field he has helped to build.

Dr. Tinashe Chikowore

Dr. Tinashe Chikowore is a distinguished scientist currently serving as an investigator in the Channing Division of Network Medicine and the Division of Genetics at Brigham and Women's Hospital and as an Instructor of Medicine at Harvard Medical School. His path to becoming a leading voice in genomic research was forged through a dedication to bridging the gap in health equity for African populations.

Dr. Chikowore holds a PhD in nutrition with a specialization in nutrigenetics. His interest in scientific inquiry was sparked during his undergraduate years at the University of Zimbabwe, where he was inspired by his lecturers in nutritional epidemiology. Initially balancing his academic pursuits with humanitarian relief work, he felt a strong calling to utilize data to address the vulnerabilities he witnessed in the field. This motivation led him to pursue advanced training, including a prestigious Wellcome Trust International Training Fellowship. During his doctoral and postdoctoral studies, he honed his expertise in laboratory techniques, statistics, and complex genetic concepts, largely within the framework of the Human Heredity and Health in Africa (H3Africa) consortium.

Dr. Chikowore's research program is at the forefront of the precision medicine revolution, specifically targeting the underrepresented populations of Africa. He focuses on the complex interplay between genetics, multi-omic profiles, diet, and environmental exposures to understand the risk of cardiometabolic diseases, such as obesity, type 2 diabetes, and hypertension.

His work integrates genomics, proteomics, and metabolomics, utilizing advanced machine learning and data-driven approaches to unravel the heterogeneous effects of lifestyle factors, including socioeconomic status, sleep, exercise, and diet, on disease risk. A core component of his research involves evaluating polygenic risk scores (PRS) and their transferability across diverse African regions. Recognizing that most global genomic studies are predominantly based on European ancestry, Dr. Chikowore emphasizes that the lack of diversity in genomic data limits the accuracy of disease prediction for people of African descent.

His research leverages expansive datasets from the H3Africa consortium, which has been critical in identifying population-enriched, clinically significant genetic variants. Dr. Chikowore's work consistently demonstrates that increasing the representation of African individuals in genomic research is not just an issue of equity but a global scientific necessity that can lead to better diagnostic tools and more effective drug design for all populations.

Dr. Chikowore's intellectual contributions have been widely recognized in the scientific community. He is a recipient of the Charles Epstein Excellence in Human Genetics Award from the American Society of Human Genetics, honored for an outstanding abstract presented during a plenary session. His pioneering work, particularly on the transferability of genetic risk scores and the development of PRS for type 2 diabetes in African populations, has been published in high-impact journals, including *Nature Medicine* and *Diabetes Care*.

He is actively supported by competitive funding, including grants from the National Institutes of Health (NIH) and the American Diabetes Association (ADA). Notably, he is an ADA Accelerator Award recipient under the "Pathway to Stop Diabetes" program, a testament to his potential to impact the future of diabetes prevention through AI and culturally sensitive, data-driven approaches.

Beyond his laboratory and clinical research, Dr. Chikowore is a committed leader dedicated to strengthening the genomic research infrastructure across Africa. He serves as the chairperson of the H3Africa Cardiovascular Working Group and holds positions on the executive steering committees of H3Africa and the Data Science for Health Discovery and Innovation in Africa (DSI-Africa) initiative. These roles place him at the center of efforts to foster international collaborations and enhance data science capacity on the continent.

Dr. Chikowore is deeply committed to mentorship, actively guiding underrepresented postdoctoral fellows and PhD students through weekly journal clubs and individual mentorship meetings. He also contributes to the global scientific community by teaching international courses led by the Wellcome Trust Connecting Science initiative in Asia and Africa. Furthermore, he serves as an ad hoc reviewer for major funding agencies, including the Wellcome Trust and the ADA, and contributes to scientific literature as a topical review editor for *Frontiers in Genetics*. Through his multifaceted roles, Dr. Chikowore continues to advocate for a more inclusive scientific landscape, ensuring that Africa's unique genetic diversity is fully harnessed to advance global health.

Chapter 1

Introductory Overview of Genomics in Africa

Tinashe Chikowore[1,2,3] *and Segun Fatumo*[4,5]

[1]Channing Division of Network Medicine, Department of Medicine, Brigham and Women's Hospital, Harvard Medical School, Boston, MA, United States
[2]SAMRC/Developmental Pathway of Health Research Unit, Faculty of Health Sciences, University of the Witwatersrand, Johannesburg, South Africa
[3]Sydney Brenner Institute of Molecular Biosciences, Faculty of Health Sciences, University of the Witwatersrand, Johannesburg, South Africa
[4]Precision Healthcare University Research Institute, Queen Mary University of London, United Kingdom
[5]MRC/UVRI and LSHTM Uganda Research Unit, Entebbe, Uganda

1.1 OVERVIEW OF GENOMICS IN AFRICA

Africa is regarded as the cradle of humankind, the continent where humans originated and spent much of their history before embarking on migrations that populated the globe.[1] This deep history has endowed the continent with an unparalleled richness of human genetic diversity.[2] African populations collectively harbour significantly more genetic variation than populations on any other continent.[3] This distinction arises primarily because the relatively small founder population groups that migrated out of Africa carried only a subset of the ancestral genetic diversity and then underwent population bottlenecks that further reduced variation.[1,4] Consequently, the genomes of individuals with African ancestry contain unique variants and combinations of variants not found elsewhere, offering a deeper, more comprehensive view into the genetic landscape of our entire species.[1] Understanding this vast, intricate tapestry of genetic diversity is not merely an academic exercise in reconstructing human origins and migration patterns, it's fundamental to advance our global knowledge of human biology, health, and disease.[3] The unique genetic architectures shaped by diverse evolutionary pressures, environmental interactions, and population histories across Africa hold immense potential for uncovering novel insights into disease susceptibility, drug response, and the fundamental workings of the human genome.[5,6]

Despite the profound significance of Africa's genetic diversity, the story of genomics research throughout much of the 20th and early 21st centuries was characterized by stark global inequities.[5] Africa, the very source of human diversity, was at the periphery of the genomics revolution. Research agendas, funding streams, and technological development were overwhelmingly

DOI: 10.1201/9781003436584-1

concentrated in high-income countries, primarily focusing on populations of European descent.[7] This resulted in global genomics databases that drastically underrepresent African populations, creating significant blind spots in our understanding of human genetic variation. The consequences of this imbalance are far-reaching, extending beyond incomplete scientific knowledge to tangible impacts on health equity.[8] Medical diagnostics, risk prediction models, and therapeutic strategies developed based on data from non-African populations often demonstrate reduced accuracy or efficacy when applied to individuals of African ancestry.[8] This disparity highlights the critical need to correct the historical imbalance and ensure that the benefits of genomics medicine are accessible to all, thereby leveraging on research that fully embraces the diversity of African people. Early efforts involving African samples were also sometimes marred by ethical oversights, including "helicopter research," where samples were collected with limited local engagement or benefit sharing, further underscoring the need for a paradigm shift towards locally led, ethically sound research.[7]

In the last few decades, however, the tide has begun to change fundamentally. The narrative of genomics in Africa is shifting slowly away from marginalization.[5] This change is being fuelled by several concurrent trends: The growing appreciation of the scientific and ethical imperatives to make research inclusive, decreasing costs of sequencing technologies, and, crucially, the determination of African scientists, governments, and institutions, aided at times by international collaborations based more explicitly on more equal terms.[5] Flagship initiatives have been at the forefront of promoting change. The Human Heredity & Health in Africa (H3Africa) consortium, launched in 2010, is the largest such coordinated effort to date.[9] Supported by the US National Institutes of Health (NIH) and the Wellcome Trust, H3Africa has facilitated the establishment of infrastructure; funded large-scale, multidisciplinary genomic studies throughout the continent with diseases of high relevance to Africa; established biorepositories; initiated the process of building up capabilities through large-scale training programs; and brought to the centre stage the crucial debate around issues of ethics and community engagement in African genomics research.[9]

Beyond H3Africa, numerous national and regional initiatives are taking root, reflecting a growing commitment to embedding genomics research within African health systems and scientific institutions.[5] Universities are establishing dedicated genomics and bioinformatics programs, local research funding agencies are increasingly recognizing the importance of the field, and a new generation of highly skilled African researchers is emerging, trained both locally and internationally.[9,10] These scientists are not only leveraging advanced genomics tools, from genome-wide association studies (GWAS) to whole-genome sequencing and multi-omics approaches. GWAS are research approaches that involve scanning the genomes of many individuals to find genetic variants associated with particular diseases or traits. Whole-genome sequencing involves determining the complete DNA sequence of an individual's genome, offering a comprehensive view of all genetic variations. Multi-omics approaches integrate diverse data types, such as genomics, transcriptomics, proteomics, and metabolomics, to provide a holistic understanding of biological systems but are also integrating this data with rich epidemiological, clinical, and environmental information available from long-standing African research cohorts. This integrated approach is vital for dissecting the complex interplay between genetic predisposition and environmental factors, which drives disease patterns on the continent.

The potential benefits stemming from this flourishing research ecosystem are immense.[5] Genomics offers powerful tools to unravel the genetic underpinnings of diseases that disproportionately affect African populations, including infectious diseases like malaria, tuberculosis, and HIV, where host genetic factors influencing susceptibility and resistance remain partly understood. Simultaneously, Africa is facing a rising burden of non-communicable diseases (NCDs), such as cardiovascular conditions, diabetes, cancers, and mental health disorders. Genomics research is key to understanding population-specific risk factors for these NCDs, potentially identifying high-risk individuals and informing prevention strategies. Furthermore, the field of pharmacogenomics—the study of how an individual's genetic makeup influences their response to drugs—allows for the customization of drug therapies to improve efficacy and minimize adverse effects and promises to tailor drug treatments based on an individual's genetic profile. For example, studies have shown that variations in the *CYP2B6* gene among African populations can significantly influence responses to the HIV drug efavirenz, affecting drug metabolism and side-effect profiles. Pharmacogenomics promises to tailor drug treatments based on an individual's genetic profile, improving efficacy and reducing adverse drug reactions. This is a particularly pertinent goal given the observed variability in drug responses among diverse African populations.[11] Beyond health, genomics data provides an unprecedented window into the intricate population histories, migrations, and admixture events that have shaped the continent's peoples.

However, the path towards fully realizing this potential is demanding, filled with significant challenges that require sustained attention and strategic investment.[7] Firstly, the *infrastructure* gap, while narrowing, remains considerable. Establishing and maintaining state-of-the-art sequencing facilities, high-capacity biorepositories with robust sample tracking systems, and the high-performance computing infrastructure needed for analysing vast genomics datasets requires substantial and continuous funding, along with reliable power and internet connectivity, elements not uniformly available across the continent. Building this infrastructure necessitates capital investment and skilled technical personnel for operation and maintenance. For instance, in some regions, researchers have reported delays of several months in data processing due to insufficient local capacity and reliance on overseas facilities.

Secondly, while growing, *funding* for genomics research in Africa is still largely dependent on international grants and donor initiatives. While these have been crucial catalysts, long-term sustainability requires increased commitment from national governments and the fostering of local philanthropic and private sector investment. Developing sustainable funding models that support discovery research and the translation of findings into tangible health benefits is a critical priority.

Thirdly, the *ethical, legal, and social implications* (ELSI) of African genomics research present unique and complex considerations. Meaningful community engagement and consent processes that are culturally appropriate and genuinely informative are required, especially given varying levels of literacy and understanding of genetics. Issues surrounding data ownership, cross-border data sharing, the potential for stigmatizing groups or individuals based on genomic findings, and ensuring equitable benefit sharing from research outcomes must be navigated with sensitivity and foresight. Efficient governance frameworks and ethical oversight capacity are essential to build trust and ensure research proceeds responsibly.

Fourthly, there remains an ongoing need, however, with the set of high-priority projects like H3Africa, to build up human capacity. It is that critical mass of skilled individuals along the entire genomics pipeline that Africa requires, from researchers proficient at wet-lab techniques and computational analysis (bioinformatics) to genetic counsellors to explain and report results to the patient and their family members, clinicians skilled at genomics medicine, ethicists skilled at the genomics interface, and policymakers skilled at translating genomic information into the health systems. That multifaceted skill set mandates extended investment into training and education programs targeting African priorities.

Confronting these interrelated challenges of infrastructure, finance, ethics, and capacities is at the heart of establishing the genomics revolution in African contexts, African led, with returns to African populations. Both local and international partnerships, based on ideals of equality and mutual regard, will be at the forefront of overcoming these challenges.

1.2 OVERVIEW OF THE BOOK

This book aims to provide a comprehensive, insightful, and forward-looking overview of the multifaceted field of genomics as it unfolds across the diverse landscape of Africa (Figure 1.1). Recognizing the rapid pace of development and the critical juncture at which African genomics currently stands, we have brought together leading voices of researchers, ethicists, and capacity-building experts actively shaping the field to share their perspectives and findings. Our collective goal is to synthesize the progress made,

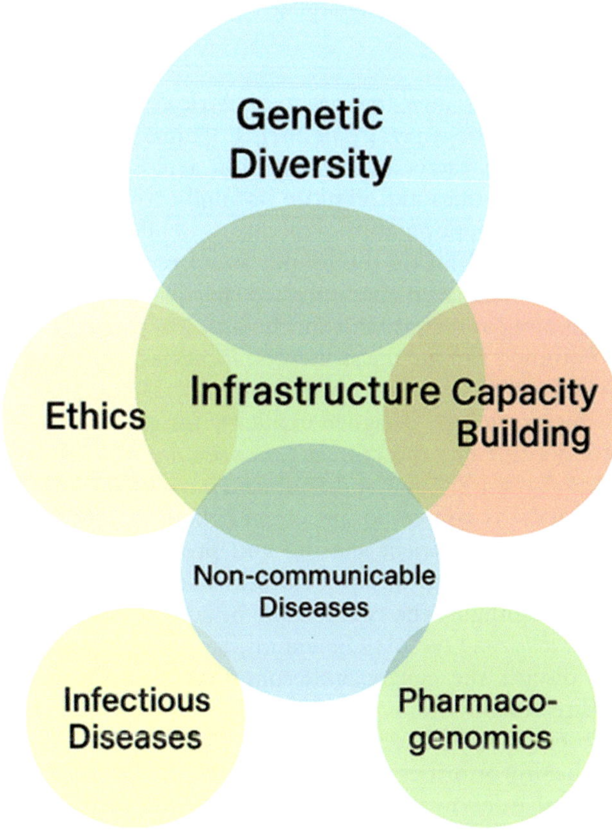

Figure 1.1 Conceptual landscape of genomics in Africa.

This figure illustrates key components: Genetic diversity, infrastructure, capacity building, ethics, and application areas such as infectious disease, non-communicable diseases, and pharmacogenomics. The overlapping circles demonstrate the interconnectedness of these domains in advancing genomics medicine in Africa.

articulate the pressing challenges, highlight the unique opportunities, and chart potential future directions. This volume is intended as a vital resource for a broad audience, including active researchers in genomics, bioinformatics, and related biomedical fields; students seeking to understand this dynamic area; clinicians interested in the potential translation of genomic findings into practice; policymakers involved in health and science strategy; ethicists grappling with the implications of genomics data; and funding agencies supporting research in Africa. We hope to foster a deeper understanding of the complexities and the immense promise of genomics on the continent.

The subsequent chapters systematically explore key dimensions of this landscape:

- **Chapter 2** delves into the heart of Africa's unique advantage: Its profound *genomic diversity in Africa*. Chimusa discusses the ongoing efforts to characterize this variation across different populations and ecological zones and explore its implications for understanding human evolution, population structure, and adaptation.
- **Chapters 3 and 4** tackle the foundational requirement for impactful genomic studies: Well-characterized participant groups. Palwendé Romuald Boua and Hermann Sorgho and Joseph Mugisha Okello and Segun Fatumo examine the strategies, successes, and challenges associated with *setting up cohorts* by leveraging existing cohort studies and establishing new ones designed explicitly for genomics research.
- **Chapter 5** confronts the critical *genomics and the ethics conundrum* in Africa. Aminu Yakubu and Paulina Tindana navigate the historical context, address contemporary ethical challenges related to consent, data sharing, and benefit sharing and propose future directions for ensuring ethically rigorous and culturally sensitive research.
- **Chapter 6** highlights the cornerstone of sustainability: *Genomics and sickle cell disease*. Oyesola Ojewunmi, Titilope Adeyemo, and Stephan Menzel discuss the genetic underpinnings of sickle cell disease and future directions.
- The subsequent block of chapters focuses on applying genomics to specific health priorities. **Chapter 7** (Amy R. Bentley, Opeyemi Soremekun, and Adebowale A. Adeyemo) examines *cardiometabolic diseases*; **Chapter 8** (Abram B. Kamiza and Solomon O. Rotimi) focuses on *cancer*; **Chapter 9** (*Allan Kalungi, Jared G. Maina, Sylvanus Toikumo, and Lukoye Atwoli*) addresses *mental health*; **Chapter 10** (Yosr Hamdi, Maroua Boujemaa, Badr-EddineAllal, Ichrak Benamri, Chiraz Mehemmai, Najla Kharrat, Fouzia Radouni, and Ahmed Rebai) provides a specific regional focus on *genomics in North Africa*, and **Chapter 11** (Gerald Mboowa) explores *infectious diseases*. Each chapter provides an overview of studies conducted, key findings, and future research directions.
- Finally, **Chapter 12** offers a visionary perspective on *genomics in Africa in the next ten years*. Nicola Mulder and Michèle Ramsay outline an ambitious future, discussing the potential impact of initiatives like the proposed Genomics Centre of Excellence and the tools needed to achieve this vision.

This book endeavours to capture genomics' dynamism, challenges, and transformative potential in Africa through these diverse contributions. It documents the remarkable progress achieved through the dedication of

African researchers and their collaborators while addressing the remaining hurdles. Ultimately, we hope this volume serves as a repository of current knowledge and as a catalyst for future research, collaboration, and innovation, contributing to a future where the power of genomics is harnessed to improve health and well-being equitably across the African continent.

REFERENCES

1. Choudhury, A., *et al.* High-depth African genomes inform human migration and health. *Nature* **586**, 741–748 (2020).
2. Gurdasani, D., *et al.* The African Genome Variation Project shapes medical genetics in Africa. *Nature* **517**, 327–332 (2015).
3. Ramsay, M., Tiemessen, C.T., Choudhury, A. & Soodyall, H. Africa: The next frontier for human disease gene discovery? *Hum Mol Genet* **20**, R214–220 (2011).
4. Sengupta, D., *et al.* Genetic substructure and complex demographic history of South African Bantu speakers. *Nat Commun* **12**, 2080 (2021).
5. Chikowore, T., Kamiza, A.B., Oduaran, O.H., Machipisa, T. & Fatumo, S. Non-communicable diseases pandemic and precision medicine: Is Africa ready? *EBioMedicine* **65**, 103260 (2021).
6. Ramsay, M. Africa: Continent of genome contrasts with implications for biomedical research and health. *FEBS Lett* **586**, 2813–2819 (2012).
7. Fatumo, S., *et al.* A roadmap to increase diversity in genomic studies. *Nat Med* **28**, 243–250 (2022).
8. Martin, A.R., *et al.* Clinical use of current polygenic risk scores may exacerbate health disparities. *Nat Genet* **51**, 584–591 (2019).
9. Dandara, C., *et al.* H3Africa and the African life sciences ecosystem: Building sustainable innovation. *Omics: J Integr Biol* **18**, 733–739 (2014).
10. Jongeneel, C.V., *et al.* Assessing computational genomics skills: Our experience in the H3ABioNet African bioinformatics network. *PLoS Comput Biol* **13**, e1005419 (2017).
11. Suhre, K., *et al.* Human metabolic individuality in biomedical and pharmaceutical research. *Nature* **477**, 54–60 (2011).

Chapter 2

A Spectrum of the Genomics Landscape of African Diversity

Emile R. Chimusa[1*]

[1]Department of Applied Sciences, Faculty of Health and Life Sciences, Northumbria University, Newcastle, Tyne and Wear, NE1 8ST, UK.

2.1 INTRODUCTION

Human migration out of, back to, and within the African continent has shaped Africans' genetic makeup due to multi-episodic admixture and adaptations to new situations, including agriculture, hunting, pastoralism, nutritional changes, climatic changes, and the emergence of new pathogens.[1] The Bantu expansion represents one of the largest migration events in African history, having started some 3,000–4,000 years ago,[2] and has been described as the most influential demographic event to have occurred in the late Holocene Africa continent.[2–4] They moved eastward and southward over the continent in waves.[3] The two main pathways of this expansion are the eastern one, which entered East Africa via the Great Lakes region, and the southern one, which travelled via the central African rainforests and ended in Southern Africa.[5,6] Early genetic research concentrated on mtDNA and Y-chromosome markers, which trace maternal and paternal lineages, respectively.[5]

Recent genomic analyses combining autosomal DNA and whole-genome sequencing have shown a more complete picture of Bantu genetic variation.[1,2,6,7–10] On the other hand, the present literature on the human signature of natural selection and phenotypic adaptation evidences a lack of global inclusiveness and approaches in polygenic adaptation and signature of selection, adversely impacting most African populations.[6,11,12] African populations have a longer history and consequently are characterized by greater diversity and differences in linkage disequilibrium (LD) architecture, with shorter haplotype blocks.[12,13] Several studies have demonstrated that the African population has undergone recent unique and complex episodic genetic mixture events and adaptation to the local environment and pathogens that shaped its genetic makeup, especially its unique traits.[1,2,6,7–10] These characteristics must be considered when designing genome-wide association studies (GWAS), selecting the appropriate technology, and analyzing the data.[14]

***Corresponding author:** emile.chimusa@.northumbria.ac.uk (Emile R. Chimusa)

 DOI: 10.1201/9781003436584-2

In addition, understanding the genomic footprints of natural selection of highly diverse and admixed communities can lead to unique discoveries. It can indicate how distinct ancestral contributions and local environmental forces have influenced the genetic makeup of these communities.

This chapter takes a more detailed look at the evidence for genetic ancestry, migration events, and natural selection posited by previous studies. We catalogue the information of existing custom-designed DNA panels and selection scans associated with humans, particularly Africans, which will benefit a large scientific community. Through data, we highlight several challenges and opportunities offered by understanding adaptation and selection through the admixture process in African ancestry populations. We conclude with discussions of research areas where further work on the signature of genomics diversity and selection in African populations is needed.

2.2 RECENT EFFORTS TO CAPTURE ADAPTIVE GENETIC DIVERSITY

In the later part of the 20th century and into the 21st century, the field of human genetics has been bombarded with innovative studies, as evidenced by the vast volumes of publications where genetic links to numerous diseases have been defined. All disease traits have a genetic component, including previously known infectious diseases.[14,15] This has in turn generated a tremendous amount of genetic data, which has swamped data scientists and bioinformaticians, who labor to connect and make sense of it. A closer scrutiny of the literature depicts not only an increased knowledge of trait heredity but also a clearer understanding of the human genome. How did we get here? Lindee (2000) suggests that this can be attributed to the integrated effort of biotechnology companies, public institutions, consortia, politicians, scientists, physicians, and society at large, which has made this knowledge useful, negotiated it into the medical realm, and further increased its value.[15]

High-throughput technologies and recent advances in statistical, data science, and computational methods have fostered the development of different techniques aimed at understanding world populations.[15] Variations in how humans relate to each other, behave, appear and in their predisposition to diseases are rooted in the differences in the genotypic make-up among individuals and interaction with their environmental aspects, arising from the different environmental exposures and lifestyle behavior.[16] These elements are primarily influenced by an individual's capability to accommodate external or internal changes, which varies among individuals.[17,18] Genetic variation is critical for the evolution process to occur and is continuously fed by the mutation process, while other micro-evolutionary factors like migration, natural selection, admixture, and random drift (Figure 2.1) are known to affect the variability patterns.[3,13] Due to population divergence,

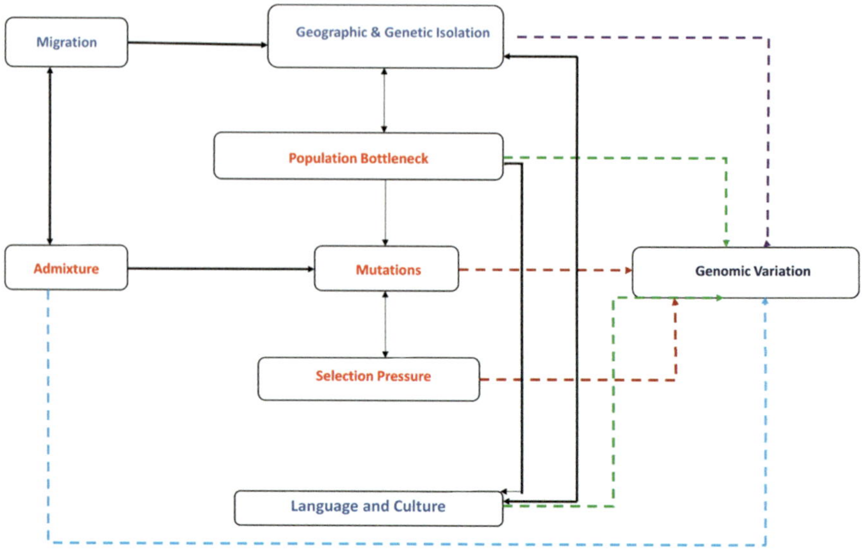

Figure 2.1 The different evolutionary forces, including migration, admixture, religion and culture, population bottleneck, natural selection, and geographic and genetic isolation, which have played a critical role in presenting human genomic variations in the genetic landscape and allele frequencies of human populations.

migration natural selection, mutation, population bottleneck, and admixture events (Figure 2.1), genetic variants differ in frequency among human populations.[19,20] This has facilitated the understanding of population history and human diseases, particularly those that result from the interaction of different genes, the environment, and their lifestyle. Nevertheless, due to their deep-rooted ancestry and environmental factors, African populations have been subjected to natural selection.[21,22] As a result, some functionally important variants that are common in African populations may be found at low frequencies or not at all in non-African genomes. Additionally, common adaptive variants that protect against infectious diseases predominant in Africa may cause susceptibility to other diseases,[21,22] and conversely, many of the markers on commercially available arrays are monomorphic in the African population, which results in a loss of usable data when using these arrays in African cohorts.

As we know, in the 19th century, Charles Darwin introduced the concept of natural selection (NS) as an evolutionary process.[23,24] It explains how certain traits become more common in successive generations because they positively impact the survival and reproduction of a species.[24] This understanding of natural selection has further enhanced research on uncovering the population admixture over time. This is because present-day humans originated in Africa and have thrived in different environments, resulting in high genetic and phenotypic diversity. In addition, studying the genetic diversity within and between African ethnic groups creates an opportunity to understand and reconstruct some of the earliest events in the history of human populations. They are essential for detecting patterns of the variation of genes that have been recently influenced by natural selection.[21,22] Hence, the research conducted by Cuadros-Espinoza et al.[22] offers dependable proof that human populations have experienced adaptive mixing in instances where genetic history is linked to periods of isolation and spatial expansions. This extension is purported to enhance the movement of genes by utilizing comprehensive computer simulations to assess various neutrality statistics to identify natural selection in the mixed population. This is done by assuming multiple mixing scenarios from the genetic material of 15 diverse mixed populations worldwide. Furthermore, it has been determined that resistance to malaria and lactase persistence have undergone adaptive mixing in South Asians, North Africans, and West Africans in Malagasy, respectively. This technique identified more adaptive combinations, such as *APOL1* in Fulani nomads and *PKN2* in East Indonesians, which had a role in the resistance to infection and metabolism, respectively. Fan et al.[25] identified evidence of local adaptation for various traits, including height, skin color, metabolic processes, and immune response. They found a variant in the lightly pigmented San population that is positively selected and affects pigmentation in vitro by regulating enhancer activity and gene expression of *PDPKI*. Caro et al.[26] found that North Africa (NA) has experienced ongoing demographic shifts due to its geographical location, which has influenced the genetic makeup of modern human populations. The study identified potential genes in NA that are associated with insulin processing (*KIF54*), immunological function (*KIF5A*, *ILIRN*, *TLR3*), and hemoglobin abnormalities (*BCL11A*). The study also observed positive selection signatures shared with European populations associated with immune function (*ILIR1*, *CD44*, *JAK1*) and skin pigmentation (*SLC24A5*, *KITLG*). Additionally, candidate genes related to insulin processing traits (*GLIS3*), other immune-related traits (*DOCK2*), and hemoglobin phenotypes (*HPSE2*, *HBE1*, *HBG2*) were found to be common among East and West African populations. The *SLC8A1* gene, which controls the sodium-calcium exchanger, was the sole candidate found as being under post-admixture selection in Western North America. However, this study on North Africa provides insights into how historical migration and selection pressure can shape the genetic makeup of an admixed group. Both populations have undergone significant admixture from different populations. An excellent example of this influence is presented in Table 2.1, which lists human genes that have undergone natural selection, specifying the associated phenotypes and the populations affected. While extensive resources such as GWAS and other variant databases provide a wealth of genetic data, no comprehensive database currently exists that specifically catalogs human genetic variants adapted through natural selection across different populations.

Table 2.1 Human Genes Adapted Through Natural Selection and Their Associated Phenotypes

Gene	*Phenotype*	*Population(s)*	*Comments*	*References*
SLC24A5	Skin pigmentation	European, South Asian, and Khoisan populations	The *SLC24A5* gene, which is involved in the pigmentation pathway, has been selected in Khoisan populations. This gene contributes to lighter skin pigmentation than that found in other African groups.	25–27
G6PD	Malaria resistance	Mediterranean, African, and South Asian populations	Variants of the *G6PD* gene, which cause glucose-6-phosphate dehydrogenase deficiency, provide a selective advantage against malaria but can lead to hemolytic anemia.	28–32
APOBEC3G	HIV resistance	Various South African populations	The *APOBEC3G* gene is part of the innate immune system and has been implicated in resistance to HIV infection.	33,34
CCR5-Δ32	Resistance to HIV infection	European and western Asian populations	Mutation confers resistance to HIV by preventing the virus from entering cells.	35–37
DARC (Duffy antigen)	Resistance to *Plasmodium vivax* malaria	Sub-Saharan African populations, including South Africa	The Duffy-null allele confers resistance to *Plasmodium vivax* malaria but is associated with increased susceptibility to certain inflammatory conditions	38–40
MCIR	Skin pigmentation (response to UV radiation)	European, East Asian populations	Variants influence pigmentation and susceptibility to UV radiation and skin cancer.	41–44

(*Continued*)

Table 2.1 (Continued)

Gene	*Phenotype*	*Population(s)*	*Comments*	*References*
LCT	Lactase persistence (ability to digest lactose in adulthood)	Europeans, some African and Asian pastoralist groups	The lactase persistence trait is due to genetic mutations that allow continued expression of lactase, the enzyme that breaks down lactose.	45–47
TPCN2	Hair color	Europeans	The *TPCN2* genes influence hair color variation; more common in European populations.	48,49
APOL1	Resistance to trypanosomiasis and kidney disease	Various African populations	The *APOL1* gene protects African trypanosomiasis but can also increase the risk of kidney disease.	50,51
(*HbS*) and *G6PD* variants	Variations in hemoglobin and glucose-6-phosphate dehydrogenase (G6PD) deficiency	Sub-Saharan Africans: Population-specific adaptation to Bantu-speaking populations	These adaptations resist malaria but come with trade-offs like sickle cell disease and hemolytic anemia.	52,53
CYP3A5	Metabolism of certain drugs	Various African populations	The *CYP3A5* gene shows high variability and affects the metabolism of several drugs, which has implications for personalized medicine.	54,55
IL10	Immune response regulation	Various South African populations	The *IL10* gene is involved in anti-inflammatory responses and has been under selection due to infectious disease pressures.	56–58, 31,32

2.3 DIVERSITY AND GENOTYPING ARRAYS AND GENOME REFERENCES

Despite the commercial availability of affordable genome-wide DNA (genotyping array) panels, custom-designed DNA panels are frequently used for high-resolution DNA studies focusing on specific genes or chromosomal regions and are now frequently used for GWAS to optimize the inference of disease scoring statistics. A genotyping array is a collection of selected probes representing single-nucleotide variants (SNVs) on an array platform and has several requirements to make it effective. Due to limited space using the existing technologies, genotyping arrays rely heavily on the association between SNVs (i.e. linkage disequilibrium). Thus, a causal variant potentially has other variants nearby that are also associated with the disease. The non-random association between SNVs means that only a "tag SNV" needs to be genotyped as associated genotypes can then be imputed. Tag SNVs and their effectiveness are thus dependent on the linkage disequilibrium (LD) structure of the population under study[59]; therefore array design requires haplotype data for the relevant populations. These reference populations are also required for reference panels for the imputation. There are several other considerations for designing a new array, including coverage of both taggable and non-taggable SNVs, the inclusion of SNVs of clinical interest, and SNVs common in the populations being used. The SNVs should be able to capture genetic diversity and at the same time have the power to detect causal variants.

Existing chip designs with imputation using large-scale sequencing panels can capture variation well relative to low-coverage sequencing designs by taking advantage of elevated levels of LD in European populations. However, the high diversity and low LD in African populations, in turn, influence the design and analysis of GWAS in these populations. Because recent strategies for finding causal variants that underlie common diseases have been based on LD, non-random association of variants at separate genetic loci and haplotype block; therefore, most current tag SNPS selection algorithms are based either on LD or haplotype block (Table 2.2). More recently, approaches that integrate cross-population LD-based, haplotype-based, and principal component analysis genome-wide tag single nucleotide polymorphism (SNP) selection have shown to efficiently identify informative variants, prioritized across multi-populations of low LD and high diversity for custom chip array.

Genetic variations that exist between populations of different ethnicities and lead to phenotypic variations are mostly due to the frequency of the genetic variation rather than its existence in either population.[13,17,21] African genomics data are significantly underrepresented in public databases, such as the GWAS catalog (https://www.ebi.ac.uk/gwas/), the European Genome-Phenome archive (https://www.ebi.ac.uk/ega/), and in the 1000 Genomes dataset.[60] This lack of appropriate reference data has the potential to impact genomic studies on African populations and the use of genomics data in clinical practice.[61] The problem is slowly being addressed by the emergence of large-scale genomics projects in Africa associated with funder requirements to submit the data to public repositories. One such initiative is the Human Health & Heredity in Africa (H3Africa)[62] funded by the National Institutes of Health and the Wellcome Trust, through the Alliance for Accelerating Excellence in Science in Africa.

Table 2.2 SNV Arrays Evaluated in the Preliminary Analysis

Array Code	*Strand File*	*Number of Single Nucleotide Polymorphisms*
AFFY6	GenomeWide_6-b37.58-v2. strand	921,570
HO1M	HumanOmni1-Quad_v1–0_C-b37.strand	1,104,250
MEGA	MEGA	1,660,493
H3Africa	–	2.5 million SNV-array
PANAF	Affy_Pan_African	2,230,182
H02p5M	HumanOmni2.5M-b37-v2.strand	2,449,626
HO2p5Ms	HumanOmni2.5S–8v1_B-b37.strand	2,014,185
HO5M	HumanOmni5–4v1_B-b37.strand	4,283,024

The African genomics characteristics need to be considered when designing new chip arrays for GWAS[11,63] and reference genome for sequence mapping[13] when selecting the appropriate technology and when analyzing the sequence data, respectively.[11,63] For example, the low genomics imputation accuracy in African populations is due to lack of large-scale African imputation reference panels.[63,64] In addition, the low performance of most tools for mapping and variant discovery in African populations is mainly due to a lack of African reference genome and underrepresentation of African populations in most of genome annotation databases.[11–13,63] Therefore, several studies suggested increasing (1) sequencing and genomics data from African and diverse populations in current imputation reference panels[11,63,64] and (2) developing pan-genome or population-specific graphs as reference genomes to increase cohort-specific information and accuracy in read alignment and variant calling.[12,13]

2.4 MAJOR AFRICAN ADMIXTURE LANDSCAPES

Regional variations in genomic diversity within the African continent are observed and driven by Bantu-speaking communities from their historical dispersal and environmental adaptations. Greater genetic diversity was found in western Bantu groups than in eastern and southern Bantu populations, based on a study that examined whole genome sequences from over 300 Bantu individuals.[7,8] The eastward expansions' successive founder effects are consistent with this gradient of diminishing variety from the west to the east. In addition to the findings reported by Choudhury et al.,[9,10] the Bantu groups residing in southern Africa exhibit a greater level of genetic diversity in comparison to those in other regions. This might be the result of mixing between the indigenous San hunter–gatherers of southern Africa and the growing Bantu populations. Together with more recent gene flow, several Bantu populations have a mixture of highly divergent ancestral components. Primary research revealed further "Southeastern Bantu" and other regional ancestry clusters, as well as a significant "Bantu" ancestry component spanning over equatorial Africa.[65] Nonetheless, subsequent analyses utilizing extensive genetic data have additionally segmented the Bantu component into unique source groups.[66]

Researchers have discovered substantial genetic mixing stemming from the Bantu expansion through genome-wide analysis of sub-Saharan African ethnicities. Skoglund et al.[67] conducted a study on Bantu-speaking people in southeastern Africa and discovered significant mixing between the migrant ancestors and the native hunter–gatherer populations living in the area. This indication of mixing was dated to 600–800 years ago. Comparably, communities in Western Africa, such as the Yoruba, exhibit evidence of Bantu gene flow, which transpired over a considerable period from 900 to 1600 CE.[68]

As stated by Skoglund et al.,[67] the Lemande of North Angola are among the most admixed Bantu-speaking communities, with about equal ancestral components from Bantu immigrants and Indigenous Kx'a populations. In agreement with Shriner and Rotimi,[69] Eastern Bantu populations, such as the Luhya, seem to have stayed mostly unadmixed throughout the previous 5,000 years. These different degrees of mixing among Bantu groups highlight the intricate genetic effects of the spread. Research shows that Bantu mixing has effects on phenotype and culture in addition to genetics. Studies in population genetics have shown that the initial dispersal of Bantu languages originates from the spread of individuals, offering insights into the genetic structure and history of Bantu speech communities.[2] Compelling evidence that the Bantu expansion was a significant demic diffusion is the comparatively low Y-chromosome diversity found in contemporary Bantu-speaking communities.[70] Li et al.[2] even go as far as to call it "one of the most dramatic demographic events in human history."

Li et al.[2] argue that the Bantu expansion ranks among the most significant demographic developments in human history. Y-chromosomal haplogroups E1b1a8 (also called E-U175) and E1b1a7a (also called E-U174) are highly prevalent among Bantu speakers and Niger-Congo-speaking populations from West Africa. These haplogroups are subgroups of E1b1a (also called E-M2) and E1b1a7 (also called E-M191), respectively.[71] The prehistory of the paternal half of a population can be inferred from Y chromosomes, which are only passed down from fathers to sons. Haplogroups are collections of similar molecules identified by a certain common mutation used for diagnosis.[71]

Since Y-chromosomes are passed down from fathers to sons, they provide insight into the prehistory of a population's paternal ancestry. According to Pakendorf,[71] haplogroups are collections of related molecules identified by a certain diagnostic common mutation. The geographic distribution of Bantu languages today does not always correspond to the initial migration of Bantu speech communities, even though Bantu speakers carried their languages with them from their ancient homeland down to Southern Africa.

Among the studies of African genome-wide data, Bantu-speaking populations within sub-Saharan regions were most studied compared to Northern African communities (Maghrebi). Maghrebi communities trace all their ancestry to autochthonous North Africans, European-related, Middle Eastern–related, and sub-Saharan African-related ancestry.[72–74]

Recent studies indicated that the Maghrebi genetic make-up is represented by 15,000-year-old Paleolithic individuals from Taforalt, Morocco, whose ancestry underwent admixture of an early Holocene Middle Eastern (63.5%), that is, Levantine Natufians, and a sub-Saharan genetic contribution.[72] These recent studies suggested that Arabization, neolithization, and sub-Saharan gene flow might lead to the dilution of Maghrebi ancestral contribution in most Northern African populations. During the Neolithization, North African populations were mixed with European Neolithic groups. This admixture is evident in ~3,000-year-old Late Neolithic individuals from Kelif el Boroud, Morocco, who are the result of admixture from European Neolithic and Ifri n'Amr or Moussa groups.[73,74] Thus recently, the Arabization introduced Middle Eastern–related ancestry ~1.4 kya, which decreases westward on a genetic cline.[74] Some scholars suggested the Arab expansion might additionally have facilitated sub-Saharan-related ancestry through the slave trade,[74,75] which is supported by sub-Saharan ancestry in North African populations that could be traced to an admixture event ~1.2 kya with a West African groups.[75] The Western Cape of Africa, at the southernmost part of South Africa, also harbors one of the most diverse admixed populations, namely, the South African colored (SAC) population, which is the largest ethnic group in this region and has its origins slightly >350 ya.[1,9–12,14]

The SAC population represents >49% of the estimated 7 million inhabitants in this province, with the vast majority being historically Afrikaan speakers (a unique South African language ancestrally linked to Dutch), although this is more recently changing.[1,9–12,14] Their complex origin of admixture is attributed to significant historical events that occurred within the last few millennia, starting ~1.7 kya, with the arrival of Bantu-speaking agro-pastoralists in South Africa.[1,9–12] During the last few centuries, European colonization of the Cape by the Dutch, Germans, and French, later followed by British seizure and rule, contributed to the complex admixture patterns at the Western Cape. During this time, slaves were traded by the Dutch East India Company from East Africa, Madagascar and surrounding islands, India, and Indonesia, leading to settler–slave admixture, including indigenous Khoe-San people.[9–12] Genomic studies of the SAC population revealed that these historical events correlate with the complex five-way admixture observed in this population, with ancestral contributions occurring predominantly from the indigenous Khoe-San, the

Bantu-speaking Africans, European-descent groups, and Southeast Asian and South Asian populations.[1,9,14] In addition, cultural and religious practices contributed to the high degree of heterogeneity in ancestral contributions among SAC individuals sampled from different regions of South Africa.[1,9,14]

Several studies have revealed a sex-biased gene flow in SAC that supports the historical records, indicating that almost all mixed marriages were between a male settler and either a free Black female (where the man bought the slave their freedom) or an indigenous Khoe-San female.[1,9–12,14] Petersen et al.[76] used uniparental markers to ascertain likely ancestral contributions using unique population-specific mtDNA and Y chromosomal haplogroup identifiers. Khoe-San-derived maternal lineage L0d had a 68% representation in the SAC group studied, while the M/N Eurasian mtDNA lineages were only represented at low frequencies. In contrast, there was a significant Eurasian paternal contribution (71.4%) defined by haplogroups R/I/G/N/O/J in the same group, and the Western European R1b haplogroup was prevalent at 44.4%. Similar distributions of mtDNA and Y haplogroups were observed from whole genome sequencing data of a small group of SAC males from the Western Cape region.[9] Overall, these findings demonstrate that recent admixture involved sex-biased gene flow.

2.5 DIVERSITY AND FUTURE DIRECTIONS

Studying admixed populations and their characteristics is especially important because they play vital roles in different specializations, including biomedical science, particularly in the fields of medical genetics and human evolution. Studying genetic variations in populations can yield historical insights, thus allowing us to understand the population history and factors that generate variations.[8,12,21] More of these studies provide the baseline study about the health inequalities or group differences in the genetic factors that are susceptible to different multifunctional diseases. Despite the advances in genotyping and sequencing platforms, comprehensive genome reference panels, and robust association approaches, challenges with GWAS persist.[77,78] As researchers grapple with the hidden heritability in GWAS and as concerns ensue over the disparity in genomic studies, the necessity to diversify GWAS is now apparent and crucial.[79] As technologies continue to advance and revolutionize the way people interact and meet their mates, the world population is becoming more heterogeneous. Extension of GWAS to non-European populations is bound to include diverse populations characterized by low LD and admixed populations, where the presence of admixture LD could lead to spurious associations. Admixture mapping has been successfully applied to these populations, though it is limited when the disease frequency is similar in the individual parental population or when the differences in the frequency of a variant between cases and controls are due to other factors, for instance, natural selection.[80,81]

As genotyping and sequencing costs decrease, association studies in admixed populations have become an appealing alternative to admixture mapping.[80,81] To account for the limitations of both GWAS and admixture mapping in these populations and also the fact that the two methods are complementary and could increase disease mapping power, joint studies of SNP and admixture associations are increasingly being explored.[79–81] However, the currently available joint SNP and admixture association methods have primarily been tailored to African American populations (two-way admixed populations), though the developers suggest that in theory these methods could be extended to more than three-way admixed populations; this has been left for future research work.[79–82] The application of these methodologies to multi-way admixed populations, like most African populations known to have high diversity within differing environments (for instance, the SAC population), remains mostly unexplored. As GWAS is being applied to multiethnic populations, there is a dire need for more powerful GWAS methods tailored for multi-way admixed populations.

1. There is an unmet need to utilize accurate knowledge of ancestry inference and disease mapping to develop joint SNP and ancestry association methods for multi-way admixed populations under a Bayesian context, implementing mixed linear model (MLM) approaches.
2. There is an urgency to diversify GWAS and the several efforts that have been put in place, to enhance its scalability to highly admixed populations and enable its clinical utility through predicting individual genetics liability.
3. While next-generation sequencing (NGS) evolution has increased our understanding of the human genome and impacted modern biological science, however, the read alignment and variant calling methods that leverage NGS data rely on the European genome reference, decreasing the effectiveness of these methods in diverse and African populations.[10–13] Therefore, pan genome or population-specific graphs of African are critically needed[10,12,83] to uncover African genetic variation and provide comprehensive and accurate analysis of genomic variation.[10]

The world population today is becoming more admixed, and though ancestry associations in admixed populations have been explored for a while now, several studies have demonstrated their limitation in locating the risk variant in the absence of ancestry deviation.[79–81] These studies[79–81] have also demonstrated that current GWAS approaches are underpowered in detecting risk variants when genotypes risk is present but weak. This could occur when the functional risk variants have become fixed in the population due to selection or when the SNPs in the study are not in LD with the SNP variant in the genotyping array. However, as genotyping and sequencing costs continue to decline, the application of associated studies to admixed populations is inevitable and becoming more appealing. To account for the limitations of both approaches and the problems of small sample sizes inherent to non-European populations, researchers are increasingly exploring joint studies of SNP and ancestry associations in admixed populations.[79–81,84,85] However, the complexity pertinent to multi-way admixed populations has limited their applications to two-way admixed populations, though the developers suggest that in theory these methods could be extended to three-way admixed populations; this work has been left for future research work in publications.[82,84,85]

Importantly, association studies have significantly contributed to medical genomics and the understanding of complex traits; however, large numbers of false positives and the small effect size of genetic risk variants have induced a need for large sample sizes and the culture of FAIR (findability, accessibility, interoperability, and reusability) data and sharing. Figure 2.2 illustrates association results from currently used GWAS tools. These results are from the simulation of different sample size GWAS case control in various simulated populations (African, European, and admixed on 6 causal variants, demonstrating that most of the tools generate significant false positive results).

Figure 2.2 shows the GWAS tools and pipelines commonly used in populations of European descent; however, high rates of false positive/negative outcomes may result if stratification is not carefully controlled, particularly in African genomes, which harbor the highest genetic diversity and are currently disproportionately underrepresented in public databases and reference panels.

We propose that even as efforts are being put in place to increase sample sizes in African populations, there is a need to explore other characteristics of African and diverse populations to ensure healthcare equity as the race to translate genetic association results gains ground. Also, there is an unmet need to acknowledge that GWAS pipelines for diverse populations should be population specific, and appropriate tools should be chosen for specific populations. Researchers could also run multiple tools and allow significance levels to recommend on the use of the right tool. It is also important to keep fostering a culture of replication to maintain the reliability of genetic association findings. Furthermore, as sequencing and thus the availability of genomic information on diverse populations increase, there are new opportunities to design appropriate novel disease-scoring statistical models that are tailored to and leverage the characteristics of African populations (with longer histories, high genetic diversity, and environmental heterogeneity) and other diverse populations. This will enable us to understand better and elucidate the genetic architecture of complex traits, variations in risk predictions, treatment responses, and disease outcomes in these populations and other populations of African descent or ancestry.

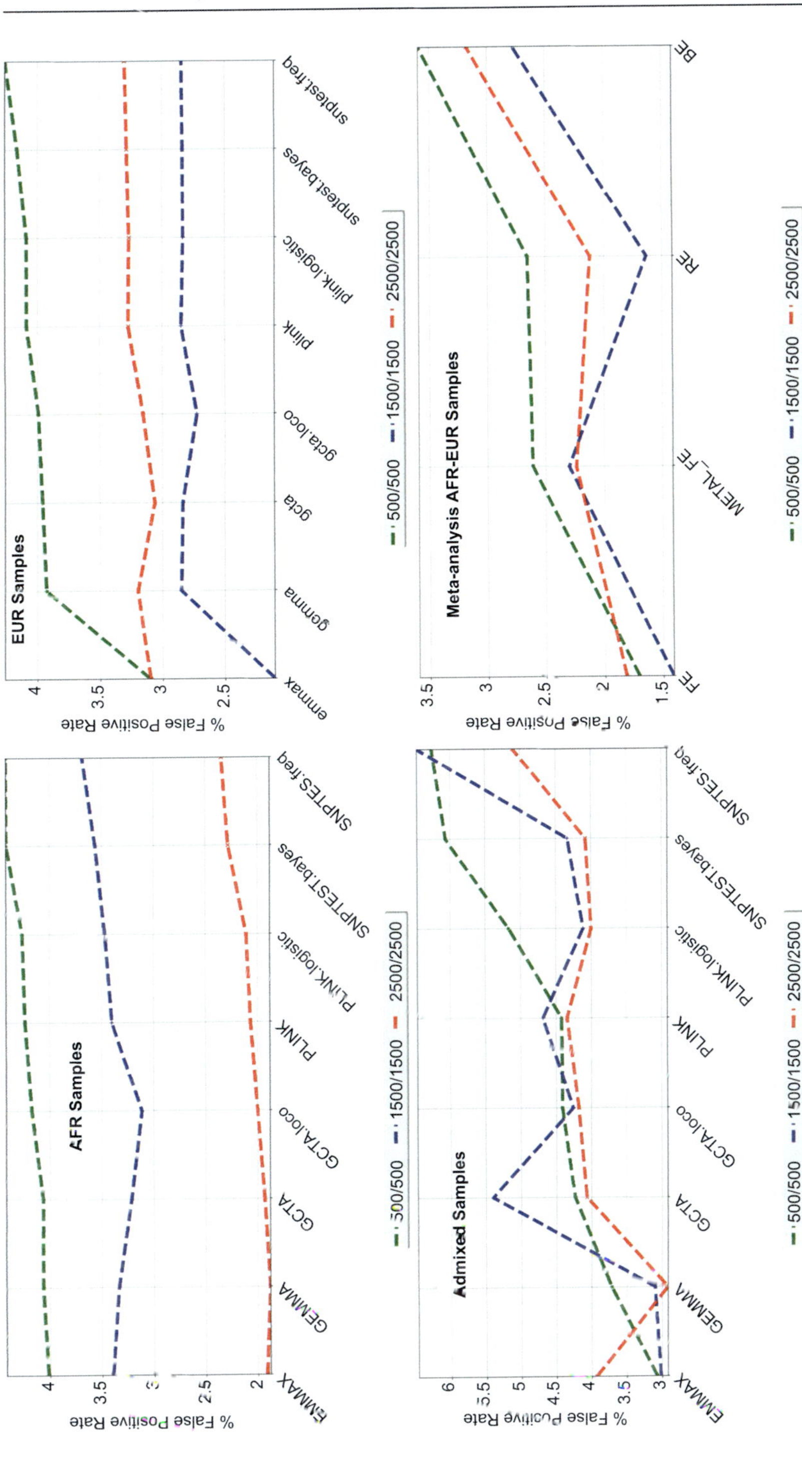

Figure 2.2 The proportion of non-causal simulated variants reaches genome-wide significance across GWAS tools: (A) from African (AFR), (B) from European (EUR), (C) admixed, and (D) cross meta-analysis of AFR and EUR simulated data sets.

Population-tailored association tools will necessitate the simulation of tools that can capture a wide range of these populations' characteristics for assessment and benchmarking.

Finally, novel African or diverse-specific disease association studies that further leverage the environmental diversity on the African and diverse continents have potentially many uses. These include a better understanding of the pathogenesis of diseases of global health relevance, new leads for studying underlying risk prediction, and advancing clinical prediction of global treatment. The findings from these African or diverse-specific disease association studies will pave the way for new, more diverse research in public health translation.

Moving forward, as we work toward increased diversification, GWAS key recommendations would be the following: (1) prioritization of funding of GWAS in diverse populations; (2) personal initiatives by researchers in recognizing the need for diversity in research; (3) training and provision of high-quality infrastructure, including good healthcare facilities that encourage genetic testing for the underrepresented populations; and (4) inclusion of local researchers with higher capacities in GWAS to both leverage contextual knowledge and encourage trust during data collection.

2.6 CONCLUSION

In this chapter, we take a detailed look at the progress of genetic disease mapping strategies. By definition, genetic disease mapping is the process of identifying gene locations that could be linked to disease phenotypes of interest without the need for prior knowledge of their biological functionality. Genetics is one of the many research approaches, including animal models and cell cultures, aiming to unravel disease etiology. However, the genome has the advantage of being consistent and finite, allowing orderly investigations of the genotype-disease trait relations that have led to numerous novel findings that reveal new and unexpected biological variants and pathways. The results also provide compelling evidence on the functionality of these genes and pathways in the disease's etiology, which in turn may eliminate or minimize the need for probable intrusive or even harmful research on human subjects. The genetics approach is, however, inadequate to untangle complex molecular mechanisms involved in the etiology of diseases and is thus considered an imperative but complementary approach in research.

REFERENCES

1. Chimusa, E. R. et al. A genomic portrait of haplotype diversity and signatures of selection in indigenous southern African populations. PLoS Genetics **11**, e1005052 (2015).

2. Li, S., Schlebusch, C. & Jakobsson, M. Genetic variation reveals large-scale population expansion and migration during the expansion of Bantu-speaking peoples. Proceedings of the Royal Society B: Biological Sciences **281**, 20141448 (2014).
3. González-Santos, M. et al. Exploring the relationships between genetic, linguistic, and geographic distances in Bantu-speaking populations. American Journal of Biological Anthropology **179**, 104–117 (2022).
4. Soares, P., Rito, T., Pereira, L. & Richards, M. B. A genetic perspective on African prehistory. Africa from MIS 6–2: Population Dynamics and Paleoenvironments 383–405 (2016).
5. Montano, V. et al. The Bantu expansion revisited: A new analysis of Y chromosome variation in Central Western Africa. Molecular Ecology **20**, 2693–2708 (2011).
6. Fortes-Lima, C. A. et al. The genetic legacy of the expansion of Bantu-speaking peoples in Africa. Nature **625**, 540–547 (2024).
7. Fan, S. et al. Whole-genome sequencing reveals a complex African population demographic history and signatures of local adaptation. Cell **186**, 923–939.e14 (2023).
8. Gurdasani, D. et al. The African Genome Variation Project shapes medical genetics in Africa. Nature **517**, 327–332 (2014).
9. Choudhury, A. et al. Whole-genome sequencing for an enhanced understanding of genetic variation among South Africans. Nature Communications **8**, 2062 (2017).
10. Choudhury, A., Aron, S., Botigué, L. R., Sengupta, D., Botha, G., Bensellak, T., Wells, G., Kumuthini, J., Shriner, D., Fakim, Y. J. & Ghoorah, A. W. High-depth African genomes inform human migration and health. Nature **586**(7831), 741–748 (2020).
11. Fan, S. et al. African evolutionary history inferred from whole genome sequence data of 44 indigenous African populations. Genome Biology **20**, (2019).
12. Tetikol, H. S., Narci, K., Turgut, D., Budak, G., Kalay, O., Arslan, E., Demirkaya-Budak, S., Dolgoborodov, A., Jain, A., Kabakci-Zorlu, D. & Brown, R. Population-specific genome graphs improve high-throughput sequencing data analysis: A case study on the Pan-African genome. bioRxiv 2021–2003 (2021).
13. Alosaimi, S., van Biljon, N., Awany, D., Thami, P. K., Defo, J., Mugo, J. W., Bope, C. D., Mazandu, G. K., Mulder, N. J. & Chimusa, E. R. Simulation of African and non-African low and high coverage whole genome sequence data to assess variant calling approaches. Briefings in Bioinformatics **22**(4), bbaa366 (2021).
14. Chimusa, E. R. et al. Genome-wide association study of ancestry-specific TB risk in the South African Coloured population. Human Molecular Genetics **23**, 796–809 (2014).
15. Buniello, A. et al. The NHGRI-EBI GWAS Catalog of published genome-wide association studies, targeted arrays and summary statistics 2019. Nucleic Acids Research **47**(D1), D1005–D1012 (2019).
16. Albert, F. W. & Kruglyak, L. The role of regulatory variation in complex traits and disease. Nature Reviews Genetics **16**(4), 197–212 (2015).
17. Pelabon, C. et al. Evolution of variation and variability under fluctuating, stabilizing, and disruptive selection. Evolution **64**(7), 1912–1925 (2010).

18. Lindee, S. Human genetics after the bomb: Archives, clinics, proving grounds and board rooms. Studies in History and Philosophy of Science Part C: Studies in History and Philosophy of Biological and Biomedical Sciences **55**, 45–53 (2016).
19. Schlebusch, C. M. et al. Human adaptation to arsenic-rich environments. Molecular Biology and Evolution **32**(6), 1544–1555 (2015).
20. Li, S., Schlebusch, C. & Jakobsson, M. Genetic variation reveals large-scale population expansion and migration during the expansion of Bantu-speaking peoples. Proceedings of the Royal Society B: Biological Sciences **281**(1793), 20141448 (2014).
21. Gomez, F., Hirbo, J. & Tishkoff, S. A. Genetic variation and adaptation in Africa: Implications for human evolution and disease. Cold Spring Harbor Perspectives in Biology **6**, a008524–a008524 (2014).
22. Cuadros-Espinoza, S., Laval, G., Quintana-Murci, L. & Patin, E. The genomic signatures of natural selection in admixed human populations. The American Journal of Human Genetics **109**, 710–726 (2022).
23. Bradley, B. Natural selection according to Darwin: Cause or effect? History and Philosophy of the Life Sciences **44**, 13 (2022).
24. Darwin, C. et al. The Correspondence of Charles Darwin: 1821–1860. Vol. 13. Cambridge University Press, 2002.
25. Fan, S. et al. Whole-genome sequencing reveals a complex African population demographic history and signatures of local adaptation. Cell **186**, 923–939. e14 (2023).
26. Caro-Consuegra, R., Lucas-Sánchez, M., Comas, D. & Bosch, E. Identifying signatures of positive selection in human populations from North Africa. Scientific Reports **13**, (2023).
27. Lamason, R. L. SLC24A5, a putative cation exchanger, affects pigmentation in zebrafish and humans. Science **310**, 1782–1786 (2005).
28. Norton, H. L. et al. Genetic evidence for the convergent evolution of light skin in Europeans and East Asians. Molecular Biology and Evolution **24**, 710–722 (2006).
29. Liu, J., Bitsue, H. K. & Yang, Z. Skin colour: A window into human phenotypic evolution and environmental adaptation. Molecular Ecology (2024). https://doi.org/10.1111/mec.17369.
30. Tishkoff, S. A. et al. Haplotype diversity and linkage disequilibrium at human G6PD: Recent origin of alleles that confer malarial resistance. Science **293**, 455–462 (2001).
31. Kumar, R., Ng, S. & Engwerda, C. The role of IL-10 in Malaria: A double edged sword. Frontiers in Immunology **10**, (2019).
32. Edwards, C. L. et al. IL-10-producing Th1 cells possess a distinct molecular signature in malaria. The Journal of Clinical Investigation **133**, (2023).
33. Ruwende, C. et al. Natural selection of hemi- and heterozygotes for G6PD deficiency in Africa by resistance to severe malaria. Nature **376**, 246–249 (1995).
34. Hill, A. V. et al. Common west African HLA antigens are associated with protection from severe malaria. Nature **352**, 595–600 (1991).
35. Belbacha, I. et al. The APOBEC3G gene rs2294367(C>G) variant is associated with HIV-1 infection in Moroccan subjects. Science Direct (2024). https://doi.org/10.1016/j.actatropica.2023.107045.

36. Hultquist, J. F. et al. Human and rhesus APOBEC3D, APOBEC3F, APOBEC3G, and APOBEC3H demonstrate a conserved capacity to restrict vif-deficient HIV-1. Journal of Virology **85**, 11220–11234 (2011).
37. Liu, R. et al. Homozygous defect in HIV-1 coreceptor accounts for resistance of some multiply-exposed individuals to HIV-1 infection. Cell **86**, 367–377 (1996).
38. Duncan, S. R. Reappraisal of the historical selective pressures for the CCR5-32 mutation. Journal of Medical Genetics **42**, 205–208 (2005).
39. Novembre, J., Galvani, A. P. & Slatkin, M. The geographic spread of the CCR5 Δ32 HIV-resistance Allele. PLoS Biology **3**, e339 (2005).
40. Tishkoff, S. A. & Williams, S. M. Genetic analysis of African populations: Human evolution and complex disease. Nature Reviews Genetics **3**, 611–621 (2002).
41. McManus, K. F. et al. Population genetic analysis of the DARC locus (Duffy) reveals adaptation from standing variation associated with malaria resistance in humans. PLoS Genetics **13**, e1006560 (2017).
42. Gartner, V. et al. Genomic insights into Plasmodium vivax population structure and diversity in central Africa. Malaria Journal **23**, (2024).
43. Nakayama, K. et al. Identification of novel functional variants of the melanocortin 1 receptor gene originated from Asians. Human Genetics **119**, 322–330 (2006).
44. Savage, S. A. et al. Nucleotide diversity and population differentiation of the Melanocortin 1 Receptor gene, MC1R. BMC Genetics **9**, (2008).
45. Swope, V. B. & Abdel-Malek, Z. A. MC1R: Front and center in the bright side of dark eumelanin and DNA repair. International Journal of Molecular Sciences **19**, (2018).
46. Feng, Y., McQuillan, M. A. & Tishkoff, S. A. Evolutionary genetics of skin pigmentation in African populations. Human Molecular Genetics **30**, R88–R97 (2021).
47. Itan, Y., Powell, A., Beaumont, M. A., Burger, J. & Thomas, M. G. The origins of lactase persistence in Europe. PLoS Computational Biology **5**, e1000491 (2009).
48. Tishkoff, S. A. et al. Convergent adaptation of human lactase persistence in Africa and Europe. Nature Genetics **39**, 31–40 (2006).
49. Segurel, L. et al. Why and when was lactase persistence selected for? Insights from Central Asian herders and ancient DNA. PLoS Biology **18**, e3000742 (2020).
50. Sulem, P. et al. Two newly identified genetic determinants of pigmentation in Europeans. Nature Genetics **40**, 835–837 (2008).
51. Böck, J. et al. Human genome diversity data reveal that L564P is the predominant TPC2 variant and a prerequisite for the blond hair associated M484L gain-of-function effect. PLoS Genetics **17**, e1009236 (2021).
52. Cooper, A. et al. APOL1 renal risk variants have contrasting resistance and susceptibility associations with African trypanosomiasis. eLife **6**, (2017).
53. Kamoto, K. et al. Association of APOL1 renal disease risk alleles with Trypanosoma brucei rhodesiense infection outcomes in the northern part of Malawi. PLoS Neglected Tropical Diseases **13**, e0007603–e0007603 (2019).
54. Esoh, K. & Wonkam, A. Evolutionary history of sickle-cell mutation: Implications for global genetic medicine. Human Molecular Genetics **30**, R119–R128 (2021).

55. Gouveia, M. H. et al. Genetic signatures of gene flow and malaria-driven natural selection in sub-Saharan populations of the 'endemic Burkitt Lymphoma belt'. PLoS Genetics **15**, e1008027 (2019).
56. Lin, M. et al. G6PD deficiency and hemoglobinopathies: Molecular epidemiological characteristics and healthy effects on Malaria Endemic Bioko Island, equatorial Guinea. PLoS One **10**, e0123991 (2015).
57. Bains, R. K. et al. Molecular diversity and population structure at the Cytochrome P450 3A5 gene in Africa. BMC Genetics **14**, 34 (2013).
58. mrr5831. Genetic Variations Influence Drug Metabolism in Patients of African Descent. News Center. https://news.feinberg.northwestern.edu/2023/02/07/genetic-variations-influence-drug-metabolism-in-patients-of-african-descent/ (2023).
59. Thanh Nguyen, D. et al. LmTag: Functional-enrichment and imputation-aware tag SNP selection for population-specific genotyping arrays. Briefings in Bioinformatics **23**(4), bbac252 (2022).
60. Popejoy, A. B. & Fullerton, S. M. Genomics is failing on diversity. Nature **538**(7624), 161–164 (2016).
61. Petrovski, S. & Goldstein, D. B. Unequal representation of genetic variation across ancestry groups creates healthcare inequality in the application of precision medicine. Genome Biology **17**, 1–3 (2016).
62. H3Africa Consortium. Enabling the genomic revolution in Africa: H3Africa is developing capacity for health-related genomics research in Africa. Science (New York, NY) **344**(6190), 1346 (2014).
63. Ibrahim, M. & Musa, M. Effective size and effectiveness: Next generation sequencing and the practice of genomics in Africa. Next Generat Sequenc & Applic **S1**, 008 (2016). https://doi.org/10.4172/2469-9853.S1-008.
64. Sengupta, D. et al. Performance and accuracy evaluation of reference panels for genotype imputation in sub-Saharan African populations. Cell Genomics 3(6), (2023).
65. Patin, E. et al. Dispersals and genetic adaptation of Bantu-speaking populations in Africa and North America. Science **356**, 543–546 (2017).
66. Micheletti, S. J. et al. Genetic consequences of the transatlantic slave trade in the Americas. The American Journal of Human Genetics **107**, 265–277 (2020).
67. Skoglund, P. et al. Reconstructing prehistoric African population structure. Cell **171**, 59–71. e21 (2017).
68. Busby, G. B. J. et al. The role of recent admixture in forming the contemporary West Eurasian genomic landscape. Current Biology **25**, 2518–2526 (2015).
69. Shriner, D. & Rotimi, C. N. Whole-genome-sequence-based haplotypes reveal single origin of the Sickle Allele during the holocene wet phase. The American Journal of Human Genetics **102**, 547–556 (2018).
70. Marks, S. J. et al. Static and moving frontiers: The genetic landscape of Southern African Bantu-speaking populations. Molecular Biology and Evolution **32**, 29–43 (2015).
71. Pakendorf, B., de Filippo, C. & Bostoen, K. Molecular perspectives on the Bantu expansion: A synthesis. Language Dynamics and Change **1**, 50–88 (2011).

72. Van de Loosdrecht, M. et al. Pleistocene North African genomes link near Eastern and sub-Saharan African human populations. Science **360**(6388), 548–552 (2018).
73. Bosch, E. et al. Population history of north Africa: Evidence from classical genetic markers. Human Biology, 295–311 (1997).
74. Lucas-Sánchez, M., Serradell, J. M. & Comas, D. Population history of North Africa based on modern and ancient genomes. Human Molecular Genetics **30**(R1), R17–R23 (2021).
75. Henn, B. M. et al. Genomic ancestry of North Africans supports back-to-Africa migrations. PLoS Genetics **8**(1), e1002397 (2012).
76. Petersen, D. C. et al. Complex patterns of genomic admixture within southern Africa. PLoS Genetics **9**(3), e1003309 (2013).
77. Uffelmann, E. et al. Genome-wide association studies. Nature Reviews Methods Primers **1**(1), 59 (2021).
78. Dennison, C. A. et al. Genome-wide association studies in schizophrenia: Recent advances, challenges and future perspective. Schizophrenia Research **217**, 4–12 (2020).
79. Peterson, R. E. et al. Genome-wide association studies in ancestrally diverse populations: Opportunities, methods, pitfalls, and recommendations. Cell **179**(3), 589–603 (2019).
80. Karczewski, K. J., Gupta, R., Kanai, M., Lu, W., Tsuo, K., Wang, Y., Walters, R. K., Turley, P., Callier, S., Shah, N. N. & Baya, N. Pan-UK Biobank genome-wide association analyses enhance discovery and resolution of ancestry-enriched effects. Nature Genetics **57**(10), 2408–2417 (2025).
81. Atkinson, E. G. et al. Reply to: On powerful GWAS in admixed populations. Nature Genetics **53**(12), 1634–1635 (2021).
82. Hou, K. et al. On powerful GWAS in admixed populations. Nature Genetics **53**(12), 1631–1633 (2021).
83. Oliva, A., Foare, R., Campbell, P., Twine, N. A., Bauer, D. C. & Johar, A. S. A pangenomic approach to improve population genetics analysis and reference bias in underrepresented middle Eastern and Horn of Africa populations. Biomolecules **15**(4), 582 (2025).
84. Geza, E. et al. A comprehensive survey of models for dissecting local ancestry deconvolution in human genome. Briefings in Bioinformatics **20**(5), 1709–1724 (2019).
85. Geza, E. et al. FRANC: A unified framework for multi-way local ancestry deconvolution with high density SNP data. Briefings in Bioinformatics **21**(5), 1837–1845 (2020).

Chapter 3

Leveraging the Health and Demographic Surveillance System to Set a Genomics Cohort

The Nanoro Experience

Palwendé Romuald Boua[1,2,3*] *and Hermann Sorgho*[1]

[1]Clinical Research Unit of Nanoro, Institut de Recherche en Sciences de la Santé, Burkina Faso

[2]MRC Unit The Gambia at London School of Hygiene and Tropical Medicine, The Gambia

[3]Sydney Brenner Institute for Molecular Biosciences (SBIMB), University of the Witwatersrand, South Africa

3.1 BACKGROUND

A genomic cohort is a group of individuals followed over time to study the relationship between genetic factors and health outcomes, including diseases. These studies aim to identify the genetic factors that influence human health and to understand their interaction with the environment and lifestyle. Genomics cohorts have been instrumental in fostering genomic discoveries, which has led to significant an understanding of biology and diseases mechanisms. Most genomics cohorts landscapes have been dominated by studies in populations of European ancestry. Establishing genomics cohorts in Africa faces significant challenges, including limited infrastructure, lack of expertise, and ethical concerns.

Many African countries lack the necessary laboratory equipment, reliable electricity, and high-speed internet access to support large-scale genomics research. Hospitals are the strategic point where non-genomics data, such as demographic, clinical, phenotypic, environmental, and imaging data, are collected. Laboratories add genomic and other layers of OMICs data to population data. Unfortunately, most African countries lack of synergy between these two players, which is most crucial for the effective setup of genomic cohort.[1] The lack of electronic health records (EHR) systems across clinics, hospitals, and public health agencies has also made difficult the setup of genomic cohorts on the continent. Genomic cohorts require data on participants across the life course, with data on diagnoses, treatments, health outcomes, and laboratory testing.

***Corresponding author:** romyboua@gmail.com (Palwendé Romuald Boua)

 DOI: 10.1201/9781003436584-3

Another challenge is the gap in expertise, due to a shortage of trained scientists, bioinformaticians, and clinicians with expertise in genomics and genetics. In an environment of limited training programs, which address this gap, the brain drain has exacerbated the gap in expertise with the absence of frameworks to retain the few trained personnel. All these factors have hindered efforts to reach the critical mass of expertise on the continent.[1,2]

A history of helicopter science has disrupted trust between communities and researchers in biomedical research. This mistrust is exacerbated in genomics research, with inadequate genomic literacy among communities and public health practitioners. This situation has also made navigation of the ethical considerations around setting up genomic cohorts difficult. Issues with data privacy, informed consent, and benefit sharing are sensitive considering diverse populations and cultural norms.

Other issues, related to lack of funding, language barriers, logistical challenges, and data management infrastructure, have added to the challenges in setting-up genomic cohort in Africa.

In this chapter, we share our experience in setting up a genomic cohort in rural Burkina Faso.

3.2 FROM THE CREATION TO THE ESTABLISHMENT OF THE CLINICAL RESEARCH UNIT OF NANORO

The Clinical Research Unit of Nanoro (CRUN) was born from the dream of a young Burkinabé researcher, Dr. Halidou Tinto, then at the end of his PhD training in parasitology, supported in his ambitions by his home institution, the Institut de Recherche en Sciences de la Santé, Regional Directorate of the West (IRSS-DRO) and his academic training institution, the Institute of Tropical Medicine in Antwerp (IMT). The origins of the CRUN trace back to 2005–2006, in collaboration with the Centre Muraz in Bobo-Dioulasso, as part of the implementation of multiple clinical trials on an antimalarial drug. It was the Saint Camille Medical District Hospital in Nanoro that welcomed the team in its facilities for recruitment and monitoring of participants. This high-standard field hospital offered adequate amenities for the implementation of clinical trials, and the low mobility of the population guaranteed adequate monitoring and a good retention rate of participants in the longitudinal studies. The excellent quality of the implementation and results of these first clinical trials conducted in Nanoro, according to international standards, despite the constraints of the moment, have aroused keen interest from partners in the research team. The team gradually established a fully equipped vaccine trial center with a clinic, a pharmacy, and a medical laboratory that meet international standards. The inauguration of this facility in 2009 by the Minister of Health at the time marked the official birth of the CRUN. The constant growth of the activity portfolio justified and led the Ministry of Research to establish the center as a regional directorate in 2016. Between 2006 and 2024, over 89 studies were conducted by the team.

From an initial interest on infectious diseases, CRUN has over time expanded its focus to include research on non-communicable diseases (NCDs) and genomics, which are becoming an increasing burden in Africa populations, with demographic and epidemiological transitions as well as lifestyles change. While infectious diseases remain a major health threat, there was a growing interest in diseases such as diabetes, hypertension, and cardiovascular disease, which are on the rise in many African countries, including Burkina Faso. CRUN's involvement in NCD and genomics research represents a significant expansion of its mission to address the full range of health challenges facing African populations.

3.3 THE HEALTH AND DEMOGRAPHIC SURVEILLANCE SYSTEM PLATFORM

The Nanoro Health and Demographic Surveillance System (HDSS) (Figure 3.1), located in the rural center of Burkina Faso, was established in 2009 by the CRUN with the aim of providing a core framework for health research and to support the Burkina Faso health authorities in generating epidemiological data that can contribute to the setup and assessment of health interventions. In the baseline of initial census, 54,781 individuals were recorded, of whom 56.1% were female.[3] By 2021, the study population reached 66,500 individuals. After the initial census, vital events such as pregnancies, births, migrations, and deaths have been monitored, and data on individuals and household characteristics are updated during regular 4-monthly household visits. The available data are categorized into demographic, cultural, socioeconomic, and health information, and are used for the monitoring and evaluation of population development issues. This platform has been instrumental in implementing studies by providing a real-time picture of the target population and potential for enrollment. The HDSS platform also facilitates the contact of potential participants through their unique identifiers and the GPS location of their households. This was leveraged to roll out the recruitment of our genomic studies.

The HDSS platform was very helpful in implementing genomic studies in the way information was leveraged. We could identify individuals from the same family, allowing family studies and/or trios for population genetics studies or cases-control studies for diseases. In the case of the AWI-Gen study, this was useful for collecting trios to generate whole genome sequences. On the other hand, for the trait association component (population-based) of the AWI-Gen study, we were interested in enrolling 2000 unrelated participants. Hence, we were able to prevent enrolment of relatives, by limiting the number of individuals recruited by compound or geographic clusters.

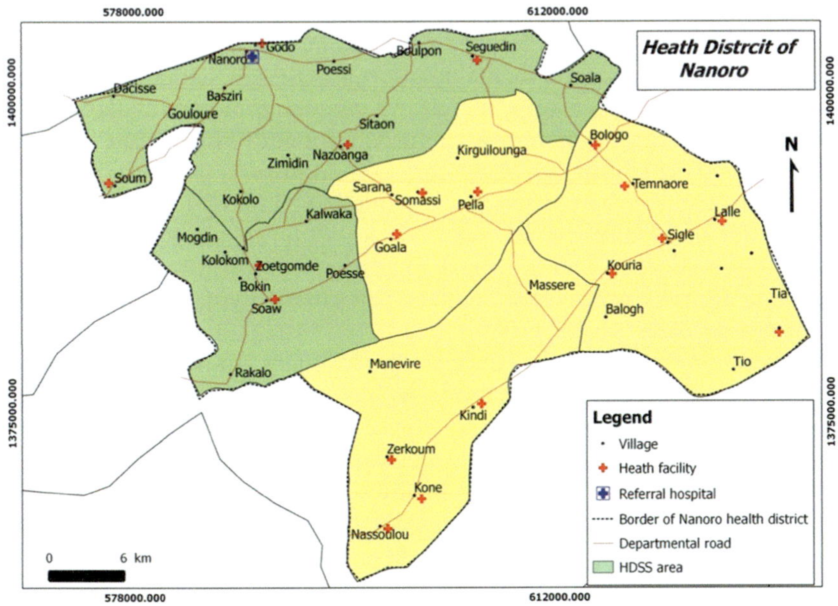

Figure 3.1 Site maps. (A) Location of Burkina Faso in Africa and the Nanoro site area in Burkina Central West region. (B) Nanoro Demographic Surveillance Area.

Located 85 kilometers from Ouagadougou, Burkina Faso, the Nanoro center serves a population of about 160,000 individuals in the Nanoro region, covering 57 villages, where malaria is prevalent during the rainy season stretching from June to November. It also has an HDSS that includes 24 of the villages, which serves as an important research platform.

3.4 A CLOSE PARTNERSHIP WITH THE COMMUNITY: THE KEY TO SUCCESSFUL GENOMICS RESEARCH

One of the characteristics of CRUN is its emphasis on community engagement and partnership. The success of its projects relies largely on the participation and trust of the local population. CRUN has established close ties with the community of Nanoro and the surrounding area, where it conducts many of its studies. These relationships are based on transparency, ethical research practices, and mutual respect. Regular contact is established with community leaders and populations through a team of social scientists and community field agents who play a key role in CRUN's research activities, as intermediaries between researchers and local populations. They implement information/awareness-raising activities, participate in participant recruitment, health education, and monitoring, ensuring that the benefits of the research are shared with the community.

CRUN's approach to ethical research goes beyond participant engagement. Indeed, CRUN has registered compliance with international standards for clinical research, including informed consent, data confidentiality and safety monitoring, a key principle. All research protocols undergo rigorous ethical and regulatory review by the relevant authorities before trials are conducted. This ethical and regulatory commitment is particularly important in a context where many participants have limited formal education and do not always understand the intricacies of clinical trials. By prioritizing clear communication and informed consent, CRUN helps ensure that participants are fully aware of what their participation entails and can make informed decisions about it.

For those reasons, when we initiated the AWI-Gen study on human genetics, we encountered less skepticism. The buy-in for the study was high and the community was happy that we moved beyond infectious disease to do research on NCDs and that the study was recruiting middle-aged adults (40–60 years old), which is how it differed from most previous studies, which had focused on children, adolescents, and pregnant women. The community was aware that genetics (heredity) was a key player of health and diseases and was keen to contribute to knowledge generation.

3.5 AWI-GEN STUDY IN A NUTSHELL

The CRUN was part of the very beginning of AWI-Gen, with its director, Dr, Halidou Tinto, indicating their interest in expanding its research to NCDs and genomics. The Nanoro HDSS area is an ethnically diverse region, with about 90% of people of Mossi ethnicity, ~7.9% Gourounsi, and ~1.7% Fulani. After recruiting 2,095 participants from Nanoro for AWI-Gen Phase 1 (2015–2016) and doing follow-up (6 years later) by recontacting and re-enrolling just over 70% of the participants (2021–2022), individual and group feedback was provided to the communities (Figure 3.2).[4,5] Partly because of the AWI-Gen experience, CRUN now has a series of genomics projects and has developed expertise in bioinformatics, ethical, societal, and legal issues around genomics research, as well as genetic counselling to participants. The AWI-Gen data and sample will be used for many projects into the future, including gene–environment interaction research, by exploring the exposome and its impact on health, which could be modified by genetic variants. The data are also suitable for machine-learning approaches in low- and middle-income countries. AWI-Gen generated an unprecedented resource of genomics in Burkina Faso, which will continue to be used for research and training.

Continued engagement has been implemented throughout the 10 years of the AWI-Gen study to ensure the success.

In 2014, before starting the field activities of the study, a thorough stepwise community and stakeholders' engagement program was implemented. With

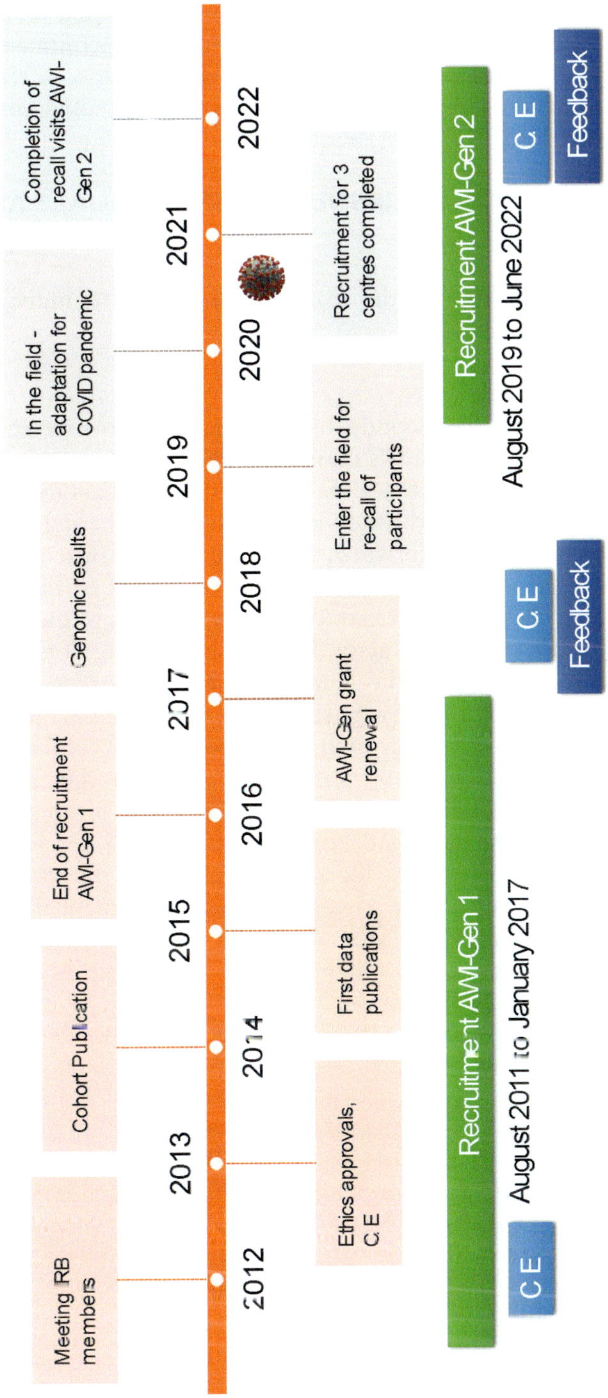

Figure 3.2 AWI-Gen study roll out over ten years, community engagement, and feedback of results.

regards to stakeholders, community leaders were consulted and informed about the upcoming study to seek their support in the recruitment. We also involved ethics committee members to sensitize them in genomic studies since most of the studies they were reviewing did not include genomic components.

3.6 THE FEEDBACK OF RESULTS TO INDIVIDUALS AND COMMUNITY

Following is a summary of the feedback of results to individuals and community.

Preparatory Meetings
Several meetings were conducted in preparation for the results feedback activities. The study's main findings were first presented to the CRUN researchers by the research coordinator. This meeting standardized the content of the individual reports and the aggregate results for the community. Afterward, a meeting was organized between the study site coordinator and the potentials actors of the study. The purpose of the meeting was to discuss the process of the activity and the implication for each actor. During the meeting, lots of concerns were raised, especially the incentives of field workers, the lack of human resources, and how participants in remote villages and those who had moved out of the area will be reached. Several propositions came out and a consensus was found. Later, another meeting was held to meet the field team to finalize the message for the community based on the aggregate results. This meeting involved the field supervisors, field workers, the study coordinator, the social scientist, and the study supervisor. Field workers made practical suggestions on how to relate information about the results and feedback to the communities.

Appointment of Staff
In preparation for the feedback activities, we took advantage of the HDSS's experience to reach the community. The deputy head of the HDSS section and the HDSS senior supervisor played the role of community facilitators because of their close collaboration with the population. In every village belonging to the HDSS catchment area, there is a community informant (a person identified by the community, to be the "eyes and ears of the HDSS" to report every demographic event). An appointment was made, and the community informants informed the village leaders about the upcoming meeting. The villages leaders were informed 2 days prior to the meeting days.

Meetings with Stakeholders (Health District)
Because the health district was a key partner in the study implementation, the aggregate results were presented to them before being

presented to the community. In that way, they could have information about the prevalence of cardiometabolic diseases in their area and act regarding the management of referred participants diagnosed with morbidity. At this meeting, all the healthcare professionals in the area were invited as well with the Health District leadership. After the presentation, a discussion session was conducted.

Meetings with Village Leaders

In every village, the supervisor introduced the team and summarized the purpose of the visit. The deputy of the HDSS and the head of the sociology team explained the purpose of the first phase of the study, followed by the presentation of the aggregate results. The CRUN team stressed that this meeting is also an opportunity for them to formulate their concerns and ask questions. The meetings offered the opportunity for people to ask questions about the study purpose and seek more clarifications concerning cardiometabolic diseases.

Individual Results

Individual feedback report-sharing sessions were scheduled by appointment in each village. Over 1 week, the participants were invited to the peripheral health center to receive their individual results. This process was done to ensure confidentiality.

Community Perspectives

The interviews conducted with the population, following their commitment, showed a satisfaction regarding communication between the research center and the community.

We noted that participants' approach to test results varied from one individual to another. For some participants, the primary interest lay in knowing the diseases they suffer from, and then having adequate free care afterward, if they tested positive. Others thought they would benefit from food for their family (the bread and drink that was offered as a token of appreciation for their participation during the recruitment). As for the reception of the results, many expected to be taken care of in one way or another (cereals and similar benefits from free care). Expectations were met for those who received negative results. and ask if they would receive care if the result was positive or if having these results will allow them to benefit from care in the event of any illness. This raises the question on how to compensate participants taking part of studies in limited-resource settings.

When asked about their willingness to be part of the follow-up visit, responses varied. Following are some comments: "if there is still blood sampling I will not participate", "with age, the human body wears out and weakens, you must take young girls and boys", "if I am still alive I will participate", "we are tired of keeping papers every time when nothing is done for us". Others noted their willingness to participate with conditions: "if they will treat us for free we will participate", or "if they will give us, for example, food that we can be shared with our families, we will participate." Others approved of the efforts that the center made to their community and population in the field of health, stating that the center cares about their health, and as long as they are alive, they will participate in the second phase and other studies.

Despite these mixed perceptions, we achieved a retention rate of 71% for the second visit, which took place 6 years later.

3.7 SCIENCE GENERATED

In recent years, our work has focused on increasing the inclusion of underrepresented populations in genomics research. To do this, we actively participated in the establishment of two research consortia, called H3Africa, or Human Heredity & Health in Africa, and AWI-Gen, or African Wits-INDEPTH Partnership for Genetic Studies.[5] This collaboration has allowed us to generate data on the one hand at the national level and to access data from the entire African continent on the other hand, to compare our data with those of European origin. Also, our work on genetic susceptibility to atherosclerosis has contributed to the knowledge of the pathophysiology of the disease, the first-ever genome-wide association study of atherosclerosis in Africa, which identified novel Africa-specific variants in atherosclerosis susceptibility genes (carotid intima-media thickness [cIMT]) with sex differences and showed the involvement of the estrogen response pathway in atherosclerosis susceptibility.[6] We reported for the first time the enrichment of the estrogen response pathway in atherosclerosis-associated genes. The study of gene-smoking interactions in West African men identified novel atherosclerosis susceptibility genes and showed that gene–environment interaction depends on the intensity of exposure.[7] Moreover, we showed through a stratified analysis that genetic susceptibility is influenced by health status and that some genetic pathways are activated by conditions such as obesity and hypertension. Furthermore, we were able to establish that LDL-C is a causal risk factor for atherosclerosis (cIMT) in individuals of European and African ancestry, following a Mendelian randomization study. This study also provided evidence that type 2 diabetes is a risk factor for cIMT in Europeans, although we found no supporting evidence in individuals of African ancestry. These data are consistent with those risk factor optimization strategies used to combat atherosclerosis in populations of European ancestry that are also applicable to populations of African ancestry.[8] Prior to our work on genetic susceptibility, we attempted to understand the epidemiology of atherosclerosis across the continent, which allowed us to understand regional differences in carotid wall thickness.[9] Our efforts to increase the representation of the African population in global genomic studies led us to engage in a global multi-ancestry analysis. Our work also includes the first genetic susceptibility to smoking,[10,11] and to other cardiometabolic traits such as lipids and arterial hypertension.[12,13] Besides, in the context of genetic susceptibility, we studied the effect of inbreeding on cardiometabolic diseases. We found that the effect of autozygosity on these phenotypes appears to be sex related, with inbreeding having a significant decreasing effect in males but a significant increasing effect in females for several traits (body mass index, subcutaneous adipose tissue, low-density lipoproteins, and total cholesterol). Overall, the effect of inbreeding is more intense in males.[14] We also worked on genetic susceptibility to kidney disease. Recessive inheritance of Africa-specific *APOL1* gene variants is associated with a higher risk of non-diabetic kidney disease, progression to kidney failure, and early albuminuria that precedes decline in estimated glomerular filtration rate. We found that *APOL1* G1 and G2 alleles and frequencies of high-risk genotypes differed between and within West and South African populations and were almost absent from the East Afrin population. *APOL1* risk variants were associated with albuminuria but not with estimated glomerular filtration rate greater than 60 ml/min per 1.73 m^2.[15,16] The Nanoro team contributed with wholes genome sequences from the trios' study, to increase representation of understudied populations, inform on demographic history, and contribute to the design of the H3Africa genotyping array.[17]

3.8 CHALLENGES

To roll out adequate recruitment with reduced time for participants at the clinic when attending study procedures, we set a turnaround time of 1 hour 45 minutes per participant. Participants were transported to the research center and back at their villages. Most of the villages included in the study were at most at a 40-minute drive from the center, leading to lesser burden for the participants.

Setting adequate laboratories for the genomic study was a challenge. The center, however, upgraded some equipment specifically to process the huge amount of sample and allow extraction of buffy coat with mobile rotor centrifuge machine.

Because of the low level of literacy in the study area, informed consent and interviews were made in local languages. For informed consent, we used an impartial witness from the community, who could certify that the information provided to the participants before signing the informed consent forms was accurately translated. Before starting the interviews, the questionnaire was assessed by a panel of clinicians and researchers, who standardized the translation of all terminologies into the two major languages spoken in the area (Mooré and Gourounsi). They also provided the team with adequate formulation of questions in local languages. In case participants and/or research assistants could not speak the same language, we identified a translator to assist during all the process.

3.9 PERSPECTIVES AND FUTURE DIRECTIONS

In line with ongoing research at our institution, we propose the following:

- Further our work on the genetic epidemiology of atherosclerosis by developing polygenic risk scores to predict cardiovascular events (stroke, myocardial infarction, etc.) based on variants identified during our previous analyses.
- Undertake the analysis of genetic susceptibility to alcoholism (alcohol consumption and cessation of consumption) to complete investigations on the alcohol–tobacco couple.
- Extend our work on the study of gene–environment interactions that influence cardiometabolic traits, the preliminary results. (These were later presented at the annual conference of the American Society of Human Genetics in 2023.)
- Carry out an in-depth analysis of the whole genome sequences generated as part of the AWI-Gen study. These analyses will focus on the fine structure of the population and on demographic history. Aware of the influence of epigenetic changes on susceptibility to cardiometabolic diseases, we plan to study DNA methylation profiles that may impact metabolic traits and/or be markers of exposure to environmental factors.
- Pursue work on the identification of genes responsible for hereditary heart diseases in Burkina Faso, using targeted sequencing (Illumina) and long-strand sequencing (Oxford Nanopore). We hope that the results will contribute to the design of diagnostic tests for hereditary heart diseases.

More generally, Africa's genetic diversity provides a unique opportunity to identify novel genes and understand diseases susceptibility. Genomic cohorts will advance research and lead to precision medicine, with more effective prevention and treatment measures for prevalent infectious and non-communicable diseases in Africa. Initiatives to build and retain skills will contribute to increasing the critical mass of researchers and foster a thriving community, along with collaboration to facilitate knowledge transfer and access to advanced technologies.

Considerations for setting up genomic cohorts should include the following:

- ***Invest in infrastructure:*** Prioritize funding for the development of genomics research centers and laboratories across Africa.
- ***Strengthen capacity building:*** Implement training programs for researchers, technicians, and clinicians in genomics and bioinformatics.
- ***Establish ethical frameworks:*** Develop clear guidelines for data management, patient consent, and benefit sharing.
- ***Promote international and South–South collaborations:*** Encourage partnerships between African countries and international research institutions.

- ***Raise awareness:*** Educate the public and policymakers about the potential benefits of genomics research.
- ***Address logistical challenges:*** Develop innovative solutions for sample collection, transportation, and data management.
- ***Foster local expertise:*** Support the development of genomics centers of excellence and encourage local scientists to lead research efforts.

3.10 CONCLUSION

Overall, setting up the genomic cohort of the AWI-Gen study in the Nanoro HDSS was a tremendous experience and a success. The study was the first of its kind in Burkina Faso and has contributed to generating valuable data, which will contribute to increase diversity in genomic studies. The resources and expertise acquired throughout the years will be used to train the next generation of African genomics researchers. With over 70 scientific publications to which the data from Nanoro have contributed, much more usage of the resources will be of benefit for the genomic community.

REFERENCES

1. Olono A, Mitesser V, Happi A, Happi C. Building genomic capacity for precision health in Africa. *Nat Med.* 2024;30(7):1856–1864. https://doi.org/10.1038/s41591-024-03081-9
2. Wonkam A, Munung NS, Dandara C, Esoh KK, Hanchard NA, Landoure G. Five Priorities of African Genomics Research: The Next Frontier. *Downloaded from www.annualreviews.org* Guest (guest). Published online 2022. https://doi.org/10.1146/annurev-genom-111521
3. Derra K, Rouamba E, Kazienga A, et al. Profile: Nanoro health and demographic surveillance system. *Int J Epidemiol.* 2012;41(5):1293–1301. https://doi.org/10.1093/ije/dys159
4. Ali SA, Soo C, Agongo G, et al. Genomic and environmental risk factors for cardiometabolic diseases in Africa: Methods used for Phase 1 of the AWI-Gen population cross-sectional study. *Glob Health Action.* 2018;11(sup2). https://doi.org/10.1080/16549716.2018.1507133
5. Ramsay M, Crowther N, Tambo E, et al. H3Africa AWI-Gen collaborative centre: A resource to study the interplay between genomic and environmental risk factors for cardiometabolic diseases in four sub-Saharan African countries. *Glob Health Epidemiol Genom.* 2016;1(e20):e20. https://doi.org/10.1017/gheg.2016.17
6. Boua PR, Brandenburg JT, Choudhury A, et al. Genetic associations with carotid intima-media thickness link to atherosclerosis with sex-specific effects in sub-Saharan Africans. *Nat Commun.* 2022;13(1):1–11. https://doi.org/10.1038/s41467-022-28276-x

7. Boua PR, Brandenburg J, Choudhury A, et al. Novel and known gene-smoking interactions with cIMT identified as potential drivers for atherosclerosis risk in West-African populations of the AWI-Gen Study. *Front Genet.* 2020;10(1):100004. https://doi.org/10.3389/fgene.2019.01354
8. Soremekun O, Slob EAW, Burgess S, Romuald Boua P, Fatumo S, Gill D. Genetically predicted lipid traits, diabetes liability, and carotid intima-media thickness in African ancestry individuals: A mendelian randomization study. *Circ Genom Precis Med.* 2022;15(6):E003910. https://doi.org/10.1161/CIRCGEN.122.003910
9. Nonterah EA, Boua PR, Klipstein-grobusch K, et al. Classical cardiovascular risk factors and HIV are associated with carotid intima-media thickness in adults from sub-Saharan Africa: Findings from the H3Africa AWI-Gen study. *J Am Heart Assoc.* 2019;(8). https://doi.org/10.1161/JAHA.118.011506
10. Boua PR, Soo CC, Debpuur C, et al. Prevalence and socio-demographic correlates of tobacco and alcohol use in four sub-Saharan African countries: A cross-sectional study of middle-aged adults. *BMC Public Health.* 2021;21(1):1–14. https://doi.org/10.1186/s12889-021-11084-1
11. Piga NN, Boua PR, Soremekun C, et al. Genetic insights into smoking behaviours in 10,558 men of African ancestry from continental Africa and the UK. *Sci Rep.* 2022;12(1). https://doi.org/10.1038/s41598-022-22218-9
12. Singh S, Choudhury A, Hazelhurst S, et al. Genome-wide association study meta-analysis of blood pressure traits and hypertension in sub-Saharan African populations: An AWI-Gen study. *Nat Commun.* 2023;14(1). https://doi.org/10.1038/s41467-023-44079-0
13. Choudhury A, Brandenburg JT, Chikowore T, et al. Meta-analysis of sub-Saharan African studies provides insights into genetic architecture of lipid traits. *Nat Commun.* 2022;13(1). https://doi.org/10.1038/s41467-022-30098-w
14. Ceballos FC, Hazelhurst S, Clark DW, et al. Autozygosity influences cardio-metabolic disease-associated traits in the AWI-Gen sub-Saharan African study. *Nat Commun.* 2020;11(1). https://doi.org/10.1038/s41467-020-19595-y
15. Brandenburg JT, Chen WC, Boua PR, et al. Genetic association and transferability for urinary albumin-creatinine ratio as a marker of kidney disease in four Sub-Saharan African populations and non-continental individuals of African ancestry. *Front Genet.* 2024;15. https://doi.org/10.3389/fgene.2024.1372042
16. Brandenburg JT, Govender MA, Winkler CA, et al. Apolipoprotein L1 high-risk genotypes and albuminuria in sub-Saharan African populations. *Clinical Journal of the American Society of Nephrology.* 2022;17(6):798–808. https://doi.org/10.2215/CJN.14321121
17. Choudhury A, Aron S, Botigué LR, et al. High-depth African genomes inform human migration and health. *Nature.* 2020;586(7831):741–748. https://doi.org/10.1038/s41586-020-2859-7

Chapter 4

Pioneering Genomics Research in Africa

Insights from a Longitudinal Ugandan Cohort

Joseph Mugisha Okello[1*] *and Segun Fatumo*[1,2]

[1]MRC/UVRI and LSHTM Uganda Research Unit, Plot 50-55, Ntiko road, Entebbe, Uganda

[2]Precision Healthcare University Research Institute, Queen Mary University of London

4.1 INTRODUCTION

Longitudinal population-based cohorts serve as a key resource for genomics research worldwide, but they are still disproportionately few in Africa. These cohorts offer researchers a unique opportunity to examine gene–environment interactions, follow disease patterns over time, and identify genetic factors influencing health in diverse populations. In low- and middle-income countries (LMICs), including those in Africa, these resources are essential for shaping public health strategies that tackle both infectious and non-communicable diseases. Creating and sustaining these cohorts is both scientifically logical and ethically necessary, considering the historical lack of representation of African populations in genomic studies.

Africa hosts the most genetically diverse populations on the planet [1], which presents a great opportunity to discover new disease-related genetic variants and rare alleles. However, genomics research often focuses mainly on individuals of European descent [2,3]. This lack of representation has resulted in skewed risk prediction models, treatment guidelines that do not apply universally, and missed chances for discoveries that could benefit all human populations [4,5]. By establishing longitudinal cohorts like the General Population Cohort (GPC) [6] in Uganda, researchers can gather extensive phenotype and environmental data over time. This data will allow for better interpretation of genetic association and the creation of fair precision medicine solutions. There is a growing recognition that genomics research in Africa is vital for understanding the unique genetic makeup of its populations and for tackling region-specific health issues [7]. With advancements in sequencing technologies and lower costs, African institutions can now undertake detailed studies that examine both rare and common variants among

*__Corresponding author:__ Joseph.mugisha@mrcuganda.org (Joseph Mugisha Okello)

DOI. 10.1201/9781003436584-4

different populations [8,9]. Conducting genomics research in Africa allows for the development of more tailored diagnostics, treatments, and preventive measures that address the distinct needs of these populations. This work is critical to advance health equity and enhances global scientific knowledge. Additionally, it builds local capacity, develops infrastructure, and encourages fair research partnerships [10]. This chapter shares insights from conducting genomic studies within the GPC. We start by outlining the study setting and methods, then detail the processes of community engagement, ethical issues, and genomics data collection. We also highlight significant scientific findings achieved with GPC data and discuss the challenges and opportunities that have arisen. This case study offers valuable lessons for other African cohorts aiming to make meaningful contributions to global genomics.

4.2 STUDY SETTING

The GPC is a landmark longitudinal research platform located in Kyamulibwa Subcounty, Kalungu District, in rural southwestern Uganda [6]. Established in 1989 through a collaboration between the UK Medical Research Council (MRC) and the Uganda Virus Research Institute (UVRI), the GPC began as an HIV-focused surveillance study and has since evolved into a robust, multi-purpose population cohort. Today, it covers 25 villages and approximately 30,000 individuals, making it one of the most enduring and expansive research platforms on the continent.

What makes the GPC especially powerful is its ability to capture rich, repeated measurements of health, behavior, and sociodemographic factors in a population living under epidemiological transition. The area faces a dual burden of disease, including persistent infectious illnesses such as malaria, HIV, and tuberculosis, and an increasing prevalence of non-communicable diseases (NCDs) like diabetes and cardiovascular conditions [6]. This dual burden renders the GPC particularly relevant for genomic studies aimed at uncovering both genetic and environmental drivers of disease.

Each village in the GPC is governed by a local council 1 chairperson and a committee of elected representatives, alongside traditional leaders and village health teams. These community-based structures have fostered deep trust and engagement between researchers and residents, which proved critical for the successful introduction of genomic studies. To further ensure this was successful, the study site is located approximately 200 kilometers southwest of Kampala, an area that is accessible via well-connected roadways. It is situated near Lake Victoria and within reach of Masaka City, a major urban hub offering medical, commercial, and logistical infrastructure that supports field operations.

Ethnically, the GPC is predominantly Baganda (about 75%), with substantial Banyarwanda and Barundi minorities, and smaller groups from Tanzania and other Ugandan ethnicities, representative of the Ugandan population. Luganda is the lingua franca, enabling streamlined communication and community mobilization throughout the research process.

Since its inception, the GPC has conducted over 30 census rounds and 27 medical surveys, making it a rich data source for longitudinal epidemiology. The cohort has been instrumental in research on HIV/AIDS, fertility, migration, and more recently, NCDs. Genomic studies, introduced in 2011, have built upon this strong infrastructure, opening new frontiers in population science.

Healthcare in the GPC area is provided through the following facilities:

1. A dedicated GPC Clinic offering free services, including antiretroviral therapy (ART)
2. Government-run health centers (with limited supplies)
3. Private clinics operating on a fee-for-service basis
4. NGOs delivering subsidized or free services

Despite this structure, public health facilities are often under-resourced, highlighting the importance of the GPC Clinic as a consistent and trusted provider.

In summary, the GPC represents a bold and successful integration of community engagement, scientific rigor, and long-term vision. Its rural yet well-connected location, trusted social structures, and epidemiological diversity make it an ideal platform for genomics research and a compelling model for similar efforts across Africa.

4.3 STUDY METHODS

Our approach to participant engagement, a critical determinant of success, became a masterclass in culturally sensitive communication. From securing the unwavering support of Local Council leaders through formal letters and megaphone announcements, to conducting open community meetings that demystified genomic concepts and pro active participant notification, every step was meticulously crafted to build trust and ensure truly informed participation

The study methods within the cohort are summarized in Figure 4.1. Study activities start with participant engagement and recruitment into the cohort. This is followed by village mapping, collection of census data, collection of survey data, sample testing and data analysis, and finally feedback of study findings to the study participants [6].

4.3.1 Participant Engagement and Recruitment into the GPC

Each village's Local Council leaders serve as representatives of their community. The main governance bodies in the study villages are Local Village

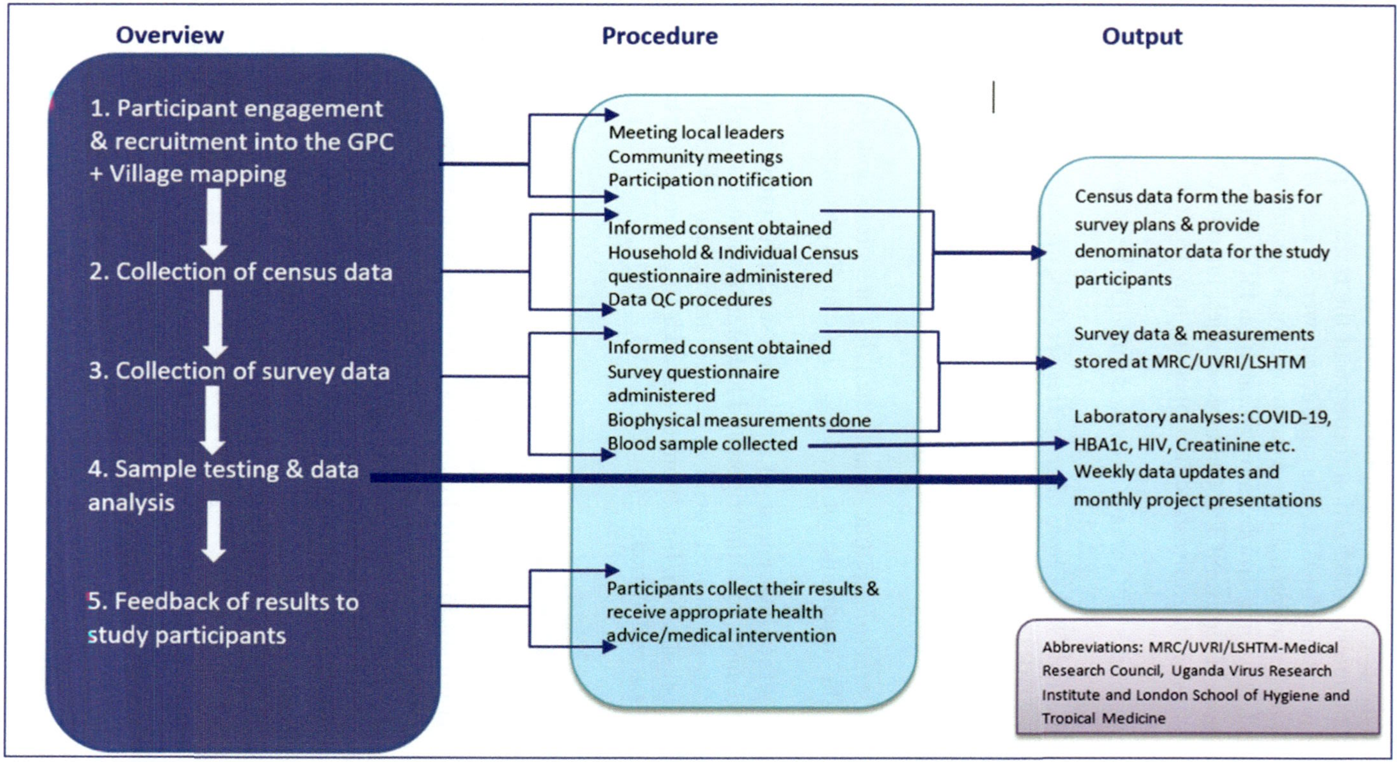

Figure 4.1 Field activities of the GPC Census and Survey

Councils, which are leadership groups usually comprising of around ten members who represent various community departments. Securing support from these groups is essential before any GPC activity is undertaken.

For the first genomic studies within the GPC [9,11], Local Council leaders played an important role in mobilizing communities in preparation for the study. Letters from the GPC program were sent to each village's Local Council executive a few days before the arrival of the census team and one month before the arrival of the survey teams in the area. The letters detailed the reason for the study, when the various study teams would arrive in that village, the duration of planned work, and served as an important channel of communication between the community and the research team.

In addition to the Local Council letters, a series of open community meetings were held to increase awareness and understanding of the study. These meetings explained to village residents the purpose of the GPC, the value of their participation in the genomic study, and the roles of the different members of the study teams. The community meetings took place in the first week of the survey team's field activities in the respective village. In conjunction with the Local Council communication officer, the GPC team members toured the village using a megaphone to publicize the meeting 1 day beforehand. Preparations for these meetings ensured that the equipment (video van, projector, PA system, generator, etc.) was ready and in operational order.

The villages considered were recruited into the GPC once their Local Councils approved participation. After this, participant engagement began directly, by first notifying them, and then visiting of their homesteads from GPC mobilization teams. These notification procedures were designed to maximize participant availability when the census teams arrived, as well as to provide an opportunity to answer any questions about the GPC or the upcoming genomic studies.

We took particular care in obtaining informed consent from study participants. For the survey, only participants aged 13 years or older were eligible to participate. Eligible adult participants provided informed consent. For participants under 18, parental consent and formal assent were obtained. Consent forms were translated from English to Luganda (the language used in the study setting) and then back to English.

4.3.2 Village Mapping

A mapper from the GPC visited each village to update its map and provide accurate geographical information for the census and survey teams. It took approximately 2 weeks to map a village. The mapper first obtained agreement from the locally elected village leader that mapping can begin. The village leader accompanied the mapper throughout the process. The mapper recorded the boundary and principal geographical features and, using global positioning system, a revised map was developed, on which buildings and key features, for example, roads, are indicated. For each building located, the mapper recorded the building use or uses, that is, residential and/or non-residential uses and assigns identifiers. For residential buildings, the mapper also recorded the unique identifier number of each household(s) member residing in that building. Updating the map involved the following activities: Adding any new house structures, removing demolished structures indicating any changes of features, for example, if a footpath was changed, and recording the head of household. Copies of the updated maps and related information were then provided for the census and survey teams.

4.3.3 Census

The GPC census consisted of two stages: Obtaining informed consent from the households and collecting questionnaire data. Every household recorded on the updated map was eligible to take part in the census. If the head of the household agreed to take part in the census, informed consent was obtained by a census team member in the presence of the village leader. If the head of the household was absent, another adult member (aged 18 years or older) was asked to provide informed consent on behalf of the household. A signed copy of the consent form was retained by the study participant. Both a paper and scanned electronic copy of the consent form were kept at the GPC field station in a secure archive. Census data were collected following standard operating procedures, using a structured standard questionnaire loaded onto a tablet. The census team administered the questionnaires by face-to-face interviews in participants' homes, accompanied by the village leader. The questionnaire comprised of household information and individual information, which was completed for every household member [6].

4.3.4 Medical Survey

The medical survey followed the census. The survey consisted of four stages. These included (1) obtaining informed assent/consent from survey participants, (2) collecting questionnaire data, (3) doing biophysical measurements, and finally (4) collecting blood or any other samples for that specific survey round.

When genomic studies were introduced in the GPC survey, blood was the only specimen obtained from the study participants. To receive consent for participation in this round, we obtained the following for each survey round: Consent to answer survey questions and biophysical measurements, consent to give blood samples, consent for a blood sample to be stored for future use, and consent for the participants' blood sample to be used for genomic studies. Both paper and scanned electronic copies of these consents were kept securely at the GPC field station, and each study participant also took their own copies.

Once the survey team staff received informed consent, they administered the survey questionnaire by face-to-face personal interviews at village central hubs. Through the questionnaire we gathered information on sociodemographic indices, lifestyle and medical history, and other risk factors. After the questionnaire interviews, they gathered biophysical measurements, including weight, height, blood pressure, waist, and hip circumference. The final stage of the survey involved the collection of random (i.e., non-fasting) venous blood samples from consenting participants. Aseptic venipuncture was performed via a closed evacuated blood collection system, using standardized procedures to minimize the risk of infection. Blood tubes were labelled with unique identification numbers, linking the samples to an individual participant.

4.3.5 Sample Handling

After drawing blood samples from study participants, the labelled blood tubes were inverted 5–6 times to ensure mixing of clot activator with blood (plain serum tube) or mixing of EDTA anticoagulant with blood to prevent clotting (EDTA whole blood tube). The blood tubes were immediately stored in cold boxes fitted with ice packs at the village hub. The cold boxes were maintained at 4–8 degrees Celsius. At the end of each day of field activities in the study villages, the cold boxes were taken back to the laboratories at the Kyamulibwa field station and logged onto a sample database. The EDTA tube was immediately stored in central refrigerators at 4–8 degrees Celsius and the serum tube was centrifuged before being stored in liquid nitrogen at −196 degrees Celsius. Samples were transported to Entebbe MRC/UVRI and LSHTM laboratories monthly [6]. For the first genomic studies within the GPC [9,11], the samples were prepared and transported to the Sanger Institute in the United Kingdom, where sequencing for the samples was performed. The handling and transportation of these samples for sequencing was done in accordance with the materials transfer agreement approved by the Uganda National Council for Science and Technology.

4.4 ETHICS

The study was approved by the Research and Ethics Committee (UVRI-REC #HS 1978), the Uganda National Council for Science and Technology (UNCST #SS 4283), and the East of England-Cambridge South (formerly Cambridgeshire 4) NHS Research Ethics Committee UK.

Genomic studies were introduced in 2011 during Survey Round 22. With recognition of the novelty and sensitivity of this research, significant emphasis was placed on community engagement. The GPC became the first platform in East Africa to integrate relatively large-scale genomic studies into an existing longitudinal cohort. Sample collection, storage, and transport adhered to stringent international standards, and sequencing was conducted at the Wellcome Sanger Institute in the UK.

Although previous small-scale efforts had extracted DNA from archived samples, this marked the first major genomics research initiative at the GPC and indeed, within the MRC Unit Uganda as a whole. The integration of genomics research into this longstanding cohort was a landmark achievement, not only for the MRC Unit but also for genomic science in the East African region.

This pioneering effort required careful navigation of complex ethical terrain, especially around informed consent, community engagement, international data sharing, and equitable benefit sharing. These are now widely recognized as central issues in responsible genomics research, especially in low-income settings

The GPC genomic studies were designed to investigate the genetic underpinnings of both infectious and chronic diseases in African populations, providing insights into disease etiology and potential therapeutic targets [11]. At the time (2010–2011), most large-scale genetic epidemiological research had focused on populations of European or Asian ancestry. Including African populations was essential to broaden the generalizability of findings and identify novel disease susceptibility loci.

In the following section, we share our experiences of implementing genomic studies within the GPC. We highlight key ethical issues and concerns, describe the study setting and methodologies, and outline the community mobilization and engagement strategies that were critical for the study's success. Finally, we reflect on the challenges encountered during the process and the strategies we used to address them.

4.5 ESTABLISHING GENOMICS RESEARCH IN THE GPC: FROM CONCEPT TO REALITY

The integration of genomics research into the GPC marked a groundbreaking milestone for biomedical science in East Africa and an important expansion of the scientific scope of the cohort. Building on the GPC's existing infrastructure and longitudinal design, genomic integration enabled the cohort to become first platform in East Africa and one of the few in sub-Saharan Africa to merge longitudinal phenotypic and socio-demographic data with genome-wide genotyping.

This achievement was possible due to the years of cumulative trust-building between researchers and the local community, investment in health infrastructure, and collaboration with international partners such as the

Wellcome Sanger Institute. The decision to collect genomics data was driven by a need to increase representation of African population in global genomics and to investigate the genetic determinants of diseases common in the region.

From its inception, this effort was built on three main pillars: Pragmatism, cultural intelligence, and a deep respect for community values. To do this, a multilayered engagement strategy was developed based on the cultural structures in place in Uganda, making use of traditional leadership, village councils, and public forums.

4.6 ETHICAL GROUNDWORK AND GOVERNANCE

The ethical landscape surrounding genomic studies in Africa is complex and evolving. At the time of implementation, questions around data sovereignty, privacy, re-consent, and the return of results were still being debated globally, making it even more critical for the GPC team to approach these issues proactively.

In our early community meetings, participants raised key questions: "Who owns my DNA?" "Will I be told if my sample reveals disease risk?" "Can I withdraw after my sample has been sequenced?" These questions reflected the broader community anxieties rooted in historical scientific exploitation. The GPC team addressed these questions openly through public dialogues during community meetings.

While the GPC already had procedures in place for informed consent and community engagement, additional safeguards were introduced to address the specific sensitivities associated with genetic data. These included detailed consent protocols for the collection, long-term storage, and international sharing of biospecimens for future genomic analyses. A landmark aspect of the GPC genomic project was its material transfer agreement, with the Wellcome Sanger Institute. This established shared ownership of data and results, ensuring that the MRC/UVRI and LSHTM Uganda Research Unit maintained stewardship over African data. Today, this database is locally stored and managed at the Unit's bioinformatics hub and remains accessible to researchers under well-regulated conditions.

4.7 COMMUNITY ADVISORY BOARD AND CAPACITY BUILDING

Community engagement and trust, built over many years, was critical to the success of the genomic program. Initially the GPC lacked a Community Advisory Board (CAB); however, this was recognized early as a critical gap. In response, a multi representative CAB was established, comprising elders, religious leaders, youth advocates, women's groups, and other key voices from the 25 participating villages.

The CAB now plays a central role in guiding research priorities, resolving ethical concerns, improving study design, and disseminating research findings. Its contributions have strengthened transparency and accountability and exemplifies what participatory science should look like. Capacity building was equally foundational to the success of the genomic program. Training went beyond lab techniques. Field staff were educated in ethical genomics, basic genetic literacy, community dialogue, and digital sample management. The GPC also served as a pipeline for building local bioinformatics expertise, through fellowships, workshops, and regional consortia (e.g., H3Africa). Importantly, the genomic project catalyzed infrastructural investments including cold-chain logistics, data servers, and improved sample tracking systems, all of which have also benefitted the broader research environment both within the GPC and beyond. It demonstrates that genomics could uplift broader scientific systems when implemented thoughtfully.

4.8 THE GPC AS A MODEL FOR GENOMICS RESEARCH IN AFRICA

The GPC's genomics journey has become a scalable model for other African countries aspiring to conduct ethical, inclusive, and impactful genomics research.

First, it shows that successful genomics begins with community trust. Genomics cannot be imported from the Global North; it must be grounded in the lived realities, values, and languages of the people.

Second, it demonstrates the value of long-standing cohorts as platforms for cutting-edge science. The GPC's ability to link genetic data with decades of phenotypic and socio-environmental information is a unique advantage.

Third, the program affirms that ethical genomics requires governance frameworks tailored to reflect local realities. This includes local data storage, African-led data access committees, and transparent benefit-sharing mechanisms.

Lastly, the GPC has proven that Africa can lead in genomic discovery. The continent is not merely a sample source but home to the world's oldest and richest human diversity. With initiatives like GPC leading the way, Africa can be at the forefront of the genomic revolution.

4.9 SCIENTIFIC CONTRIBUTIONS OF THE GPC IN GENOMICS

The scientific outputs emerging from the GPC's genomic program have been transformative, not only for understanding health in African populations, but also for advancing global genomic science. Leveraging high-quality phenotypic, demographic, and environmental data, GPC-based studies have produced some of the most extensive genomic analyses in sub-Saharan Africa to date.

One major contribution was the genome-wide association study (GWAS) of 34 cardiometabolic traits including lipid levels, body mass indices, blood cell counts, and HbA1c conducted across multiple African populations (Uganda, Ghana, Kenya, Nigeria, and South Africa) [9]. This study identified 43 novel genetic loci and proposed a revised genome-wide significance threshold ($p < 5 \times 10^{-9}$) more appropriate for African populations due to lower linkage disequilibrium and greater genetic diversity [9]. This work demonstrated the power of pan-African collaboration and the importance of ancestry-specific analysis strategies.

The GPC also enabled the first-ever GWAS of kidney function in continental Africa. This study validated two loci *GATM* and *HBB* previously reported in other populations, thereby confirming their relevance across ancestries and reinforcing the utility of African cohorts in validating global findings [12,13].

Beyond single-trait studies, the GPC pioneered multivariate GWAS approaches, particularly for blood cell traits [14]. These analyses uncovered pleiotropic loci genetic variants influencing multiple traits and showcased how multivariate models can outperform traditional univariate methods in detecting novel associations. The implications are profound: In populations with complex disease burdens, multivariate methods offer a pathway to uncover shared biological pathways.

In partnership with a South African Zulu cohort, GPC data supported the first GWAS of body shape phenotypes among Africans [15]. Novel loci such as *FGF12*, *GRM8*, *TLX1NB*, and *TRAP1* were identified, with findings showing associations between body shape and cardiometabolic markers, including blood pressure and lipid profiles.

GPC data have also contributed to the development and evaluation of polygenic risk scores (PRS) for lipid levels and type 2 diabetes [16]. With the integration of genetic information from African American cohorts, the study demonstrated enhanced prediction accuracy for sub-Saharan African populations, though regional variation remains a challenge [16,17]. This work highlighted both the promise and the limitations of cross-ancestry PRS translation, emphasizing the need for more African-ancestry-specific discovery datasets.

Lastly, Mendelian randomization studies using GPC data have shown causal relationships between lipid traits and both type 2 diabetes, kidney function, and stroke [18–20]. These findings, including other bioinformatic work [21–30], move beyond association toward inference of causality, reinforcing the translational potential of genomic studies for precision public health. Together, these contributions position the GPC not only as a local asset but as a globally impactful genomics research platform. They also affirm the central thesis of this chapter: That sustained investment in African cohorts can produce world-leading science, reshape biomedical paradigms, and ensure that the genomics revolution is truly global in scope.

4.10 OPPORTUNITIES AND CHALLENGES

Conducting genomics research in rural African settings highlighted several systemic, cultural, infrastructural, and scientific hurdles and, however, unveiled transformative opportunities. A key challenge was the lack of terminology in local languages, for many genomic concepts, which initially complicated informed consent processes. This led to interdisciplinary collaborations with linguists at Makerere University, who co-developed locally resonant translations and analogies. These are now reference points for other genomic efforts across the region.

Managing expectations around individual feedback also presented another challenge. In previous studies, GPC participants had routinely received medical test results. Genomics research, however, often doesn't yield directly actionable results at the individual level. Addressing this required clear communication strategies to ensure participants understood the objectives and limitations of discovery-based science.

Ensuring equitable data governance emerged as another major hurdle. In the past, African-generated genomics data have often been housed and analyzed in high-income countries. The GPC disrupted this pattern by establishing a pioneering data-sharing and joint data custodianship model with the Wellcome Sanger Institute. This framework enabled equitable access and local leadership in data analysis. It also encouraged investment in the local bioinformatics infrastructure and research capacity, which now benefits numerous studies.

Operational constraints, such as unreliable electricity, intermittent internet outages, and gaps in laboratory logistics, were practical as well as persistent constraints. These were addressed through infrastructural upgrades, including solar-powered freezers, improved sample tracking systems, and robust data management pipelines, which now serve multiple health research platforms.

There were regulatory and bureaucratic challenges too. Aligning the standards and timelines of international ethics boards with national regulations required diplomacy, patience, and adaptability. These early efforts helped harmonize ethical review processes and opened new channels of trust between Ugandan and international oversight bodies.

Human resource gaps in areas such as bioinformatics and genomics interpretation posed another barrier. This gap was mitigated through postgraduate training, mentorship programs, regional workshops, and collaborations with global consortia such as H3Africa. Today, GPC alumni and staff are active contributors to global genomic science.

Finally, community trust had to be continually earned. The creation of a diverse and representative CAB has since strengthened the legitimacy, transparency, and responsiveness of all ongoing genomic work.

The GPC genomic study confronted and transformed a wide spectrum of challenges into sustainable scientific and structural opportunities. These experiences should inform and inspire other initiatives striving to advance equitable genomic science across the African continent.

4.11 CONCLUSION

This chapter has traced the GPC's remarkable journey from its origins as an HIV cohort to its evolution into a flagship site for large-scale genomics research in East Africa.

The GPC's genomic program provides a valuable example of how African-led population cohorts can meaningfully contribute to the global genomics agenda. It shows that with appropriate investment in infrastructure, governance, and local capacity, it is possible to conduct high-quality genomics research that is both scientifically rigorous and ethically governed.

The impact of this work is already evident in globally relevant scientific discoveries, which has strengthened institutional capacity, empowered African scientists, and engaged communities.

As precision medicine continues to expand globally, the GPC model offers a path forward for equitable, inclusive, and scientifically rigorous genomics. It proves that the future of genomics should reflect and benefit the global population and be a combination of both scientific excellence and community empowerment.

In the years ahead, cohorts like the GPC will shape the trajectory of genomic science through the data they generate and the values they uphold. The call to action is clear: Build ethically, engage deeply, share equitably, and lead boldly.

REFERENCES

1. Tishkoff, S. A., & Verrelli, B. C. (2003). Patterns of human genetic diversity: Implications for human evolutionary history and disease. *Annual Review of Genomics and Human Genetics*, *4*(1), 293–340. https://doi.org/10.1146/annurev.genom.4.070802.110226
2. Popejoy, A. B., & Fullerton, S. M. (2016). Genomics is failing on diversity. *Nature*, *538*(7624), 161–164. https://doi.org/10.1038/538161a
3. Fatumo, S., Chikowore, T., Choudhury, A., Ayub, M., & Martin, A. R. (2022). A roadmap to increase diversity in genomic studies. *Nature Medicine*, *28*(2), 243–250. https://doi.org/10.1038/s41591-021-01672-4

4. Martin, A. R., Kanai, M., Kamatani, Y., Okada, Y., Neale, B. M., & Daly, M. J. (2019). Clinical use of current polygenic risk scores may exacerbate health disparities. *Nature Genetics*, *51*(4), 584–591. https://doi.org/10.1038/s41588-019-0379-x
5. Petrovski, S., & Goldstein, D. B. (2016). Unequal representation of genetic variation across ancestry groups creates healthcare inequality in the application of precision medicine. *Genome Biology*, *17*, 157. https://doi.org/10.1186/s13059-016-1016-y
6. Asiki, G., Murphy, G., Nakiyingi-Miiro, J., Seeley, J., Nsubuga, R. N., Karabarinde, A., . . . & Kaleebu, P. (2013). The general population cohort in rural south-western Uganda: A platform for communicable and non-communicable disease studies. *International Journal of Epidemiology*, *42*(1), 129–141. https://doi.org/10.1093/ije/dys234
7. Mulder, N. J., Adebiyi, E., Adebiyi, M., Adeyemi, S., Ahmed, A., Ahmed, R., . . . & as members of the H3Africa Consortium. (2017). Development of bioinformatics infrastructure for genomics research. *Global Heart*, *12*(2), 91–98.
8. Fatumo, S., Yakubu, A., Oyedele, O., Popoola, J., Attipoe, D. A., Eze-Echesi, G., . . . & Ene-Obong, A. (2022). Promoting the genomic revolution in Africa through the Nigerian 100K Genome Project. *Nature Genetics*, *54*(5), 531–536.
9. Gurdasani, D., Carstensen, T., Tekola-Ayele, F., Pagani, L., Tachmazidou, I., Hatzikotoulas, K., . . . & Sandhu, M. S. (2015). The African genome variation project shapes medical genetics in Africa. *Nature*, *517*(7534), 327–332.
10. Munung, N. S. (2025). Science and society: Pathways to equitable access and delivery of genomics medicine in Africa. *Current Genetic Medicine Reports*, *13*(1), 1.
11. Fatumo, S., Mugisha, J., Soremekun, O. S., Kalungi, A., Mayanja, R., Kintu, C., . . . & Kaleebu, P. (2022). Uganda genome resource: A rich research database for genomic studies of communicable and non-communicable diseases in Africa. *Cell Genomics*, *2*(11).
12. Fatumo, S., Chikowore, T., Kalyesubula, R., Nsubuga, R. N., Asiki, G., Nashiru, O., . . . & Newton, R. (2021). Discovery and fine-mapping of kidney function loci in first genome-wide association study in Africans. *Human Molecular Genetics*, *30*(16), 1559–1568.
13. Mayanja, R., Machipisa, T., Soremekun, O., Kamiza, A. B., Kintu, C., Kalungi, A., . . . & Fatumo, S. (2023). Genome-wide association analysis of cystatin-C kidney function in continental Africa. *EBioMedicine*, *95*.
14. Fatumo, S., Carstensen, T., Nashiru, O., Gurdasani, D., Sandhu, M., & Kaleebu, P. (2019). Complimentary methods for multivariate genome-wide association study identify new susceptibility genes for blood cell traits. *Frontiers of Genetics*, *10*, 334. https://doi.org/10.3389/fgene.2019.00334
15. Nakabuye, M., Kamiza, A. B., Soremekun, O., Machipisa, T., Cohen, E., Pirie, F., Nashiru, O., Young, E., Sandhu, M. S., Motala, A. A., et al. (2022). Genetic loci implicated in meta-analysis of body shape in Africans. *Nutrition Metabolism and Cardiovascular Diseases.*, *32*, 1511–1518. https://doi.org/10.1016/j.numecd.2022.03.010
16. Chikowore, T., Ekoru, K., Vujkovi, M., Gill, D., Pirie, F., Young, E., Sandhu, M. S., McCarthy, M., Rotimi, C., Adeyemo, A., et al. (2022). Polygenic

prediction of type 2 diabetes in Africa. *Diabetes Care*, *45*, 717–723. https://doi.org/10.2337/dc21-0365

17. Kamiza, A. B., Toure, S. M., Vujkovic, M., Machipisa, T., Soremekun, O. S., Kintu, C., Corpas, M., Pirie, F., Young, E., Gill, D., et al. Transferability of genetic risk scores in African populations. *Nature Medicine*, *28*, 1163–1166. https://doi.org/10.1038/s41591-022-01835-x
18. Soremekun, O., Karhunen, V., He, Y., Rajasundaram, S., Liu, B., Gkatzionis, A., Soremekun, C., Udosen, B., Musa, H., Silva, S., et al. (2022). Lipid traits and type 2 diabetes risk in African ancestry individuals: A Mendelian Randomization study. *EBioMedicine*, *78*. https://doi.org/10.1016/j.ebiom.2022.103953
19. Kintu, C., Soremekun, O., Kamiza, A. B., Kalungi, A., Mayanja, R., Kalyesubula, R., . . . & Fatumo, S. (2023). The causal effects of lipid traits on kidney function in Africans: Bidirectional and multivariable Mendelian-randomization study. *EBioMedicine*, *90*.
20. Fatumo, S., Karhunen, V., Chikowore, T., Sounkou, T., Udosen, B., Ezenwa, C., Nakabuye, M., Soremekun, O., Daghlas, I., Ryan, D. K., et al. (2021). Metabolic traits and stroke risk in individuals of African ancestry: Mendelian randomization analysis. *Stroke*, *52*, 2680–2684. https://doi.org/10.1161/STROKEAHA.121.034747
21. Soremekun, O., Soremekun, C., Machipisa, T., Soliman, M., Nashiru, O., Chikowore, T., & Fatumo, S. (2021). Genome-wide association and Mendelian randomization analysis reveal the causal relationship between white blood cell subtypes and asthma in Africans. *Frontiers Genetics*, *12*. https://doi.org/10.3389/fgene.2021.749415
22. Soremekun, O., Mhlongwe, T., Kirabo, G., Akinyele, C., Chikowore, T., Fatumo, S., & Gill, D. (2024). Genetically proxied lipid-lowering drug target perturbation and ischemic stroke risk in European and African ancestry individuals: Mendelian randomization study. *Stroke*, *55*(7), e185–e186.
23. Kalungi, A., Kinyanda, E., Akena, D. H., Gelaye, B., Ssembajjwe, W., Mpango, R. S., . . . & Fatumo, S. (2024). Prevalence and correlates of common mental disorders among participants of the Uganda Genome Resource: Opportunities for psychiatric genetics research. *Molecular Psychiatry*, 1–9.
24. Soremekun, C., Jjingo, D., Kateete, D., Nash, O., Grallert, H., Peters, A., . . . & Fatumo, S. (2024). Structural insights into conformational stability of both wild-type and mutant Insulin Receptor Gene. *Next Research*, *1*(2), 100041.
25. Kamiza, A. B., Touré, S. M., Zhou, F., Soremekun, O., Cissé, C., Wélé, M., . . . & Fatumo, S. (2023). Multi-trait discovery and fine-mapping of lipid loci in 125,000 individuals of African ancestry. *Nature Communications*, *14*(1), 5403.
26. Kintu, C., Soremekun, O., Machipisa, T., Mayanja, R., Kalyesubula, R., Bagaya, B. S., . . . & Fatumo, S. (2023). Meta-analysis of African ancestry genome wide association studies identified novel locus and validates multiple loci associated with kidney function. *BMC Genomics*, *24*(1), 1–9.
27. Soremekun, O., Musanabaganwa, C., Uwineza, A., Ardissino, M., Rajasundaram, S., Wani, A. H., . . . & Fatumo, S. (2023). A Mendelian randomization study of genetic liability to post-traumatic stress disorder and risk of ischemic stroke. *Translational Psychiatry*, *13*(1), 237.

28. Udosen, B., Soremekun, O., Kamiza, A., Machipisa, T., Cheickna, C., Omotuyi, O., . . . & Fatumo, S. (2023). Meta-analysis and multivariate GWAS analyses in 77,850 individuals of African ancestry identify novel variants associated with blood pressure traits. *International Journal of Molecular Sciences*, *24*(3), 2164.
29. Soremekun, C., Machipisa, T., Soremekun, O., Pirie, F., Oyekanmi, N., Motala, A. A., . . . & Fatumo, S. (2023). Multivariate GWAS analysis reveals loci associated with liver functions in continental African populations. *Plos One*, *18*(2), e0280344.
30. Traore, M., Sangare, H., Diabate, O., Diawara, A., Cissé, C., Nashiru, O., . . . & Fatumo, S. (2025). Causal effect of severe and non-severe malaria on dyslipidemia in African Ancestry individuals: A Mendelian randomization study. *Annals of Human Genetics*, *89*(4), 178–187.

Chapter 5

Genomics and the Ethics Conundrum in Africa

Aminu Yakubu[1,2] *and Paulina Tindana*[3*]
[1]Center for Bioethics and Research, Ibadan, Nigeria
[2]Federal Ministry of Health, Abuja, Nigeria
[3]Department of Health Policy, Planning and Management, School of Public Health, College of Health Sciences, University of Ghana, Legon, Ghana

5.1 HISTORICAL ETHICAL CHALLENGES IN AFRICAN GENOMIC STUDIES

Contemporary discourse and considerations of the ethical, legal, and social implications (ELSI) of human genomics research can be traced to the successes of the human genome project (Collins and McKusick 2001). Building on the work of James Watson and Francis Crick in describing the double helix structure of the DNA (Watson and Crick 1953), the human genome project, launched in 1990, generated the first complete set of DNA and its arrangement that defines who we are as human beings (International Human Genome Sequencing Consortium, 2004). The scientific value of this development can only be imagined, encompassing human biology, the genetic basis of diseases, differential responses to drugs between different populations, personalized medicine, and human evolution studies (Hood and Rowen 2013). The ELSI of these applications of our newfound knowledge of the human genome were also considered a priority by the initiators and leaders of the project, as well as other ELSI researchers (Dolan et al. 2022). Over the years, thousands of materials from manuscripts, books, book chapters, and others have been produced on the ELSI of genomics research with increased funding in empirical ethics projects (Morrissey and Walker 2012).

Discourse on the ELSI of genomics research often focused on moral considerations at the individual and the societal level. At the individual level, questions frequently deliberated concerned informed consent models, the extent of individual participant' rights to the genomics data generated from research, risks to privacy and confidentiality breaches, and risks from knowledge of genomic risk markers to individuals and their families. At the societal level, ELSI concerns focused on benefits to society, equity in the distribution of benefits, and the avoidance of wrongful use of genomics research data and products.

*__Corresponding author:__ ptindana@gmail.com (Paulina Tindana)

DOI: 10.1201/9781003436584-5

The literature suggests that most genomics research and its benefits to human health have been explored with data from Europe and the Americas, leaving out Africa, where some of the most genetically diverse populations live (Corpas et al. 2025). It is estimated that on average, a genome from African populations has about one million more genetic variants than that from non-African populations (Sibomana 2024). Thus, without adequate representation of African and other diverse populations in genomics research, the world would fall significantly short of realizing the full potential of the extant knowledge and capabilities to understand the human genome, explore its relationship to diseases, and improve human health and well-being.

Several efforts have increased the representation of African genomics data in the global genomics research landscape. This includes the Malaria Genomic Epidemiology Network (MalariaGen), the Human Hereditary & Health in Africa initiative (H3Africa) (J. de Vries et al. 2015), and African Human Variation Project (Gurdasani et al. 2014), arguably the pioneer pan-African genomics initiative. There are also recent country-driven initiatives such as the 100k Nigeria Genetic Heritage (Fatumo et al. 2022), Egyptian Genome (Wohlers et al. 2020), and the Senegalese Genome (Gaye et al. 2024) studies, among others.

Heightened attention to ELSI in genomic studies in Africa has resulted from these initiatives, especially the H3Africa, Africa's most far-reaching and ambitious genomic initiative. Considering the novelty of genomics research at a large scale as envisaged in the H3Africa, an ethics working group, which was later consolidated into an Ethics and Community Engagement Working Group, was established (Tindana et al. 2015, 2019b). The working group guided investigators on ethics and community engagement issues in implementing the various H3A research projects. It also developed guidance documents on the governance of genomics research in Africa (Yakubu et al. 2018), informed consent, community engagement (Tindana et al. 2017), and feedback of results (Matimba et al. 2022). These have become reference materials for individuals and institutions in Africa conducting genomics research. Other seminal events that further heightened ELSI concerns include the Ebola virus disease epidemic in West Africa (Nichol and Antierens 2021) and, most recently, the allegations of DNA misuse against the Sanger Institute from a collaboration with some South African universities (Stokstad 2019).

A cursory look at the contemporary ELSI issues of concern in genomics research in Africa shows that they are not new or novel from those raised in the context of research in developed countries. Instead, the question is mostly about how these ELSI issues are seen from an African perspective, given the communitarian nature of most African societies (Maqoma 2020), history of mistrust and exploitation (de Vries et al. 2015; Fernando et al. 2019; Mohamed et al. 2022), and deep-rooted cultural and religious belief systems. These include ethical concerns related to the appropriate consent model that allows the collection, storage (Munung et al. 2016), and sharing of biospecimens and data (Amoakoh-Coleman et al. 2023; Staunton et al. 2019); ownership of biospecimens and data in both primary and secondary research (Nielsen and Nicol 2024); return of primary and incidental genomic results (Mulder et al. 2017; Matimba et al. 2022; de Vries et al. 2017); use of samples and data by the industry in discovery research that may involve some commercial gains (Marshall et al. 2022); and responsible and fair international genomics research collaborations (de Vries et al. 2015; Munung et al. 2021).

5.1.1 Responsible and Fair International Genomics Research Collaborations

Researchers and regulators in African institutions have long decried irresponsible and unfair practices as the bane of international collaborative research. In genomics research, this often applies to two key practices—unfair collaboration terms involving African researchers and unethical practices in sample and data storage, sharing, and use (Tiffin 2024).

Unfair collaboration practices involving African researchers concern tokenistic research partnerships in which the African researchers are not treated as partners with equal intellectual contributions. This is often seen when African researchers are only included in the research collaboration when the research idea has been fully developed, and are only required to help implement the research in their countries. Some researchers have described this phenomenon as treating the African researchers as mere "data collectors," with no regard for their intellectual achievements and contributions to the research (Kwachou 2025; Nolte 2019).

A related perspective is where African researchers are used to legitimize sample and data collection led by the collaborators from high-income countries (HICs), who, in turn, ship out the samples to their institutions without recourse to any regulatory regime in the host African countries (Moyi et al. 2015; Munung et al. 2017). This concern has been termed "parachute research" and was most recently heightened during the Ebola virus disease outbreak in West Africa (Odeny and Bosurgi 2022; The Lancet Global Health 2018). Scientists from diverse HICs, in the name of further research, took samples to their countries with no accountability to the workers and the governments of the African countries from where the samples were taken (Freudenthal 2019).

Due to limited infrastructure, samples are often shipped out, and data are also stored in databases largely controlled by collaborators from the HICs (de Vries et al. 2011; Tindana et al. 2014). While this may not be a problem, the central issue has often been the lack of transparent and clear terms indicating the rights of African researchers and their institutions to the samples and data. Further, despite global acknowledgement of the limited capacity among African researchers in genomics and bioinformatics, such collaborations are not accompanied by robust proposals to help bridge these gaps.

Thus, African researchers are left handicapped, only involved in the primary research outcomes generation and dissemination, and often excluded from downstream secondary sample and data uses (Skelly and Chiware 2022).

5.1.2 National Sovereignty, Sharing, and Ownership of Samples and Data

Sharing stored samples and data from primary genomics research studies is considered most valuable in the genomics research field. Against the practice of collecting samples for one-off use, which are discarded thereafter, genomics research emphasized the need to store samples for use in the future (Wolf et al. 2010). Such storage and sharing of samples and data allows researchers to meet sample size requirements without having to collect new data and to increase the diversity of their datasets to improve the reliability and generalizability of their findings, especially considering the diversity of the human genome (Trinidad et al. 2010). This is also expected to lead to equitable access to the benefits of genomics research, given the applicability of the findings to many different population groups. The initiators of the modern genomics research era had called for an open data access regime, supported by other international declarations such as the Bermuda and Fort Lauderdale declarations (Marshall 2001; Kaye et al. 2009; The Wellcome Trust 2003). The open data access regime has been extended to apply to stored samples in genomics research. Therefore, as with data, allowing for in-country and cross-border sharing of samples is considered a global best practice in genomics research (Ali et al. 2021). However, rising concerns about the privacy of individual genomics data in open databases with unrestricted access, use of shared samples against approved causes (Hubaux et al. 2017; Remes Lenicov and Fink 2023), and unfair distribution of benefits of genomics research (Lemke et al. 2022; Atutornu et al. 2022), including data and sample commercialization concerns (Marshall et al.2022), have led many organizations to resort to controlled data access regimes. With controlled access databases, some level of authorization is required to allow access to genomics databases with further limitations on the data elements that are accessible.

In African countries, the concern of sharing samples and data is beyond privacy alone. It also includes concerns about inequitable access to available data and the capacity to utilize the data and samples optimally. Due to poor access to funding, inadequate genomics data generation capacity, and inadequate capacity for genomics data insights generation, a truly open data and sample access regime would result in researchers from HICs nearly exhausting the utility of any stored samples or data well before their African counterparts (Ramsay 2022; Barnes et al. 2019).

African researchers and authorities are also increasingly raising concerns about sample and data sovereignty. This may not be unconnected with recent experiences of exploitative research, in which samples were taken from some African countries without accountability, as in the experience during the Ebola virus disease outbreak in West Africa or when some African countries were feared to have received unfair benefits from the use of samples from their countries, as in the case of vaccines for COVID-19. During the COVID-19 era, African and other LMICs had decried the cost of COVID-19 vaccines to their countries, when they believed that samples and data from their citizens had contributed to the discovery of the vaccines, and thus deserved some benefit, in this case, low-cost vaccines (Eccleston-Turner and Rourke 2021). For these concerns, some African countries now require that data and sample storage for future use be within their jurisdictions, with restrictive policies on the export of samples and sharing of genomics data. Termed "genomic sovereignty," this practice of restricting the storage and use of samples and genomics data is arguably a protectionist approach by the respective African governments to protect their natural heritage and harness the benefits achievable from their use optimally (Staunton et al. 2025). It is a strategic approach to addressing concerns about historical exploitation in international research collaborations between African countries (including other LMICs) and HICs. Although the reasons for such restrictive policies are clear, a cursory look at their impact on research discoveries that could benefit African people and general humanity would suggest a more pragmatic approach to addressing the concerns raised by the African governments (Staunton and de Vries 2020; Staunton et al. 2025).

Some African nation-states' pursuit of genomic sovereignty reflects one arm of the debate about ownership of samples and data from genomics research. At the individual level, empirical studies have reported that individual research participants have indicated some interest in "owning" their genomics data (Kutyabami et al. 2024). By claiming such a right, it is thought that the participants would be better empowered to have some control over the use of their samples and data, including the ability to claim benefits that may accrue through such uses. Considering their effort in collecting and processing samples and data, there are arguments that researchers and, in other instances, their institutions could also claim some ownership rights over the samples and data from their research studies (Kabanda et al. 2023) This shows the multiplicity of claimants of ownership of samples and data from genomic studies, making the concept of ownership complex, with different owners based on jurisdiction (Nielsen and Nicol 2024). This multiplicity and complexity are seen in a survey of regulations from 22 African countries. In some countries (Malawi and Zambia), the Ministry of Health implicitly or explicitly owns the samples. However, in other countries, such as Botswana, Ghana, Ethiopia, Rwanda, and Uganda, the institutions that collected the samples are said to be "custodians," holding the samples in trust on behalf of the research participants and the government (de Vries et al. 2017). The Alliance for accelerating excellence in Africa (AESA), an initiative of the African Academy of Sciences also preferred this concept of custodianship in its recommendations for data and biospecimen governance in Africa. AESA posited that the "custodianship" concept implies a "care-taking and governance responsibility by the data [and samples] community, which allows for fair and transparent sample and data use practices based on ethical rather than strictly legal principles" (AESA 2020, p. 3).

5.2 EMERGING ETHICAL CONCERNS IN GENOMIC STUDIES IN AFRICA

The ethical issues examined in the preceding discussion, such as informed consent, community engagement, privacy issues, data sovereignty, and ownership, are still relevant and continue to inform the review and implementation of genomics research and biobanking in Africa. However, in recent years, other contentious issues have emerged. These include benefit sharing, commercialization, and feedback of findings.

5.2.1 Benefit Sharing

Benefit sharing as an ethical requirement is not exclusive but has gained greater prominence with increased genomics research and its applications in medicine, including drug discovery, and precision medicine. It is considered the act of giving some advantage or profit from using samples or data from human genomics research (Schroeder 2007). Such benefit sharing is said to be particularly important for individuals with limited access to healthcare products or services, which is a key characteristic of most African countries, where only 17% of the population has any form of health insurance (Odipo et al. 2024; Schroeder 2007). Over the years, researchers struggled to agree on what constitutes benefit, benefit sharing, and how it should be implemented. Benefit sharing is sparsely discussed in international research ethics guidelines, and where it is, the provisions therein lack the specificity for application in practice. One of the earliest guidelines for benefit sharing in human genetic research was issued by the Human Genetics Organization ethics committee in 2000 (Hugo ethics committee 2000), albeit in anticipation of the launch of the Human Genome Project, which started a year later. This document recommends that benefits to be shared should seek to promote the welfare of an individual and/or of a community. It also cautioned about considering benefits as profit in the monetary or economic sense, perhaps to avoid the commodification of science. To guard against such concerns, it was recommended that benefit sharing be based on specific ethical principles, norms, and values.

In Africa, benefit sharing in scholarly discourse and practice is only just emerging. They include studies that discuss the normative principles that ought to guide benefit sharing in Africa and those on the experiences and preferences of researchers and research participants (Sudoi et al. 2021). The principles of fairness, solidarity, and reciprocity, as well as those of humanness, friendliness, and partiality, were considered reflective of the African communitarian structure and best fit for benefit-sharing design in Africa. In some empirical studies, researchers in Africa reported a low awareness of benefit sharing but supported the practice based on the principles of solidarity and reciprocity (Munung et al. 2020; Sudoi et al. 2021; Selormey et al. 2025). In another study, researchers and ethics committee members in Uganda were more concerned about the equitable sharing of benefits and the need for researchers, communities, and participants to benefit directly from genomics research. To address these concerns and limited knowledge, a team of researchers proposed a framework for ethical benefit sharing in genomics research (Bedeker et al. 2022). This framework suggests that benefit sharing might benefit many stakeholders in research, including nation-states (macro level), institutions within nation-states (meso level), and people and their communities (micro level). Rather than focusing on monetary benefit sharing or capacity building, as advocated by the H3A, the framework provided other categories of benefits that researchers, sponsors, and communities should consider when negotiating benefit-sharing agreements in a specific study (Bedeker et al. 2022). Some potential advantages include money gains, equipment acquisition, information gain, infrastructure establishment, healthy well-being, skill advancement, career growth, and recognition of efforts (Bedeker et al. 2022). While this paradigm is valuable, its application in policy and practice in African nations still needs some adaptation to reflect the social norms and values of the people of these countries.

5.2.2 Commercialization of Genomic Resources

An ethical concern closely related to benefit sharing is the issue of commercialization. This arises when data or samples from genomics research are used to generate products of economic value for one party or the other. Samples or data may be used to develop a DNA microarray chip that is then sold to researchers across the globe for genetic imputation. In other cases, samples or data may be used to identify and validate a drug target for a disease of interest, which is of economic value to pharma partners in developing new drugs in the precision medicine era (Marshall et al. 2022). It is important to point out that it is not the commercialization that is a concern in genomics research, since, of course, one important aim of medical science is to generate products that improve human health. In almost all instances, where there is a product, someone would have to pay to use it. The central issue in commercialization in genomics research, however, is whether commercialization intent is contained in the informed consent for the data or sample used to develop a product with economic value; whether there is a fair agreement negotiated between those that provided the sample or data and the party that used same to develop the product; and whether in such an agreement, there is an explicit provision to ensure that some of the economic value that may be earned would benefit the research participants from whom the samples or data was derived initially, either directly or indirectly (Bedeker et al. 2022; Schultz et al. 2024). Instances where such considerations are not defined or adhered to are the reason for the concerns about commercialization in genomics research in Africa. Some of such instances include the alleged development of a DNA microarray chip using samples from some South African institutions without authorization (Stokstad 2019), equity concerns with the new gene therapy for sickle cell disease, where Black Africans most affected by the disease are unable to pay for the therapy (Stokstad 2019), as well as concerns by some African countries about the unfairly high cost of the SARS-COV2 vaccine to their countries despite contributing virus sequence data that facilitated the vaccine's development (Eccleston-Turner and Rourke 2021).

Contrary to some traditional perceptions, commercialization in genomics research does not include the direct sale of samples between parties or profit sharing with research participants. Most genomics research products use aggregate data from diverse populations, making it challenging to attribute discoveries to individual participant samples (Bentley et al. 2017). Further engagements on these issues would be key to allaying participants fears about the potential for commercialization of their biosamples and associated data.

5.2.3 Feedback of Findings

An arrangement to return the results from genomic studies to research participants is considered an important ethical consideration in genomics research (Wonkam and de Vries 2020). Normatively, the return of results from genomics research is rooted in the principle of autonomy on the one hand and reciprocity and beneficence on the other (Shalowitz and Miller 2008; Ralefala et al. 2020; Kasule et al. 2023). Whether research participants should be given their results is a question that squarely borders on respecting individual choice. The importance of deliberation on whether to return the results to the participants is anchored on two key considerations: (1) clinical utility and (2) clinical actionability of the genomic results (Matimba et al. 2022). Researchers have argued against returning results that have no clinical utility or are not clinically actionable, that is, results that cannot be used towards improving the health condition of the research participant. Clinical utility is when there is potential to use a test result to improve the health outcome of a person, through established prevention, diagnosis, or treatment modalities, while clinical actionability is about whether there is a clear medical course of action that can be taken considering the state of the health system in a place (Fabsitz et al. 2010; Goddard et al. 2022). A clinical test result may be considered clinically actionable in theory but not actionable in practice because the technology does not yet exist on what to do, or it is unavailable to the participant due to health system challenges or poverty, among other reasons (Lerner et al. 2017; Goddard et al. 2022).

In both instances, the primary concern is to avoid psycho-social harm that research participants may face if they become aware of their risk for a debilitating or highly stigmatized health condition for which clinical treatment cannot be rendered. Some ethics experts have, however, added personal utility as a third condition for considering the return of genomics research results to participants. Regarding personal utility, research participants may wish to know the outcome out of curiosity or want to be helped so they may make some life adjustments that could allow them to live with their condition, barring the ability to improve it clinically (Lemke et al. 2013). Given all these considerations, however, when researchers choose to return results from genetics research to participants, the information given to the participants must be in a comprehensible way and manner that reduces the risks of harm from such disclosure (Appelbaum et al. 2014). This can be achieved through genetics counsellors. However, there is a dearth of such professionals in Africa, making it challenging to achieve high standards in returning genomics results to research participants. In the interim, health professionals who have received some abridged training in genetics counselling are assigned to undertake this task. Thus, even where genomics results are returned, there is the concern that it is done through a less desirable standard, due to the limited number of genetics counsellors on the continent (Matimba et al. 2022).

5.2.4 Gene Therapy

Diseases caused by gene changes can now increasingly be prevented, managed, or treated by replacing or repairing the genes causing such diseases. This approach is called gene therapy. After several years of research, gene therapy is now a reality for many genetic and malignant diseases (Adair et al. 2023). With many clinical trials for other therapies already underway, genetic therapies for many more diseases are expected in the coming decade (High 2020). However, nearly all genetic therapies are developed from clinical trials in HICs (Adair et al. 2023). For example, whereas two thirds of the people living with HIV (World Health Organization 2023) and 300,000 children born annually with sickle cell anaemia (Aygun and Odame 2012) are in Africa, only 4 of more than 800 clinical trials on these diseases involved sites in Africa (Adair et al. 2023). This has several implications. First, given the higher genetic variation among African populations, there is the risk that the therapies developed from populations residing in HICs lack the requisite representativeness and may thus not have the same efficacy among Africans. Second, when developed, these therapies are costly, with the per-patient cost of gene therapy for sickle cell disease estimated at $2 million (Nikitin et al. 2023). Thus, even when gene therapies are available for diseases of priority to Africans, they may not be readily accessible to people in African countries where the people's ability to pay for healthcare services is extremely low due to high levels of poverty. The poor health insurance coverage in most African countries further complicates this. There are also infrastructure, human, and regulatory capacity concerns in African gene therapy studies. Too few institutions and experts have the requisite infrastructure and skills to host gene therapy studies in Africa (Arbuthnot et al. 2017). Similarly, research regulatory bodies are inexperienced in overseeing gene therapy studies (Arbuthnot et al. 2017; Munung et al. 2024).

Beyond concerns of equity, which may be worsened by the high cost and the potential low efficacy of gene therapies due to the limited representation of African populations, other classical ethical issues are also of concern. One key ethical concern is the model and process for obtaining informed consent from prospective gene therapy research participants in Africa (Munung et al. 2024; Buedo et al. 2023a). Obtaining informed consent for genomics research is still an ongoing pain point in Africa, with debates ongoing between broad consent and tiered consent. Whereas broad consent, which is preferred in most HICs and is also seen as a good model that protects the rights and welfare of research participants when accompanied by a good governance model (Tindana and de Vries 2016; Tindana et al. 2019a; Mweemba et al. 2019), has benefits, some African researchers and, indeed, the African Academy of Sciences have indicated a preference for tiered consent, as one that empowers and respects the participants' rights to autonomous choice better (Nembaware et al. 2019; AESA 2020). In the case of gene therapy studies, regardless of which model between broad or tiered consent is chosen, the key concerns raised are the appropriate presentation of risks and ensuring comprehension among the research participants (Woollard et al. 2021). Some of the dangers in gene therapy include insertional mutagenesis from retroviral vectors, immunotoxicity from some nanoparticle gene therapy delivery systems, off-target effects, environmental contamination, and inadvertent infection of others from viral shedding, and unknown long-term effects of both somatic and germline gene editing research among others (Kohn et al. 2023). Because gene therapies introduce permanent changes, such studies also raise the concern about voluntary withdrawal, which is a hallmark of traditional informed consent (Munung et al. 2024; Buedo et al. 2023a). In addition to their impact on the informed consent process, these concerns have implications for the ethical review of gene therapy studies in African countries. Studies have reported the limited capacity of African research ethics committees to provide adequate oversight for studies, including genomic studies, generally.

5.3 ETHICS AND GENOMICS: FUTURE DIRECTIONS IN AFRICA

The past two decades have witnessed growth in capacity development and scholarship addressing the ethical, legal, and socio-cultural issues in genomics and biobanking in Africa. As discussed in earlier sections of this chapter, significant progress has been made in addressing issues related to consent, community engagement, and feedback of findings. Additionally, funding opportunities have been increased from key research funders such as the National Institutes of Health and the Wellcome Trust to support research focusing on these issues. However, some of these issues continue to persist and are likely to influence the conduct of genomics research across the continent. Recent developments in research funding policies, particularly from the United States, are likely to affect progress made so far. There are also re-emerging concerns that would require further discussion.

5.3.1 From Consent to Deliberative Engagement

Although the issue of consent has received considerable attention, there is still a lack of consensus on the practical implementation and ethical acceptability of consent models such as broad and tiered consent. Despite these challenges, there has been a growing recognition that a governance framework and meaningful community engagement must support any consent model that is adopted for genomics research. Several articles based on empirical ethics projects from African scholars have addressed both the intrinsic and the instrumental values as well as the practical implementation of community engagement in Africa (Tindana et al. 2017, 2015; Nankya et al. 2024) and there is a growing body of resources that is now available to support research teams to implement community engagement strategies in specific contexts.

Recently, the World Health Organization recommended that "the use of human genome data has implications beyond an individual, and the views of communities should be ascertained on access to, use, and sharing of these data throughout its lifecycle" (WHO 2024). Questions, however, remain on how participants and communities can be engaged meaningfully in genomics research beyond the sample and data collection stage. This has led to proposals for deliberative engagement approaches as an alternative to traditional engagement models. While this approach continues to receive attention, empirical research on the practical application of deliberative engagement and its effect on participation in genomics research remains limited (Ralefala et al. 2021; Wand et al. 2023; Tsey et al. 2025). We therefore call for further research to assess this approach, identify innovative ways of engaging communities meaningfully in genomics research on the continent, promote genomic literacy, and ensure that local voices and concerns are considered in decisions regarding sample and data sharing practices.

5.3.2 The Ethics and Legal Status of Benefit Sharing and Commercialization

Although there has been an established practice, a modest literature, and some frameworks proposed, benefit sharing in genomics research in Africa remains one of the unsettled ethical issues. While there is growing scholarship on this subject and a few frameworks have been proposed, there are minimal ethics and legal frameworks to guide its implementation. Some scholars have highlighted complexities in the legal status of benefit sharing in some African countries (Kamau et al. 2025), and others have suggested that this practice may even be considered an offence in some jurisdictions, particularly in South Africa (Thaldar and Shozi 2023). These uncertainties have implications for the conduct of genomics research, particularly on the ethics and legal tensions relating to research practices such as benefit sharing. While research ethics committees are beginning to highlight benefit sharing as an ethical imperative, legal requirements may hinder its implementation. We therefore recommend further analysis on the legal status of benefit sharing across the continent and its implications for genomics research. We also recommend more empirical studies to elicit the views of key stakeholders, including research participants, on what, how, and when benefits from genomics research should be shared.

As previously discussed, concerns regarding the use of genomics data to develop commercially viable products or of scientists in for-profit organizations as part of their discovery process—both commonly referred to as commercialization—are directly linked to the subject of benefit sharing and will continue to be an important area influencing the ethics review and implementation of genomics research in Africa. The lack of empirical evidence and perspectives of key stakeholders, including research participants, further complicates this issue and would require empirical research to inform the development of specific ethics and legal frameworks governing the commercialization of genomics data.

To address some of the concerns related to commercialization, it is important to adopt a consent model suitable to the population involved in genomic studies. Where commercialization is envisaged, a clear description of the nature of the commercialization should be provided in the consent form, supported by effective community engagement strategies using deliberative approaches.

As recommended by other authors, rather than using the term "commercialization," which some consider synonymous with buying and selling, a typology of partnerships that may lead to products with economic value or involve financial exchange is proposed. This may include terms like "product development partnership" and "drug discovery research" (Marshall et al. 2022). An agreement about the nature of the benefit that may accrue to researchers, their institutions, or the country at large from commercially potential genomics research should be negotiated in good faith. These would help to build trust and allay reservations about genomics research that some individuals and countries may raise due to commercialization concerns.

5.3.3 Data Ethics and Data Protection Considerations

The increase in the collection, use, and sharing of data has also led to greater attention on data ethics, defined as "a new branch of ethics that studies and evaluates moral problems related to data (including generation, recording, curation, processing, dissemination, sharing, and use" (Floridi and Taddeo 2016). Building on initial efforts by H3Africa, the Data Science for Health Discovery and Innovation in Africa initiative has advanced the understanding of the ethical, legal, and social implications of data science and has made efforts to develop specific guidance on the responsible and equitable use of data (Mistry et al. 2023). This initiative has highlighted the need for genomics researchers to remain both ethically and legally compliant.

Despite some progress, there remains a significant lack of data regarding specific guidelines for data sharing and use on the continent. The principles for genomics data collection, access, use, and sharing recently published by the World Health Organization serves as a helpful starting point. Of the eight principles outlined in the guideline (Figure 5.1), particular emphasis is placed on the first principle, which advocates for the respect of both

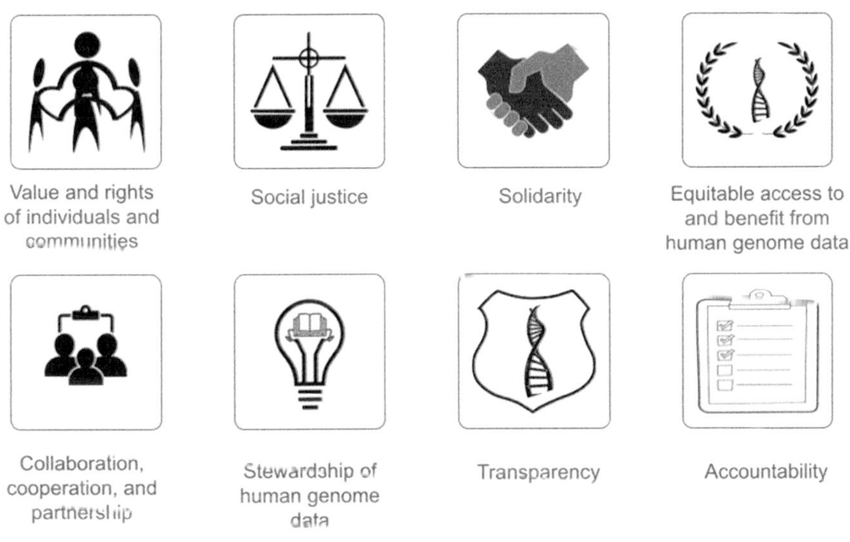

Figure 5.1 Principles for human genome collection, access, use, and sharing. (Source: WHO 2024; World Health Organization, 2024).

individual and community values, as well as the principles of solidarity and equitable access to and benefit from genomics data (WHO 2024). These principles support similar arguments made by ethicists and genomics experts in Africa.

The introduction of data protection laws across the African continent is likely to affect how personal data from genomic studies are shared (Staunton et al. 2025). Data from the online platform Data Protection Africa (https://dataprotection.africa) indicate that as of July 2025, 36 out of the 55 African countries have enacted data protection laws. It is worth noting that almost all African countries have specific constitutional provisions that address privacy issues. However, among these 36 countries, only 26 have established specific regulatory agencies to implement and enforce compliance of the provisions in data protection laws.

The variations in legal instruments among African countries have resulted in tensions between ethical standards and legal frameworks concerning genomics data sharing. These tensions also extend to the authority of research ethics committees and data protection commissions regarding approval of these research practices. For instance, while most of the data protection laws focus primarily on personal data, they lack explicit provisions regarding the handling of personal data collected for research purposes, although some instruments do make exceptions for research (Staunton et al. 2025). The emphasis on personal data in these laws does not adequately address the specific use of anonymous data. This ambiguity raises questions about the applicability of these laws to genomics research and presents potential challenges for researchers navigating both the ethical and legal requirements in cross-border studies. Clarifying these legal frameworks is important to ensuring that genomics researchers are able to conduct their research responsibly while adhering to ethical standards.

There is also the need for clarity on how existing data protection laws intersect with research ethics, particularly in the context of genomics and the remit of research ethics committees regarding data use and cross-border data sharing. This calls for further empirical studies and engagements with policy makers and other key stakeholders to address the responsibilities of researchers regarding data sharing and protection. These efforts could ultimately promote trust and accountability in genomics research.

5.3.4 Strengthening and Sustaining Bioethics Capacity in Africa

Addressing the ELSI issues in genomics requires a cadre of skilled bioethicists to support both the review and the implementation of genomics research. This support should take into account the key ethical principles and frameworks guiding these initiatives. For over a decade, the H3Africa Ethics and Community Engagement Working Group has been instrumental in supporting capacity development and scholarships focused on ELSI of genomics research. Several other groups have also emerged to promote discourse on ethical issues in research and clinical care across Africa.

In addition, there is a growing number of bioethics scholars—individuals who have received PhD-level training in bioethics from foreign and African institutions and who have a visibly growing scholarly portfolio. Collectively, the groups, organizations, and individual bioethics scholars provide a valuable resource for advancing discussions on bioethical issues in Africa, including unresolved and emergent challenges in genomics research on the continent.

However, barriers to African scholarship persist, including challenging working environments and limited access to research tools. For example, anecdotal reports suggest that despite the progress made, many research ethics committees in Africa lack bioethics experts, there is limited faculty positions available for bioethics scholars in African universities, and very few research consortia and centres actively engage bioethics experts in the planning, review, and implementation of genomics research and biobanking on the continent. To address these challenges, universities, research institutions, and research ethics committees should leverage the growing bioethics expertise on the continent to facilitate the review of genomics research proposals and promote scholarly discourse on contemporary and emergent bioethical issues in medical practice and research on the continent. This collaborative effort is essential to foster a more ethically informed research environment that is responsive to the unique challenges faced in the sub-region.

With funding from the Fogarty Centre of the National Institutes of Health, various universities in Africa, such as the University of Ibadan, Nigeria; University of Ghana School of Public Health; University of Cape Town, South Africa; University of Addis Ababa; Makerere University in Uganda; among others, are now offering graduate-level degrees in bioethics. Some of these programmes include specific courses on the ethics of genomics and biobanking as part of its core curriculum. This emphasis on genomics ethics within advanced academic programmes is key for preparing professionals who can effectively navigate the complexities of ethical issues arising in genomics research and its applications in clinical settings and support the implementation of genomics research across the continent. It is important for these programmes to incorporate legal analysis as a fundamental component of the core competences required of trainees and to better equip bioethics scholars to comprehend the implications of data protection laws, consent regulations, and other legal frameworks related to genomics research.

Bioethics networks are also emerging across the continent and have the potential to support the career growth of bioethics scholars. However, targeted initiatives are needed to support early-career bioethicists. Opportunities such as those provided by the Brocher Institute, which provides enabling spaces for bioethicists to work on special projects, such as manuscripts, workshops, and policy development, would be beneficial to advance ethical enquiry. We encourage the establishment of a network of bioethics scholars on the continent to provide a platform for collaborative research and mentorship. This network could also facilitate knowledge sharing and translations, resources, and best practices and contribute to strengthening ethical standards and practices in the region.

5.4 CONCLUSION

The ethical, legal, and social implications of genomics in Africa have evolved significantly over the past two decades. Scholarship on issues such as consent, data and sample sharing, benefit sharing, and feedback of findings has advanced overtime. There is now a wealth of resources—ethics guidelines and frameworks as well as human resources—to support the review and implementation of genomics research on the continent.

Given the current challenges and uncertainties surrounding global funding for research, it is imperative to secure sustainable financial support for ELSI-related activities in genomics with funding. It is encouraging that several countries in Africa are establishing national research funds to support locally driven research efforts. This shift towards national funding is a positive step towards building a sustainable research ecosystem in Africa. These funds could also provide an opportunity to enhance local research capacity for addressing ELSI issues in genomics while promoting research that is relevant to local contexts.

We call on organizations such as Africa Union, Africa CDC and the WHO Regional Office for Africa to play a key role in facilitating engagements with policy makers and other key stakeholders, including bioethicists, to harmonize ethical standards and legal frameworks across the continent. Initiatives to promote genomic literacy across the continent would also be key to ensure that no community is left behind. These efforts could contribute to a more robust and integrated approach to genomics research in Africa and ensure that ethical and legal standards are upheld while advancing scientific research across the continent.

REFERENCES

Adair, Jennifer E., Lindsay Androski, Lois Bayigga, Deus Bazira, Eugene Brandon, Lynda Dee, Steven Deeks, et al. 2023. "Towards Access for All: 1st Working Group Report for the Global Gene Therapy Initiative (GGTI)." *Gene Therapy* 30 (3–4): 216–221.

AESA. 2020. "Setting Scientific Priorities for Data and Biospecimen Governance in Africa: Outcome Document for Experts Consultative Roundtable Held in

March 2020." Nairobi, Kenya: Alliance for Accelerating Excellence in Science in Africa. https://portal.aasciences.ac.ke/storage/publications/31102022110023Recommendations%20for%20Data%20and%20Biospecimen%20Governance%20in%20Africa.pdf.

Agamah, F. E., A. Mistry, T. Muzambi, G. Dankyi, L. Povlich, and M. Skelton. 2025. "How the DS-I Africa Consortium Is Harnessing the Power of Partnerships for Data Science in Africa." *Data Science Journal* 24 (32): 1–11.

Ali, J., B. Cohn, E. Mwaka, J. M. Bollinger, B. Kwagala, J. Barugahare, N. K. Sewankambo, and J. Ochieng. 2021. "A Scoping Review of Genetics and Genomics Research Ethics Policies and Guidelines for Africa." *BMC Medical Ethics* 22 (1): 39.

Amoakoh-Coleman, Mary, Dorice Vieira, and James Abugri. 2023. "Ethical Considerations for Biobanking and Use of Genomics Data in Africa: A Narrative Review." *BMC Medical Ethics* 24 (December): 108.

Appelbaum, Paul S., Erik Parens, Cameron R. Waldman, Robert Klitzman, Abby Fyer, Josue Martinez, W. Nicholson Price 2nd, and Wendy K. Chung. 2014. "Models of Consent to Return of Incidental Findings in Genomic Research." *The Hastings Center Report* 44 (4): 22–32.

Arbuthnot, P., M. B. Maepa, A. Ely, and M. S. Pepper. 2017. "The State of Gene Therapy Research in Africa, Its Significance and Implications for the Future." *Gene Therapy* 24 (9): 581–589.

Atutornu, Jerome, Richard Milne, Alesia Costa, Christine Patch, and Anna Middleton. 2022. "Towards Equitable and Trustworthy Genomics Research." *eBioMedicine* 76 (February): 103879.

Aygun, Banu, and Isaac Odame. 2012. "A Global Perspective on Sickle Cell Disease." *Pediatric Blood & Cancer* 59 (2): 386–390.

Barnes, Karen I., Julio Arturo Canario, Sunil S. Vernekar, Shivaprasad S. Goudar, Roberto Espinal, Laura Merson, and Phaik Yeong Cheah. 2019. "Equitable Data Sharing: Challenges and Suggestions for Ways Forward." *Wellcome Open Research* 4 (November): 172.

Bedeker, Anja, Michelle Nichols, Taryn Allie, Tsaone Tamuhla, Peter van Heusden, Olorunyomi Olorunsogbon, Nicki Tiffin, and PHA4GE Ethics and Data-Sharing Working Group. 2022. "A Framework for the Promotion of Ethical Benefit Sharing in Health Research." *BMJ Global Health* 7 (2). https://doi.org/10.1136/bmjgh-2021-008096.

Bentley, Amy R., Shawneequa Callier, and Charles N. Rotimi. 2017. "Diversity and Inclusion in Genomic Research: Why the Uneven Progress?" *Journal of Community Genetics* 8 (4): 255.

Buedo, Paola, Alahi Bianchini, Katarzyna Klas, and Marcin Waligora. 2023a. "Bioethics of Somatic Gene Therapy: What Do We Know so Far?" *Current Medical Research and Opinion* (October): 1355–1365.

Collins, F. S., and V. A. McKusick. 2001. "Implications of the Human Genome Project for Medical Science." *JAMA: The Journal of the American Medical Association* 285 (5): 540–544.

Corpas, Manuel, Mkpouto Pius, Marie Poburennaya, Heinner Guio, Miriam Dwek, Shivashankar Nagaraj, Catalina Lopez-Correa, Alice Popejoy, and Segun Fatumo. 2025. "Bridging Genomics' Greatest Challenge: The Diversity Gap." *Cell Genomics* 5 (1): 100724.

Data Protection Africa. https://dataprotection.africa. Accessed on 19th May 2025.
de Vries, Jantina, Susan J. Bull, Ogobara Doumbo, Muntaser Ibrahim, Odile Mercereau-Puijalon, Dominic Kwiatkowski, and Michael Parker. 2011. "Ethical Issues in Human Genomics Research in Developing Countries." *BMC Medical Ethics* 12 (March): 5.
de Vries, Jantina, Syntia Nchangwi Munung, Alice Matimba, Sheryl McCurdy, Odile Ouwe Missi Oukem-Boyer, Ciara Staunton, Aminu Yakubu, Paulina Tindana, and H3Africa Consortium. 2017. "Regulation of Genomic and Biobanking Research in Africa: A Content Analysis of Ethics Guidelines, Policies and Procedures from 22 African Countries." *BMC Medical Ethics* 18 (1): 8.
de Vries, Jantina, P. Tindana, K. Littler, M. Ramsay, C. Rotimi, A. Abayomi, N. Mulder, and B. M. Mayosi. 2015. "The H3Africa Policy Framework: Negotiating Fairness in Genomics." *Trends in Genetics: TIG* 31 (3): 117–119.
Dolan, Deanne Dunbar, Sandra Soo-Jin Lee, and Mildred K. Cho. 2022. "Three Decades of Ethical, Legal, and Social Implications Research: Looking Back to Chart a Path Forward." *Cell Genomics* 2 (7): 100150.
Eccleston-Turner, Mark, and Michelle Rourke. 2021. "Arguments Against the Inequitable Distribution of Vaccines Using the Access and Benefit Sharing Transaction." *International & Comparative Law Quarterly* 70 (4): 825–858.
Fabsitz, Richard R., Amy McGuire, Richard R. Sharp, Mona Puggal, Laura M. Beskow, Leslie G. Biesecker, Ebony Bookman, et al. 2010. "Ethical and Practical Guidelines for Reporting Genetic Research Results to Study Participants." *Circulation: Cardiovascular Genetics* (December). https://doi.org/10.1161/CIRCGENETICS.110.958827.
Fatumo, Segun, Aminu Yakubu, Olubukunola Oyedele, Jumi Popoola, Delali Attiogbe Attipoe, Golibe Eze-Echesi, Fatima Z. Modibbo, et al. 2022. "Promoting the Genomic Revolution in Africa through the Nigerian 100K Genome Project." *Nature Genetics* 54 (5): 531–536.
Fernando, B., M. King, and A. Sumathipala. 2019. "Advancing Good Governance in Data Sharing and Biobanking—International Aspects." *Wellcome Open Research* 4: 184.
Floridi, Luciano, and Mariarosaria Taddeo. 2016. "What Is Data Ethics?" *Philosophical Transactions of the Royal Society A: Mathematical, Physical and Engineering Sciences* (December). https://doi.org/10.1098/rsta.2016.0360.
Freudenthal, Emmanuel. 2019. "Ebola's Lost Blood: Row over Samples Flown out of Africa as 'Big Pharma' Set to Cash in." *The Sunday Telegraph*, February 6. https://www.telegraph.co.uk/global-health/science-and-disease/ebolas-lost-blood-row-samples-flown-africa-big-pharma-set-cash/.
Gaye, Amadou, Andrea Regina G. Sene, Macoura Gadji, Alioune Deme, Aynina Cisse, and Rokhaya Ndiaye. 2024. "Toward Building a Comprehensive Human Pan-Genome: The SEN-GENOME Project." *American Journal of Human Genetics* 111 (10): 2074–2078.
Goddard, Katrina A. B., Kristy Lee, Adam H. Buchanan, Bradford C. Powell, and Jessica Ezzell Hunter. 2022. "Establishing the Medical Actionability of Genomic Variants." *Annual Review of Genomics and Human Genetics* 23 (April): 173.
Gurdasani, Deepti, Tommy Carstensen, Fasil Tekola-Ayele, Luca Pagani, Ioanna Tachmazidou, Konstantinos Hatzikotoulas, Savita Karthikeyan, et al. 2014.

"The African Genome Variation Project Shapes Medical Genetics in Africa." *Nature* 517 (7534): 327–332.
High, Katherine A. 2020. "Turning Genes into Medicines-What Have We Learned from Gene Therapy Drug Development in the Past Decade?" *Nature Communications* 11 (1): 5821.
Hood, Leroy, and Lee Rowen. 2013. "The Human Genome Project: Big Science Transforms Biology and Medicine." *Genome Medicine* 5 (9): 1–8.
Hubaux, Jean-Pierre, Stefan Katzenbeisser, and Bradley Malin. 2017. "Genomic Data Privacy and Security: Where We Stand and Where We Are Heading." *IEEE Security & Privacy* 15 (5): 10–12.
Hugo ethics committee. 2000. "Hugo Ethics Committee Statement on Benefit Sharing: April 9, 2000." *Clinical Genetics* 58 (5): 364–366.
International Human Genome Sequencing Consortium. 2004. "Finishing the Euchromatic Sequence of the Human Genome." *Nature* 431 (7011): 931–945.
Kabanda, Siti M., Nezerith Cengiz, Kanshukan Rajaratnam, Bruce W. Watson, Qunita Brown, Tonya M. Esterhuizen, and Keymanthri Moodley. 2023. "Data Sharing and Data Governance in Sub-Saharan Africa: Perspectives from Researchers and Scientists Engaged in Data-Intensive Research." *South African Journal of Science* 119 (5–6): 15129.
Kamau, Claude, Larisse Prinsen, and Donrich Thaldar. 2025. "Benefit-Sharing with Human Participants in Health Research in South Africa: A Call for Clarity." *Developing World Bioethics* 25 (2): 144–153.
Kasule, Mary, Mogomotsi Matshaba, Ambroise Wonkam, and Jantina de Vries. 2023. "Feeding Back of Individual Genetic Results in Botswana: Mapping Opportunities and Challenges." *BMC Medical Ethics* 24 (1): 37.
Kaye, Jane, Catherine Heeney, Naomi Hawkins, Jantina de Vries, and Paula Boddington. 2009. "Data Sharing in Genomics—Re-Shaping Scientific Practice." *Nature Reviews. Genetics* 10 (5): 331.
Kohn, Donald B., Yvonne Y. Chen, and Melissa J. Spencer. 2023. "Successes and Challenges in Clinical Gene Therapy." *Gene Therapy* 30 (10–11): 738–746.
Kutyabami, Paul, Herbert Muyinda, Moses Mukuru, Erisa Mwaka, Kamba Pakoyo, Joan Kalyango, and Nelson K. Sewankambo. 2024. "Unmasking the Ethical Dimensions of Data-Sharing in Health Research: Perspectives from Researchers at a Public University in Uganda." *Research Square* (October), rs.3.rs—5204585.
Kwachou, Monique. 2025. "Pathways for Pragmatic Decolonisation in Research." *Open Research Europe* 5 (April): 112.
The Lancet Global Health. 2018. "Closing the Door on Parachutes and Parasites." *The Lancet Global Health* 6 (6): e593.
Lemke, Amy A., D. Bick, D. Dimmock, P. Simpson, and R. Veith. 2013. "Perspectives of Clinical Genetics Professionals toward Genome Sequencing and Incidental Findings: A Survey Study." *Clinical Genetics* 84 (3): 230–236.
Lemke, Amy A., Edward D. Esplin, Aaron J. Goldenberg, Claudia Gonzaga-Jauregui, Neil A. Hanchard, Julie Harris-Wai, Justin E. Ideozu, et al. 2022. "Addressing Underrepresentation in Genomics Research through Community Engagement." *American Journal of Human Genetics* 109 (9): 1563–1571.

Lerner, Barbara, Nell Marshall, Sabine Oishi, Andrew Lanto, Martin Lee, Alison B. Hamilton, Elizabeth M. Yano, and Maren T. Scheuner. 2017. "The Value of Genetic Testing: Beyond Clinical Utility." *Genetics in Medicine: Official Journal of the American College of Medical Genetics* 19 (7): 763–771.

Maqoma, Walter P. 2020. "In Defence of Communitarianism Philosophy: The Contribution of Moderate Communitarianism to the Formation of an African Identity." *Verbum et Ecclesia* 41 (1). https://doi.org/10.4102/ve.v41i1.1984.

Marshall, Eliot. 2001. "Bermuda Rules: Community Spirit, with Teeth." *Science* (February). https://doi.org/10.1126/science.291.5507.1192.

Marshall, Patricia, Charmaine D. M. Royal, and Ruth Chadwick. 2022. "Translational Science, DNA Commercialization, and Informed Consent: The Need for Specific Terminology, Insights from a Review of H3Africa Projects." *Public Health Genomics* 25 (3–4): 112.

Matimba, A., S. Ali, K. Littler, E. Madden, P. Marshall, S. McCurdy, V. Nembaware, et al. 2022. "Guideline for Feedback of Individual Genetic Research Findings for Genomics Research in Africa." *BMJ Global Health* 7 (1). https://doi.org/10.1136/bmjgh-2021-007184.

Mohamed, Farah Nabil Abdelnaim, Mercury Shitindo, and Angel Gasch. 2022. "African History of Research Ethics Violations: Lessons for the Future." *University of Zaragoza*. https://doi.org/10.13140/RG.2.2.24967.68004.

Morrissey, Clair, and Rebecca L. Walker. 2012. "Funding and Forums for ELSI Research: Who (or What) Is Setting the Agenda?" *AJOB Primary Research* 3 (3): 51–60.

Moyi Okwaro, Ferdinand, and P. W. Geissler. 2015. "In/Dependent Collaborations: Perceptions and Experiences of African Scientists in Transnational HIV Research." *Medical Anthropology Quarterly* 29 (4): 492–511.

Mulder, Nicola, Clement A. Adebamowo, Sally N. Adebamowo, Oladimeji Adebayo, Osimhiarherhuo Adeleye, Mohamed Alibi, Shakuntala Baichoo, et al. 2017. "Genomic Research Data Generation, Analysis and Sharing—Challenges in the African Setting." *Data Science Journal* 16 (49): 1–15.

Munung, Nchangwi Syntia, and Jantina de Vries. 2020. "Benefit Sharing for Human Genomics Research: Awareness and Expectations of Genomics Researchers in Sub-Saharan Africa." *Ethics & Human Research* 42 (6): 14–20.

Munung, Nchangwi Syntia, Jantina de Vries, and Bridget Pratt. 2021. "Genomics Governance: Advancing Justice, Fairness and Equity through the Lens of the African Communitarian Ethic of Ubuntu." *Medicine, Health Care, and Philosophy* 24 (3): 377.

Munung, Nchangwi Syntia, Patricia Marshall, Megan Campbell, Katherine Littler, Francis Masiye, Odile Ouwe-Missi-Oukem-Boyer, Janet Seeley, D. J. Stein, Paulina Tindana, and Jantina de Vries. 2016. "Obtaining Informed Consent for Genomics Research in Africa: Analysis of H3Africa Consent Documents." *Journal of Medical Ethics* 42 (2): 132–137.

Munung, Nchangwi Syntia, Bongani M. Mayosi, and Jantina de Vries. 2017. "Equity in International Health Research Collaborations in Africa: Perceptions and Expectations of African Researchers." *PLoS One* 12 (10): e0186237.

Munung, Nchangwi Syntia, Obiageli E. Nnodu, Patrick Ohiani Moru, Akpaka A. Kalu, Benido Impouma, Marsha J. Treadwell, and Ambroise Wonkam. 2024. "Looking Ahead: Ethical and Social Challenges of Somatic Gene Therapy for Sickle Cell Disease in Africa." *Gene Therapy* 31 (5–6): 202–208.

Mweemba, Oliver, John Musuku, Bongani M. Mayosi, Michael Parker, Rwamahe Rutakumwa, Janet Seeley, Paulina Tindana, and Jantina de Vries. 2019. "Use of Broad Consent and Related Procedures in Genomics Research: Perspectives from Research Participants in the Genetics of Rheumatic Heart Disease (RHDGen) Study in a University Teaching Hospital in Zambia." *Global Bioethics* (March): 184–199.

Nankya, Harriet, Vincent P. Alibu, Enock Matovu, Edward Wamala, and John Barugahare. 2024. "Community Engagement Conduct for Genetics and Genomics Research: A Qualitative Study of the Experiences and Perspectives of Key Stakeholders in Uganda." *BMC Medical Ethics* 25 (1): 1–16.

Nembaware, Victoria, African Genomic Medicine Training Initiative, and Nicola Mulder. 2019. "The African Genomic Medicine Training Initiative (AGMT): Showcasing a Community and Framework Driven Genomic Medicine Training for Nurses in Africa." *Frontiers in Genetics* 10 (December): 1209.

Nichol, Ariadne A., and Annick Antierens. 2021. "Ethics of Emerging Infectious Disease Outbreak Responses: Using Ebola Virus Disease as a Case Study of Limited Resource Allocation." *PloS One* 16 (2): e0246320.

Nielsen, Jane, and Dianne Nicol. 2024. "Data Ownership in Genomic Research Consortia." *Journal of Law and the Biosciences* 11 (2): lsae024.

Nikitin, Dmitriy, Francesca L. Beaudoin, Praveen Thokala, Avery McKenna, Emily Nhan, David M. Rind, and Steven D. Pearson. 2023. "Gene Therapies for Sickle Cell Disease: Effectiveness and Value: A Summary from the Institute for Clinical and Economic Review's California Technology Assessment Forum." *Journal of Managed Care & Specialty Pharmacy* 29 (11): 1253–1259.

Nolte, Insa. 2019. "The Future of African Studies: What We Can Do to Keep Africa at the Heart of Our Research." *Journal of African Cultural Studies* (March): 296–313.

Odeny, Beryne, and Raffaella Bosurgi. 2022. "Time to End Parachute Science." *PLoS Medicine* 19 (9): e1004099.

Odipo, Emily, Prashant Jarhyan, Jacinta Nzinga, Dorairaj Prabhakaran, Amit Aryal, Emma Clarke-Deelder, Sailesh Mohan, et al. 2024. "The Path to Universal Health Coverage in Five African and Asian Countries: Examining the Association between Insurance Status and Health-Care Use." *The Lancet. Global Health* 12 (1): e123–133.

Ralefala, Dimpho, Mary Kasule, Olivia P. Matshabane, Ambroise Wonkam, Mogomotsi Matshaba, and Jantina de Vries. 2021. "Participants' Preferences and Reasons for Wanting Feedback of Individual Genetic Research Results From an HIV-TB Genomic Study: A Case Study From Botswana." *Journal of Empirical Research on Human Research Ethics* (December). https://doi.org/10.1177/15562646211043985.

Ralefala, Dimpho, Mary Kasule, Ambroise Wonkam, Mogomotsi Matshaba, and Jantina de Vries. 2020. "Do Solidarity and Reciprocity Obligations Compel African Researchers to Feedback Individual Genetic Results in Genomics Research?" *BMC Medical Ethics* 21 (1): 112.

Ramsay, Michèle. 2022. "African Genomic Data Sharing and the Struggle for Equitable Benefit." *Patterns (New York, N.Y.)* 3 (1): 100412.

Remes Lenicov, Federico, and Nilda E. Fink. 2023. "Ethical Issues in the Use of Leftover Samples and Associated Personal Data Obtained from Diagnostic

Laboratories." *Clinica Chimica Acta; International Journal of Clinical Chemistry* 548 (August): 117442.

Schroeder, D. 2007. "Benefit Sharing: It's Time for a Definition." *Journal of Medical Ethics* 33 (4): 205–209.

Schultz, Baergen, Francis E. Agamah, Cornelius Ewuoso, Ebony B. Madden, Jennifer Troyer, Michelle Skelton, and Erisa Mwaka. 2024. "Webinar Report: Stakeholder Perspectives on Informed Consent for the Use of Genomic Data by Commercial Entities." *Journal of Medical Ethics* 50 (1): 57–61.

Selormey, P. E., I. H. Tsey, J. Ganle, P. Akweongo, and P. Tindana. 2025. "Developing a Contextually and Culturally Relevant Benefit-Sharing Framework for Pathogen Genomic Research and Biobanking in Africa: A Deliberative Expert Approach." *BMC Medical Ethics* 26: 77.

Shalowitz, David I., and Franklin G. Miller. 2008. "Communicating the Results of Clinical Research to Participants: Attitudes, Practices, and Future Directions." *PLoS Medicine* 5 (5): e91.

Sibomana, Olivier. 2024. "Genetic Diversity Landscape in African Population: A Review of Implications for Personalized and Precision Medicine." *Pharmacogenomics and Personalized Medicine* 17 (November): 487.

Skelly, Lara, and Elisha R. T. Chiware. 2022. "African Researchers Do Not Think Differently about Open Data." *Frontiers in Research Metrics and Analytics* 7 (July): 950212.

Staunton, C., R. Adams, E. S. Dove, N. Harriman, L. Horn, M. Labuschaigne, N. Mulder, et al. 2019. "Ethical and Practical Issues to Consider in the Governance of Genomic and Human Research Data and Data Sharing in South Africa: A Meeting Report." *AAS Open Research* 2: 15.

Staunton, Ciara, and Jantina de Vries. 2020. "The Governance of Genomic Biobank Research in Africa: Reframing the Regulatory Tilt." *Journal of Law and the Biosciences* 7 (1): lsz018.

Staunton, Ciara, Aliki Edgcumbe, Lukman Abdulrauf, Amy Gooden, Paul Ogendi, and Donrich Thaldar. 2025. "Cross-Border Data Sharing for Research in Africa: An Analysis of the Data Protection and Research Ethics Requirements in 12 Jurisdictions." *Journal of Law and the Biosciences* 12 (1): lsaf002.

Stokstad, Erik. 2019. "Major U.K. Genetics Lab Accused of Misusing African DNA." October 30. https://www.science.org/content/article/major-uk-genetics-lab-accused-misusing-african-dna.

Sudoi, Allan, Jantina de Vries, and Dorcas Kamuya. 2021. "A Scoping Review of Considerations and Practices for Benefit Sharing in Biobanking." *BMC Medical Ethics* 22 (1): 1–16.

Thaldar, Donrich, and Bonginkosi Shozi. 2023. "Is Benefit Sharing with Research Participants Lawful in South Africa? An Unexplored Question in the Governance of Genomics Research." *Journal of Law and the Biosciences* 10 (1): lsad018.

Tiffin, Nicki. 2024. "Colonial Legacies Fuel Unfair Practices in African Research." SciDev.Net. February 16. https://www.scidev.net/global/opinions/colonial-legacies-fuel-unfair-practices-in-african-research/.

Tindana, P., and J. de Vries. 2016. "Broad consent for Genomic Research: Perspectives from Low and Middle in Come Countries." *Annual Review of Genomics and Human Genetics* 22.

Tindana, Paulina, M. Campbell, P. Marshall, K. Littler, R. Vincent, J. Seeley, J. de Vries, D. Kamuya, and H3Africa Community Engagement Working Group. 2017. "Developing the Science and Methods of Community Engagement for Genomic Research and Biobanking in Africa." *Global Health, Epidemiology and Genomics* 2: e13.

Tindana, Paulina, J. de Vries, M. Campbell, K. Littler, J. Seeley, P. Marshall, J. Troyer, et al. 2015. "Community Engagement Strategies for Genomic Studies in Africa: A Review of the Literature." *BMC Medical Ethics* 16: 24.

Tindana, Paulina, Catherine S. Molyneux, Susan Bull, and Michael Parker. 2014. "Ethical Issues in the Export, Storage and Reuse of Human Biological Samples in Biomedical Research: Perspectives of Key Stakeholders in Ghana and Kenya." *BMC Medical Ethics* 15 (1): 1–11.

Tindana, Paulina, Sassy Molyneux, Susan Bull, and Michael Parker. 2019a. "'It Is an Entrustment': Broad Consent for Genomic Research and Biobanks in Sub-Saharan Africa." *Developing World Bioethics* 19 (1): 9–17.

Tindana, Paulina, Aminu Yakubu, Ciara Staunton, Alice Matimba, Katherine Littler, Ebony Madden, Nchangwi Syntia Munung, and Jantina de Vries. 2019b. "Engaging Research Ethics Committees to Develop an Ethics and Governance Framework for Best Practices in Genomic Research and Biobanking in Africa: The H3Africa Model." *BMC Medical Ethics* 20 (1): 1–7.

Trinidad, Susan Brown, Stephanie M. Fullerton, Julie M. Bares, Gail P. Jarvik, Eric B. Larson, and Wylie Burke. 2010. "Genomic Research and Wide Data Sharing: Views of Prospective Participants." *Genetics in Medicine: Official Journal of the American College of Medical Genetics* 12 (8): 486–495.

Wand, Hannah, Daphne O. Martschenko, Annamaria Smitherman, Sheryl Michelson, Ting Pun, John S. Witte, Stuart A. Scott, Mildred K. Cho, and Euan A. Ashley. 2023. "Re-Envisioning Community Genetics: Community Empowerment in Preventive Genomics." *Journal of Community Genetics* 14 (5): 459–469.

Watson, J. D., and F. H. C. Crick. 1953. "Molecular Structure of Nucleic Acids: A Structure for Deoxyribose Nucleic Acid." *Nature* 171 (4356): 737–738.

The Wellcome Trust. 2003. "Sharing Data from Large-Scale Biological Research Projects: A System of Tripartite Responsibility. Report of a Meeting Organized by the Wellcome Trust and Held on 14–15 January 2003 at Fort Lauderdale, USA." Fort Lauderdale, USA. https://www.genome.gov/Pages/Research/WellcomeReport0303.pdf.

WHO. 2024. "Guidance for Human Genome Data Collection, Access, Use and Sharing." Geneva: World Health Organization. https://iris.who.int/bitstream/handle/10665/379554/9789240102149-eng.pdf?sequence=1.

Wohlers, Inken, Axel Künstner, Matthias Munz, Michael Olbrich, Anke Fähnrich, Verónica Calonga-Solís, Caixia Ma, et al. 2020. "An Integrated Personal and Population-Based Egyptian Genome Reference." *Nature Communications* 11 (1): 4719.

Wolf, Leslie E., Timothy A. Bouley, and Charles E. McCulloch. 2010. "Genetic Research with Stored Biological Materials: Ethics and Practice." *IRB* 32 (2): 7.

Wonkam, Ambroise, and Jantina de Vries. 2020. "Returning Incidental Findings in African Genomics Research." *Nature Genetics* 52 (1): 17.

Woollard, Laurence, Richard Gorman, and Dakota J. Rosenfelt. 2021. "Improving Patient Informed Consent for Haemophilia Gene Therapy: The Case for Change." *Therapeutic Advances in Rare Disease*. https://doi.org/10.1177/26330040211047244.

World Health Organization. 2023. "HIV Statistics, Globally and by WHO Region." Geneva: World Health Organization. July 17. https://www.who.int/publications/i/item/WHO-UCN-HHS-SIA-2023-01.

World Health Organization. 2024. *Guidance for Human Genome Data Collection, Access, Use and Sharing*. World Health Organization.

Yakubu, A., P. Tindana, A. Matimba, K. Littler, N. S. Munung, E. Madden, C. Staunton, and J. De Vries. 2018. "Model Framework for Governance of Genomic Research and Biobanking in Africa: A Content Description." *AAS Open Research* 1: 13.

Chapter 6

The Genomics of Sickle Cell Disease in Africa

Oyesola Ojewunmi[1*], *Titilope Adeyemo*[2], *and Stephan Menzel*[3]

[1]Precision Healthcare University Research Institute, Queen Mary University of London, London, UK.
[2]Department of Haematology, University of Lagos, Lagos, Nigeria.
[3]Red Cell Biology, Comprehensive Cancer Centre, King's College London, London, UK.

6.1 INTRODUCTION

6.1.1 Overview and Pathophysiology of Sickle Cell Disease

Globally, approximately 300,000 children are born with sickle cell disease every year; 75% of these are in sub-Saharan Africa.[1] Mortality due to sickle cell disease (SCD) is still alarming in Africa, where 50% of children with the disorder do not live up to their 5th birthday, in contrast to high-income countries, where up to 99% of children with SCD live into adulthood with a median survival of 60 years.[2–7]

Sickle cell anaemia (SCA) is the first known molecular disease and one of the most common single-gene disorders globally. It is an autosomal recessive blood disorder caused by a single nucleotide substitution (a point mutation replacing glutamic acid: GAG for valine: GTG) in the 6th position of the β-chain of haemoglobin, resulting in a structurally abnormal β-haemoglobin (HbS). The term "sickle cell disease" encompasses homozygosity for HbS, known as sickle cell anaemia (HbSS), compound heterozygosity for HbS and other structural haemoglobin variants like HbC (HbSC disease), or with thalassaemia (HbSβ-thalassaemia occurring as $HbS\beta^+$ or $HbS\beta^0$). HbSS is the most common phenotype for SCD, followed by HbSC.

Haemoglobin is a polypeptide tetramer composed of four globin subunits—two α-like and two β-like globin chains—responsible for oxygen transport. The α-globin chain—consisting of 141 amino acid residues—is encoded by duplicated genes on the short arm of chromosome 16. These genes, HBA2: α2 and HBA1: α1, are separated by approximately 4kb of DNA. The β-globin chain, consisting of 146 amino acid residues, is mapped to chromosome 11 along with δ and γ globin chains. A diploid cell, therefore,

***Corresponding author:** o.ojewunmi@qmul.ac.uk (Oyesola Ojewunmi)

DOI: 10.1201/9781003436584-6

has four α-globin genes, two β-globin genes δ, γ (Aγ and Gγ), and ε globin genes, alongside a β-globin pseudogene (ψβ) (Figure 6.1).

The primary driver for the pathophysiological events in SCD is the polymerization of HbS. This occurs under low oxygen tension when HbS

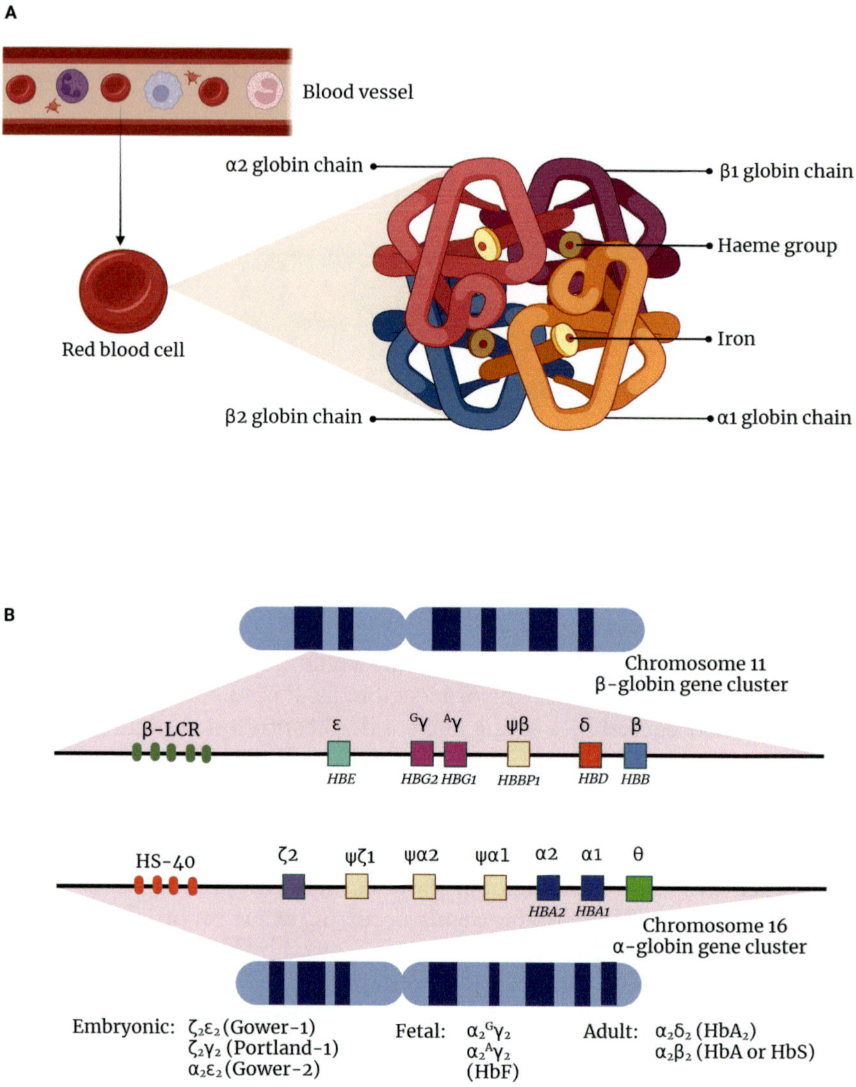

Figure 6.1 Haemoglobin structure: α- and β-globin genes. (A) Haemoglobin is composed of four globin subunits—two α-like and two β-like globin chains—responsible for oxygen transport. (B) The α-globin chain comprises 141 amino acid residues, which is encoded by duplicated genes on the short arm of chromosome 16. The β-globin chain, which consists of 146 amino acid residues, is mapped to chromosome 11 alongside δ and γ globin chains. A diploid cell has four α-globin genes, two β-globin genes δ, γ (Aγ and Gγ), and ε globin genes, along with a β-globin pseudogene (ψβ).

becomes deoxygenated (and forms polymers through hydrophobic interactions between valine on one chain and other amino acids, like alanine and phenylalanine, on the other chain). This process distorts the normal bi-concave red blood cell shape to give rise to a deformed, "sickled" red blood cell with altered properties and functions.[8] A key effect is the blockage of small blood vessels and impaired oxygen delivery to tissues, cascading into major clinical complications such as episodes of pain, multi-organ damage, and early mortality. These complications are mediated by a set of pathogenic processes, including vaso-occlusion, haemolytic anaemia, oxidative stress, hypercoagulability, inflammation, endothelial dysfunction, and ischaemia-reperfusion injury.[9,10]

6.1.2 Historical Perspectives of SCD and Distribution in Africa

Sickled red blood cells were first described in the United States in a dental student from Grenada, West Indies, in 1910 by James B. Herrick,[11] and, in 1949, Pauling and colleagues[12] coined the term "molecular disease" as they identified and explained the electrophoretic abnormalities in sickle haemoglobin (HbS).

The sickle mutation (β^S) originated in the Sahara or western central Africa around 7,000 years ago,[13] subsequently spreading across the continent and expanding to reach high frequencies within distinct geographical regions that overlap with historical malaria endemicity: Three in West Africa, two in Central/South-West Africa, and the third covering the Arabian peninsula and parts of India.[14] Today, effects of such focussed regional expansion and later migration can still be seen in patients' genotypes: Genetic variation within the β-globin gene cluster has shown that the sickle mutation (HbS: rs334—c.20A>T [p.Glu7Val] in *HBB*) is mapped to five typical β-globin haplotypes based on the patterns (presence or absence) of restriction fragment length polymorphism sites (Figures 6.2 and 6.3), named after the geographic or ethnolinguistic regions where they mostly occur: Senegal, Cameroon, Benin, Bantu/Central Africa Republic, and Arab-Indian).[15] Generally, these haplotypes can be inferred using four restriction sites: 5′ ε HincII, G_γ HindIII, A_γ HindIII, and ψβ HincII.[13] Other haplotypes that do not fit into these classes are referred to as atypical haplotypes.

6.1.3 Epidemiology and Burden of SCD in Africa

Nigeria, the Democratic Republic of Congo, Uganda, and Tanzania are the countries most affected in Africa due to the high prevalence of the sickle cell trait (SCT, HbAS phenotype) (Figure 6.4). The prevalence of SCT is below 1% in South Africa, while it is about 2% in North Africa. In contrast, it ranges between 15% and 30% across West African countries like Nigeria and the Democratic Republic of Congo.[16] These differences reflect the

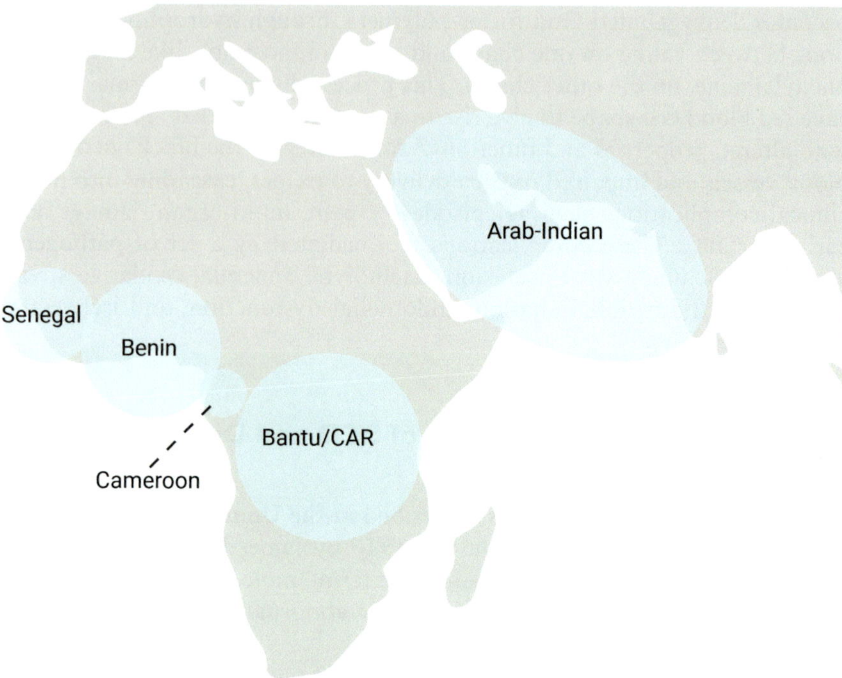

Figure 6.2 Origins of classical β-globin haplotypes. Five sickle haplotypes reflect historical population movements and contribute to differences in SCD severity and clinical outcomes. Individuals with Senegal and Arab-Indian haplotypes usually are characterized by higher foetal haemoglobin levels and milder disease severity, whereas the Bantu haplotype is associated with more severe complications.

protective role of SCT against malaria, as regions with higher prevalence correspond with historically malaria-endemic areas. The selection pressure exerted by malaria led to the elevated frequencies of the mutant gene seen in these regions. Conversely, homozygote HbS (sickle cell anaemia) is not protective against malaria infection.

In Nigeria, the frequency of SCT is 24%. With an incidence of Hb SS of 20 per 1,000 births, Nigeria has an annual birth of approximately 150,000 affected babies. It is the country with the largest burden of SCD globally.[16] HbSC distribution in Africa is commonly concentrated in West Africa, with the highest prevalence in Ghana (40% of those attending a sickle cell clinic).[17] In Nigeria, the HbC allele is predominant in the southwest (3.4%), and the estimated birth prevalence is 0.83%, compared to 0.01% in the South east.[7] HbS is sometimes co-inherited with β-thal0/β-thal+ in Africa, although not very frequently. SCD also occurs in individuals of African descent in America and the Caribbean, in the Mediterranean, and in regional Indian populations.

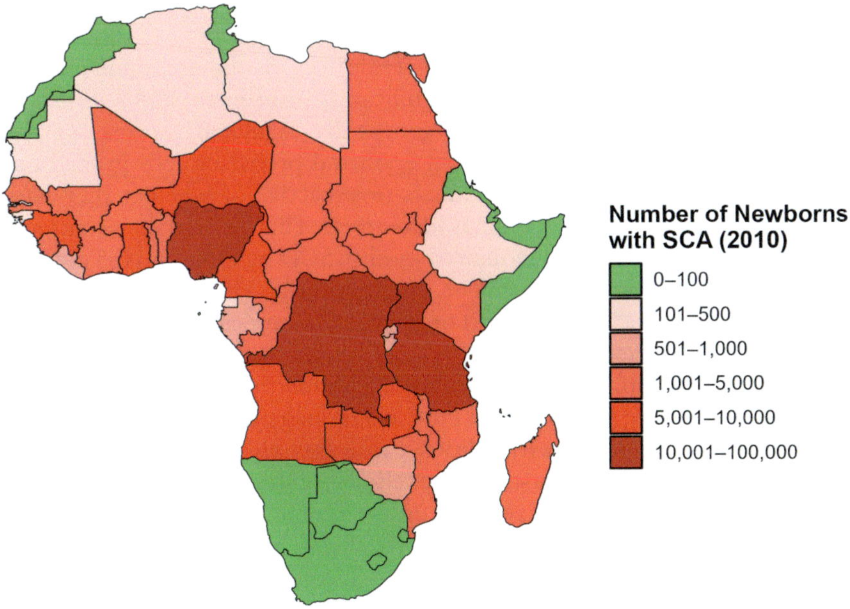

Figure 6.3 The annual newborns of SCA in Africa. Nigeria, Democratic Republic of Congo, Uganda, and Tanzania are the countries with the largest number of newborns of SCD in Africa and globally every year. (We used published data from Piel et al. for the map.[1])

Epidemiological data for most African countries are still sparse due to the lack of universal newborn screening. However, the crucial, practical effect of this is the failure to intervene early, for instance, with pneumococcal vaccination. Overall, inadequate healthcare (including absence of primary stroke prevention, hydroxyurea treatment, and blood transfusion), along with malaria infection and malnutrition, dampen the health-related quality of life and life expectancy of SCD individuals.[10,18]

6.2 GENETIC MODIFIERS AND HETEROGENEITY OF SCD IN AFRICA

In addition to the actual mutational genotype at the β-globin locus, proven genetic SCD modifiers fall roughly into three categories: Modifiers of foetal haemoglobin (HbF) production, modifiers of cellular globin concentration (mainly α-thalassaemia), and variants that act in specific organ systems impacted by SCD pathology.

6.2.1 Foetal Haemoglobin ($\alpha_2\gamma_2$: HbF)

During gestation, foetal haemoglobin (HbF: $\alpha_2\gamma_2$) is the predominant haemoglobin (up to 75%). It declines significantly after birth, and a switch to adult haemoglobin (HbA: $\alpha_2\beta_2$), a genetically regulated process (known as haemoglobin switching), occurs in erythroid precursor cells in the bone marrow. However, in some individuals (irrespective of whether they have SCD), the haemoglobin switching is incomplete. This leads to the persistence of HbF that is pronounced enough to modify SCD by diluting HbS concentration and inhibiting sickling. The distribution of HbF across red blood cells, whether even ("pancellular") or uneven ("heterocellular"), also affects the clinical disease expression.[19]

Deletional and non-deletional mutations within the β-globin gene cluster can cause hereditary persistence of HbF in a monogenic fashion, though they have little impact on SCD variability because they are caused by allelic variants of the β-globin gene cluster that exclude the sickle mutation. However, such variants have inspired research into γ-globin gene regulation, which has led to novel gene-therapeutic approaches. Common genetic variants throughout the genome have more muted effects on HbF levels, but in combination, they have the potential to alter clinical presentation and severity.

Foetal haemoglobin levels themselves have been shown to influence the clinical course of SCD considerably. In a study by Platt et al., it was shown that SCD patients with HbF $\geq$ 8.6% had a higher survival rate.[20] Many studies have also reported the beneficial effect of increased HbF on SCD clinical manifestations, such as splenic function, pain, and leg ulcers.[21–23]

6.2.2 β-Globin Haplotypes and HbF

The five β-globin haplotypes surrounding the sickle cell mutation are associated with different HbF levels and, consequently, different clinical outcomes. Individuals with SCD carrying the Arab-Indian (AI) and the Senegal haplotype are known to have higher levels of HbF (up to 22% and 10%, respectively), and a milder clinical expression of SCD compared to those carrying the Bantu haplotype, which is associated with more severe complications. However, avascular necrosis has been reported in Kuwaiti patients carrying AI haplotype.[24] Despite this haplotype being common in the Middle East and India, evidence suggests that it originated from eastern Africa to the Middle East and India through the slave trade or migration.[13,25,26] The XmnI (*HBG2*) polymorphism characterizing Senegal and Arab-Indian haplotypes is thought to be one of the causal variants conferring their HbF-boosting advantage. The Senegal haplotype is the predominant haplotype in Senegal.[27] Central African Republic (Bantu) haplotypes usually have the lowest HbF level and are associated with the most severe clinical presentation of SCD, while Cameroon and Benin haplotypes have intermediate HbF levels. The Bantu haplotype is the prevalent haplotype in Congo, Uganda, Kenya, and Tanzania.[28–31] Benin haplotype is the common haplotype in Nigeria (87–97%) and Cameroon (74%), with an additional 19% Cameroon haplotype in SCD patients in Cameroon.[32–35]

Renewed study of the five (and other) haplotypes, using modern genomic technology and focusing on African populations, creates opportunities to unlock additional insight with potential therapeutic relevance. So far, studies have not been able to differentiate between the effect of the haplotypes themselves and that of the genetic/ethnic background associated with them. A mainstay of quality control for genetic studies, the Hardy-Weinberg test, often flags such analysis as unreliable, due to population stratification. Modern-day genetic research involves principal component analysis or the inclusion of relationship matrices based on genome-wide data, which can adjust for this and help to correctly identify local (e.g., β-globin haplotype-derived) effects. Furthermore, studying specific African populations where atypical haplotypes are prevalent will enable resolving and fine-mapping HbF-modifying genetic variation in and around the β-globin gene cluster and identify new regions with therapeutic potential.

6.2.3 α-Thalassaemia

The co-inheritance of α-thalassaemia has been recognized as an alleviating factor for SCD. The commonest form of α-thalassaemia in sub-Saharan Africa is the 3.7 kb α-globin chain deletion ($\alpha^{3.7}$ kb deletion), characterized by the loss of a single (heterozygous: –α/αα) or of two copies of the gene (homozygous: –α/–α), which reduces α-globin chain production. The frequency of the $\alpha^{3.7}$ kb deletion (both heterozygous and homozygous) in Africa ranges from 30% to 50%.[36–40] The prevalence of α-thalassaemia in Africa is linked to its selective advantage through survival of malaria and sickle cell anaemia.[28,41,42] α-Thalassaemia reduces HbS concentration and polymerization and, in turn, lowers haemolytic rates.[43] It is associated with a lower mean corpuscular volume, mean corpuscular haemoglobin concentration, red blood cell counts, reticulocyte counts, haemoglobin levels, HbF, and HbA2.[41–46] The effect of α-thalassaemia depends on whether the deletion is homozygous or heterozygous, with stronger clinical and laboratory effects of the heterozygous deletion.[44,46] Studies involving individuals with SCD in continental Africa have shown that co-inheritance of α-thalassaemia with SCA is associated with later onset of complications and prolonged survival, although another study among HbSS in the UK reported that α-thalassaemia did not impact survival.[42,47] Furthermore, α-thalassaemia improves haematological indices and lowers hospital consultations; it also provides protection against abnormal transcranial Doppler and hepatomegaly[38,48–50] but does not protect against splenomegaly. In addition, it reduces the risk of leg ulcerations, acute chest syndrome, nephropathy, splenic sequestration, and cholelithiasis.[21,23,46,47,51–53]

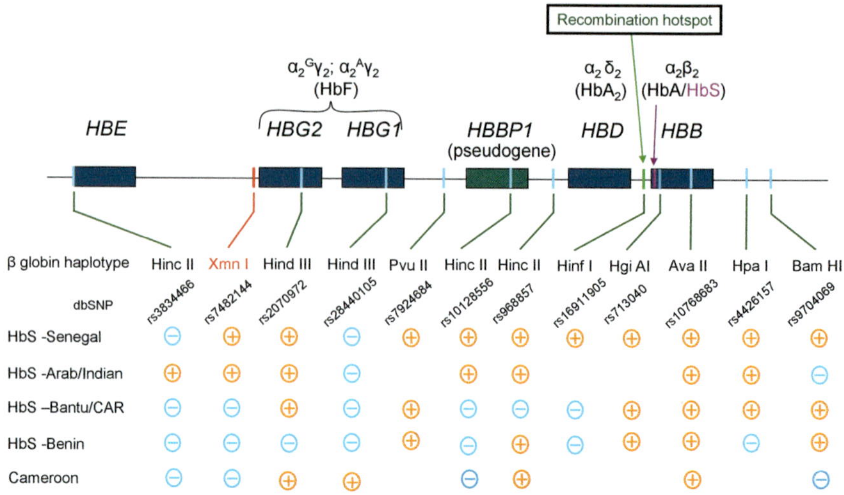

Figure 6.4 β-Globin haplotypes around the sickle mutation. Genetic variation within the β-globin gene cluster has shown that the sickle mutation is mapped to five typical β-globin haplotypes. The haplotypes are named after the geographic or ethnolinguistic regions where they mostly occur (namely, Senegal, Cameroon, Benin, Bantu/Central Africa Republic, and the Arab-Indian).

6.3 GENETICS OF HBF CONDUCTED IN AFRICA AND ELSEWHERE

Three major HbF-modifier loci have been shown in healthy individuals and persons with SCD.[54,55] These loci are *BCL11A* on chromosome 2, *HBS1L-MYB* (*HBS1L-MYB* intergenic polymorphism: *HMIP*) on chromosome 6, and *XmnI-HBG2* on chromosome 11. While the first two loci have been replicated in genome-wide association studies (GWAS) of HbF conducted in Africa among SCD individuals,[56,57] the third locus, XmnI-*HBG2*, seen in GWAS of African American patients, was not detected. In the Nigerian cohorts, for instance, less than 1% (7 out of 1006) of individuals with SCD had *XmnI-HBG2* polymorphism similar to a MAF of 0.5% (in 596 individuals) reported in a candidate gene study in Cameroonians.[58] This reflects the rarity of the Senegal haplotype in Nigeria and Cameroon.[57]

Association of HbF with ten SNPs in *BCL11A*, *HMIP*, *XmnI-HBG2*, and *OR51B5/6* in Cameroonian SCD patients showed that rs4671393 in *BCL11A*, rs9399137, and rs9494142 in *HBS1L-MYB* are HbF-boosting SNPs while rs28384513 in *HBS1L-MYB* is associated with lower levels of HbF.[58] The lead SNP at *BCL11A* in the Tanzanian cohort was rs1427407 with rs6545816 as an independent signal. Meanwhile, rs7606173 was identified in the Nigerian cohort and rs1427407 as the secondary signal.[56,57] The Nigerian cohort further refined the fine-mapping resolution of

BCL11A and *HBS1L-MYB (HMIP)* loci. At *BCL11A*, two HbF-boosting haplotypes were identified (defined by rs1427407 T>G, rs7565301 G>A, and rs7606173 G>A). One haplotype (TGG) is tagged by rs1427407-T, a known functional SNP, while the second HbF-increasing haplotype harbours rs7565301-A, indicating that a causative variant linked to rs7565301 may influence this haplotype's effect. At *HBS1L-MYB*, two haplotypes (defined by rs61028892 G>C and rs9399137 T>C) also have a strong HbF boosting effect; one haplotype was characterized by rs9399137-C and carries rs66650391, a known functional variant. The second haplotype, tagged rs61028892-C (lies within the HMIP-2A region), shares the same haplotype block as the *HMIP-2B* sub-locus previously identified in the Tanzanian cohort. This suggests that *HMIP-2A* and *HMIP-2B* form a single sub-locus characterized by the presence of two HbF-boosting alleles: rs66650371-del and rs61028892-C. The differences observed between these two African cohorts highlight haplotype diversity in African populations and how variations in linkage disequilibrium (LD) structure and allele frequencies across populations can affect fine-mapping efforts. Studying more African SCD populations could provide novel insight into the genetic architecture of SCD, aiding the discovery of disease-/trait-associated variants for therapeutic development.

Recent studies reported novel loci, including *BACH2*, *ABCC1*, *NFIX*, *ZBTB7A*, *KLF1*, and *FLT1*, to be associated with HbF levels in individuals with SCD and without SCD.[59–61] A replication study will be necessary to validate these new loci and three novel loci (*SLC28A3*, *TICRR/KIF7*, and *PIEZO2*) identified in the Nigerian cohort.[57] The heritability of HbF was estimated to be 89%, obtained in a study of non-anaemic twins.[62] Phenotypic variance attributed to SNPs at the main known HbF loci explained only 20–25% of the overall trait variance in Cameroonian, Nigerian, and Tanzanian SCD patients, but 31% in African American patients.[57,63,64] The presence of the *XmnI-HBG2* site in African American patients increases the impact of genetic factors on HbF variability in this population. Still, at least half of the heritability of HbF levels in SCD has not been accounted for.

The discovery of the *BCL11A* gene (HbF repressor gene) through GWAS led to the functional experiment that showed that *BCL11A* harbours erythroid enhancer elements with three DNase I hypersensitive sites, with rs1427407 being causal as an Hb-F boosting SNP.[65] This functional insight and many validation experiments enabled the development of a gene therapy to target *BCL11A*, and this has been approved for use in clinical care in the UK and the US for individuals with SCD.[66]

Looking at the history of HbF-genetic research, it is quite revealing that solidly confirmed modifier loci have arisen almost exclusively from the study of non-anaemic individuals. However, a confirmation that such genes are indeed disease relevant always requires their study in patient cohorts of sufficient size, which will likely be available only in high-prevalence regions and countries. On the other hand, to fully utilize the African genetic landscape for disease-relevant discoveries, with its diversity, haplotype structure, and population-genetic history, will require its thorough characterization based on non-anaemic population cohorts with recorded haematological data.

6.4 CHALLENGES IN DIAGNOSING AND MANAGING SCD

The diagnosis of sickle cell disease depends mainly on the laboratory identification and confirmation of the haemoglobin variant through a combination of biochemical and molecular tests. However, in many African countries, there are practically few, if any, facilities capable of providing definitive diagnosis and management of SCD. Molecular tests are virtually inaccessible, while conventional techniques—such as full blood count, peripheral blood smears, sickling solubility test, Hb electrophoresis, and high-performance liquid chromatography—are considered the gold standard in the continent. These methods are plagued by low sensitivity and specificity and poor accuracy, leading to a high incidence of misdiagnosis.

A study of 182 parents with a child with SCD found that 59 (32.4%) had previously received a haemoglobin phenotype misdiagnosis.[67] Similarly, another study reported that 35% of all parents had received incorrect results on the risk of SCD in their offspring based on phenotyping.[68]

Managing SCD in Africa is fraught with significant challenges, primarily due to a lack of comprehensive care programmes. In most cases, children are not diagnosed until there are clinical manifestations, delaying interventions. Infection remains a leading cause of mortality among children under 5 with SCD in sub-Saharan Africa. The lack of national newborn screening programmes accounts for high mortality rates, which would typically include pneumococcal vaccination against infection, prophylactic penicillin, and counselling for caregivers. Comprehensive management of SCD, such as stroke prevention programmes, hydroxyurea therapy, and blood transfusions, is also limited across the region.

While newborn screening pilot studies have been conducted and some are ongoing in some African countries, the lack of governmental policies and infrastructural support introduces major gaps in scaling up this type of programme.[69–75] At the same time, some health facilities have initiated newborn screening programmes for SCD as part of their standard of care.[76]

Hydroxyurea (HU) was the first and until 2017 the only US Food and Drug Administration–approved drug (in 1998) for SCD management. Three additional drugs have been approved since then: L-glutamine and crizanlizumab for the management of vaso-occlusive pain in SCD and voxelotor to increase haemoglobin levels.[77–79] However, voxelotor (Oxbryta) was withdrawn in September 2024 due to safety concerns. HU primarily boosts HbF (reducing intracellular HbS polymerization—a primary pathophysiological process in SCD) while improving clinical and laboratory outcomes. For instance, it reduces neutrophil, reticulocyte, platelet counts, haemolysis, and rates of acute chest syndrome, transfusion, and vaso-occlusive crises while enhancing quality of life and survival.[80–82] Since the response to hydroxyurea is genetically determined, pharmacogenomics research is essential to optimize its use in African populations.

A study involving 606 SCA children from Angola, the Democratic Republic of Congo, Kenya, and Uganda demonstrated that HU led to a reduction in clinical events such as malaria, vaso-occlusion, transfusion needs, and mortality.[83] Similarly, a multicentre trial in Nigeria demonstrated lower hospitalization rates among children on HU therapy.[84]

HU is an established protocol for stroke prevention in SCD, and is essential, especially for difficulties associated with regular blood transfusion in Africa. Studies in Africa have shown that HU lowers abnormal transcranial Doppler velocities (stroke risk) in children.[85,86]

Despite its proven benefits and safety in children and adults with SCD and being a cost-effective and relatively affordable treatment option for SCD in Africa, access to HU remains alarmingly low.[83,87] Financial barriers are a significant obstacle to HU usage, as caregivers often bear the cost of treatment. Ensuring its widespread access and use is critical for reducing morbidity and improving the quality of life of individuals affected by this condition.

Blood transfusion is a core part of the standard of care for SCD in high-income settings. However, in many parts of Africa, there are substantial challenges to accessing it. Firstly, there is an insufficient supply of blood and blood components due to inefficient blood banking services. Additionally, the financial burden of blood transfusion therapy often falls on patients and their families, especially where health insurance schemes are underdeveloped and out-of-pocket payments are required for healthcare services.

Hematopoietic stem cell transplantation for SCD has up to a 93% success rate for those who meet the eligibility criteria.[88] However, once again, the high cost limits its accessibility in sub-Saharan Africa, despite its availability in three countries (Nigeria, Tanzania, and South Africa).[89,90]

If basic interventions like newborn screening programmes and HU therapy cannot be successfully implemented at a national scale in African countries, the prospects for comprehensive management of SCD will remain bleak. In such a context, access to advanced treatments such as stem cell transplantation and gene therapy is likely to remain out of reach for the majority of those affected by this condition in Africa.

6.5 SCD GENOMICS RESEARCH IN AFRICA

While the sheer amount of suffering due to SCD in Africa is unparalleled, there is also hope due to the unique opportunities for genomic research arising from African patient populations and from the exceptional experience and insight of local clinical and non-clinical researchers, which eventually will lead to better treatment and prevention.

With the increased availability of newborn screening[91] and early intervention, SCD is being transformed from one of a paediatric disease with a marked childhood mortality to one of adult disease characterized by end-organ complications. A rigorous evaluation of different forms of traditional and modern treatments and of different environments and lifestyles, malaria exposure, nutritional status, and other factors can disentangle the complexities of the clinical spectrum of SCD, supported by the potential to assemble large patient research cohorts in African countries.

Few genotype–phenotype correlation studies in SCD have been conducted in Africa, limiting our understanding of disease-causing variants. Beyond the genetic studies on HbF and the effects of α-thalassaemia mentioned here, there is a pressing need for more genomic studies focusing on the clinical phenotypes of SCD, such as stroke (especially in children), nephropathy, retinopathy, cholelithiasis, priapism, leg ulceration, and vaso-occlusive pains. The limited availability of SCD in African genomics data reflects a broader issue of the underrepresentation of African populations in global genomic research efforts. Many landmark SCD genomic studies have been conducted on African American populations, which may not represent the diversity of African populations due to differences in ancestry, environments, lifestyle, and other socio-economic factors. Hence, there is an urgent need to prioritize SCD genomic research within the African continent. While most studies and funding have focused on epidemiological and clinical data, there remain significant gaps in well-characterized longitudinal phenotypes and biorepositories that could support SCD genomic studies in Africa. These resources will facilitate the integration of environmental, clinical, and social determinants into genomics research, advancing our understanding of SCD clinical phenotypes across diverse African populations.

Apolipoprotein L1 (*APOL1*) is an established risk factor for kidney disease in the general population of African descent and has also been linked to nephropathy in SCD.[92–94] Two studies involving individuals with SCD in Cameroon and the Democratic Republic of Congo have also investigated the relationship between *APOL1* variants and albuminuria.[95,96] Additionally, the *CLCN6* (a voltage-dependent chloride channel) and *OGHDL* (implicated in arginine metabolism) genes have been identified as being associated with survival in Cameroonian SCD patients.[97] To better understand the genetic determinants of stroke, kidney disease, and other phenotypes in SCD, larger studies across many sub-Saharan African countries using whole-genome sequencing are necessary. These studies would also benefit from integrating epidemiological and environmental factors to help explain the heterogeneity of SCD manifestations in Africa. Such genomic findings could revolutionize the management of children identified through newborn screening for preventive measures and offer opportunities for risk prediction, early diagnosis, and therapeutics.

6.6 GENE THERAPY IN AFRICA

Whilst the African continent has not demonstrated much-needed commitment to healthcare and research, high-income countries have made a significant impact in advancing healthcare, including gene therapy research and its clinical utility. Many African countries are still facing challenges such as a lack of political will and funding, inadequate infrastructure for research, and a shortage of specialized research skills. Gene therapy for SCD has been approved in the UK and the US. In the UK, the world's first CRISPR (Clustered Regularly Interspaced Short Palindromic Repeats) gene therapy (Casegevy, also known as Exa-cel) was approved in 2025 for individuals with SCD and β-thalassaemia aged 12 and above. This therapy is estimated to cost over USD1 million per individual, and it is planned for 50 individuals annually. This high cost currently prevents widespread use in Africa, where the majority of the affected individuals live. However, African-led therapies could be a game-changer if African countries invest in building research infrastructure and develop capacity and regulatory frameworks.[98] The Muhimbili Sickle Cell Programme has kick-started international collaborations and is currently developing research infrastructure in preparation for gene therapy for sickle cell disease in Tanzania.[99]

6.7 SCD GENOMICS AND ETHICAL CONSIDERATIONS IN AFRICA

6.7.1 Access to Genetic Testing

Worldwide, millions of people live with genetic diseases, with a likely higher birth prevalence in sub-Saharan Africa. Sadly, access to genetic testing remains inequitable. In high-income countries, a baby with a genetic disease can be diagnosed within hours to days after birth, allowing for timely intervention. In sub-Saharan countries, diagnosis for individuals with genetic diseases is missed for months or even years, if a diagnosis is made at all.

6.7.2 Informed Consent

Ethical issues related to genetic testing, including informed consent, have been recognized for some time. Ethicists have long argued that genetic testing should be treated uniquely and be subjected to a more rigorous informed consent process than medical tests. Unlike standard medical tests, genetic testing provides information not only about the diagnosis of an existing disease but also foretells the possibility of a future disease. This knowledge can expose individuals to psychological distress, stigmatization, and discrimination based on a predisposing gene. Additionally, the discovery of a particular genetic trait has implications not only for the individual tested but also for their family members and may be a cause of family discord.

Informed consent ensures that individuals undergoing genetic testing are apprised that the process could reveal whether they have, are at risk for, or are carriers of a specific disease. It also informs them that the results may have significant consequences or require difficult decisions regarding their current or future health, career, marriage, or reproductive options.

6.7.3 Genetic Counselling

Premarital counselling and testing can reduce the prevalence of inherited disorders of haemoglobin by identifying and offering counselling to intending couples with high-risk pregnancies. In the USA, genetic counselling and testing, established as early as 1970, have also demonstrated positive outcomes in reducing the likelihood of at-risk couples giving birth to affected children. Similarly, antenatal screening and counselling programmes for haemoglobin disorders based on ethnicity have been introduced in many European countries. In countries with universal genetic counselling and screening for genetic disorders, the information collected provides valuable insight into the epidemiology of these conditions and serves as a crucial resource for health planning.

Genetic counselling is central to any effective SCD control programme. However, despite sub-Saharan Africa's significant burden of SCD, dedicated genetic counselling and testing services are largely unavailable in these countries due to a lack of trained genetic counsellors. Sickle Cell Foundation Nigeria has been effective in this area by organizing an intensive 2-week genetic counselling training course on SCD to healthcare professionals. In 2022, a formal, accredited genetic counselling training programme commenced at the West African Genetic Medicine Centre, University of Ghana. They welcomed the first cohort of students into their flagship master of science degree programme—the first of its kind in Ghana and sub-Saharan Africa, and only the third such programme on the continent.

To build capacity for genetic counselling in sub-Saharan Africa, it has been suggested that governments should establish accredited genetic counselling degree-training programs; create national, regional, and continental regulatory bodies; integrate genetic counselling into mainstream care for all patients; organize specialized training for nurses; and promote genetic and genomic research and policy development.

6.8 FUTURE DIRECTIONS

Integrating newborn screening and access to preventive care for SCD into national health systems across Africa will change the narrative of the outcomes for people with this disease. These programmes would not only improve clinical care but also serve as a powerful resource for genomic and translational research. By enabling the investigation of SCD across the life course, they can also inform the development of context-specific and targeted interventions.

With significant gaps remaining in SCD genomics and without dedicated resources to bridge these gaps, health disparities will continue to widen. Additionally, the novel insight that African genomics could offer, which may not be observable in other populations, will be missed. To bridge these gaps, it is essential to prioritize establishing SCD registries in Africa with longitudinal cohorts and genomics data, invest in biorepositories, and build capacity in statistical genetics and bioinformatics. One cohort addressing this gap is the Sickle Cell Disease Genomics of Africa Network, established in 2018, as part of H3Africa (an initiative jointly funded by the US National Institutes of Health and the UK-based Wellcome Trust).[100] This study has recruited 7,000 individuals with SCD from Ghana, Nigeria, Tanzania, and South Africa to investigate the genomics of intravascular haemolysis and the development of organ damage.

Expanding on such initiatives in SCD genomics will enhance research and unlock the potential of genomics to transform the lives of millions affected by this disorder across Africa.

ACKNOWLEDGEMENTS

We are grateful to Julian Hoffmann Anton of Precision Healthcare University Research Institute, Queen Mary University of London, for his assistance with Figure 6.4.

REFERENCES

1. Piel, F.B. *et al.* Global epidemiology of sickle haemoglobin in neonates: A contemporary geostatistical model-based map and population estimates. *Lancet* **381**, 142–51 (2013).
2. Gardner, K. *et al.* Survival in adults with sickle cell disease in a high-income setting. *Blood* **128**, 1436–8 (2016).
3. Couque, N. *et al.* Improvement of medical care in a cohort of newborns with sickle-cell disease in North Paris: Impact of national guidelines. *Br J Haematol* 173, 927–37 (2016).
4. Quinn, C.T., Rogers, Z.R., McCavit, T.L. & Buchanan, G.R. Improved survival of children and adolescents with sickle cell disease. *Blood* **115**, 3447–52 (2010).

5. Telfer, P. *et al.* Clinical outcomes in children with sickle cell disease living in England: A neonatal cohort in East London. *Haematologica* **92**, 905–12 (2007).
6. Grosse, S.D. *et al.* Sickle cell disease in Africa: A neglected cause of early childhood mortality. *Am J Prev Med* **41**, S398–405 (2011).
7. Nnodu, O.E. *et al.* Child mortality from sickle cell disease in Nigeria: A model-estimated, population-level analysis of data from the 2018 Demographic and Health Survey. *Lancet Haematol* **8**, e723–31 (2021).
8. Odievre, M.H., Verger, E., Silva-Pinto, A.C. & Elion, J. Pathophysiological insights in sickle cell disease. *Indian J Med Res* **134**, 532–7 (2011).
9. Rees, D.C., Williams, T.N. & Gladwin, M.T. Sickle-cell disease. *Lancet* **376**, 2018–31 (2010).
10. Piel, F.B., Steinberg, M.H. & Rees, D.C. Sickle cell disease. *N Engl J Med* **376**, 1561–73 (2017).
11. Herrick, J.B. Peculiar elongated and sickle-shaped red blood corpuscles in a case of severe anemia. *Arch Intern Med* **83** (1910).
12. Pauling, L., Itano, H.A. & et al. Sickle cell anemia a molecular disease. *Sci* **110**, 543–8 (1949).
13. Shriner, D. & Rotimi, C.N. Whole-genome-sequence-based haplotypes reveal single origin of the Sickle Allele during the holocene wet phase. *Am J Hum Genet* **102**, 547–56 (2018).
14. Bodmer, W. & Cavalli-Sforza, L.L. *Genetics, Evolution and Man*, 307–28 (W H Freeman & Co, San Francisco, 1976).
15. Nagel, R.L. & Steinberg, M.H. Genetics of the beta-S gene: Origins, genetic epidemiology, and epistasis in sickle cell anemia. In *Disorders of Hemoglobin: Genetics, Pathophysiology, and Clinical Management* (ed. Steinberg, M.H.F.B., Higgs, D.R., & Nagel, R.L.) 711–55 (Cambridge University Press, Cambridge, UK, 2001).
16. WHO. *World Health Organisation Sickle-Cell Anaemia Report by the Secretariat* (Fifty Ninth World Health Assembly, 2006).
17. Asare, E.V. *et al.* Burden of sickle cell disease in Ghana: The Korle-Bu experience. *Adv Hematol* **2018**, 6161270 (2018).
18. Piel, F.B. *et al.* Defining global strategies to improve outcomes in sickle cell disease: A Lancet haematology commission. *Lancet Haematol* **10**, e633–86 (2023).
19. Thein, S.L., Menzel, S., Lathrop, M. & Garner, C. Control of fetal hemoglobin: New insights emerging from genomics and clinical implications. *Hum Mol Genet* **18**, R216–23 (2009).
20. Platt, O.S. *et al.* Mortality in sickle cell disease: Life expectancy and risk factors for early death. *N Engl J Med* **330**, 1639–44 (1994).
21. Koshy, M. *et al.* Leg ulcers in patients with sickle cell disease. *Blood* **74**, 1403–8 (1989).
22. Platt, O.S. *et al.* Pain in sickle cell disease: Rates and risk factors. *N Engl J Med* **325**, 11–6 (1991).
23. Wali, Y.A. *et al.* Splenic function in Omani children with sickle cell disease: Correlation with severity index, hemoglobin phenotype, iron status, and alpha-thalassemia trait. *Pediatr Hematol Oncol* **19**, 491–500 (2002).

24. Adekile, A.D. Limitations of Hb F as a phenotypic modifier in sickle cell disease: Study of Kuwaiti Arab patients. *Hemoglobin* **35**, 607–17 (2011).
25. Gelpi, A.P. Migrant populations and the diffusion of the sickle-cell gene. *Ann Intern Med* **79**, 258–64 (1973).
26. Bhagat, S., Patra, P.K. & Thakur, A.S. Fetal haemoglobin and beta-globin gene cluster haplotypes among sickle cell patients in Chhattisgarh. *J Clin Diagn Res* **7**, 269–72 (2013).
27. Currat, M. *et al.* Molecular analysis of the beta-globin gene cluster in the Niokholo Mandenka population reveals a recent origin of the beta(S) Senegal mutation. *Am J Hum Genet* **70**, 207–23 (2002).
28. Mouele, R., Pambou, O., Feingold, J. & Galacteros, F. Alpha-thalassemia in Bantu population from Congo-Brazzaville: Its interaction with sickle cell anemia. *Hum Hered* **50**, 118–25 (2000).
29. Ndugwa, C. *et al.* Homozygous sickle cell disease in Uganda and Jamaica a comparison of Bantu and Benin haplotypes. *West Indian Med J* **61**, 684–91 (2012).
30. Ojwang, P.J. *et al.* Haplotypes and alpha globin gene analyses in sickle cell anaemia patients from Kenya. *Br J Haematol* **65**, 211–5 (1987).
31. Oner, C. *et al.* Beta S haplotypes in various world populations. *Hum Genet* **89**, 99–104 (1992).
32. Adabale, A. *et al.* Frequency of beta S globin gene haplotypes among sickle cell patients in Nigeria. *J Int Med Res* **49**, 3000605211019918 (2021).
33. Adekile, A.D. *et al.* Haplotypes in SS patients from Nigeria; characterization of one atypical beta S haplotype no. 19 (Benin) associated with elevated HB F and high G gamma levels. *Ann Hematol* **65**, 41–5 (1992).
34. Falusi, A.G. & Kulozik, A.E. Relationship of foetal haemoglobin levels and beta s haplotypes in homozygous sickle cell disease. *Eur J Haematol* **45**, 1–4 (1990).
35. Bitoungui, V.J., Pule, G.D., Hanchard, N., Ngogang, J. & Wonkam, A. Beta-globin gene haplotypes among Cameroonians and review of the global distribution: Is there a case for a single sickle mutation origin in Africa? *OMICS* **19**, 171–9 (2015).
36. Adeyemo, T.A., Ojewunmi, O.O., Oyetunji, I.A., Kalejaiye, O.O. & Menzel, S. Fetal-haemoglobin enhancing genotype at BCL11A reduces HbA(2) levels in patients with sickle cell anaemia. *EJHaem* **2**, 459–61 (2021).
37. Saraf, S.L. *et al.* Associations of alpha-thalassemia and BCL11A with stroke in Nigerian, United States, and United Kingdom sickle cell anemia cohorts. *Blood Adv* **1**, 693–8 (2017).
38. Lubega, I., Ndugwa, C.M., Mworozi, E.A. & Tumwine, J.K. Alpha thalassemia among sickle cell anaemia patients in Kampala, Uganda. *Afr Health Sci* **15**, 682–9 (2015).
39. Mikobi, T.M. *et al.* Association between sickle cell anemia and alpha thalassemia reveals a high prevalence of the alpha(3.7) triplication in Congolese patients than in worldwide series. *J Clin Lab Anal* **32**(2018).
40. Wonkam, A. *et al.* Coinheritance of sickle cell anemia and alpha-thalassemia delays disease onset and could improve survival in Cameroonian's patients (Sub-Saharan Africa). *Am J Hematol* **89**, 664–5 (2014).

41. Enevold, A. *et al.* Associations between alpha+-thalassemia and Plasmodium falciparum malarial infection in northeastern Tanzania. *J Infect Dis* **196**, 451–9 (2007).
42. Mears, J.G., Lachman, H.M., Labie, D. & Nagel, R.L. Alpha-thalassemia is related to prolonged survival in sickle cell anemia. *Blood* **62**, 286–90 (1983).
43. de Ceulaer, K. *et al.* alpha-Thalassemia reduces the hemolytic rate in homozygous sickle-cell disease. *N Engl J Med* **309**, 189–90 (1983).
44. Steinberg, M.H. & Embury, S.H. Alpha-thalassemia in blacks: Genetic and clinical aspects and interactions with the sickle hemoglobin gene. *Blood* **68**, 985–90 (1986).
45. Steinberg, M.H. Predicting clinical severity in sickle cell anaemia. *Br J Haematol* **129**, 465–81 (2005).
46. Higgs, D.R. *et al.* The interaction of alpha-thalassemia and homozygous sickle-cell disease. *N Engl J Med* **306**, 1441–6 (1982).
47. Brewin, J.N. *et al.* The pleiotropic effects of alpha-thalassemia on HbSS and HbSC sickle cell disease: Reduced erythrocyte cation co-transport activity, serum erythropoietin, and transfusion burden, do not translate into increased survival. *Am J Hematol* **97**, 1275–85 (2022).
48. Rumaney, M.B. *et al.* The co-inheritance of alpha-thalassemia and sickle cell anemia is associated with better hematological indices and lower consultations rate in Cameroonian patients and could improve their survival. *PLoS One* **9**, e100516 (2014).
49. Ojewunmi, O.O. *et al.* Association of alpha-thalassemia and Glucose-6-Phosphate Dehydrogenase deficiency with transcranial Doppler ultrasonography in Nigerian children with sickle cell anemia. *J Clin Lab Anal* **35**, e23802 (2021).
50. Cox, S.E. *et al.* Haptoglobin, alpha-thalassaemia and glucose-6-phosphate dehydrogenase polymorphisms and risk of abnormal transcranial Doppler among patients with sickle cell anaemia in Tanzania. *Br J Haematol* **165**, 699–706 (2014).
51. Adekile, A.D. & Haider, M.Z. Morbidity, beta S haplotype and alpha-globin gene patterns among sickle cell anemia patients in Kuwait. *Acta Haematol* **96**, 150–4 (1996).
52. Adekile, A.D. *et al.* Influence of alpha-thalassemia trait on spleen function in sickle cell anemia patients with high HbF. *Am J Hematol* **53**, 1–5 (1996).
53. Vasavda, N. *et al.* The linear effects of alpha-thalassaemia, the UGT1A1 and HMOX1 polymorphisms on cholelithiasis in sickle cell disease. *Br J Haematol* **138**, 263–70 (2007).
54. Menzel, S. *et al.* A QTL influencing F cell production maps to a gene encoding a zinc-finger protein on chromosome 2p15. *Nat Genet* **39**, 1197–9 (2007).
55. Uda, M. *et al.* Genome-wide association study shows BCL11A associated with persistent fetal hemoglobin and amelioration of the phenotype of beta-thalassemia. *Proc Natl Acad Sci U S A* **105**, 1620–5 (2008).
56. Mtatiro, S.N. *et al.* Genome wide association study of fetal hemoglobin in sickle cell anemia in Tanzania. *PLoS One* **9**, e111464 (2014).
57. Ojewunmi, O.O. *et al.* The genetic dissection of fetal haemoglobin persistence in sickle cell disease in Nigeria. *Hum Mol Genet* **33**, 919–29 (2024).
58. Wonkam, A. *et al.* Association of variants at BCL11A and HBS1L-MYB with hemoglobin F and hospitalization rates among sickle cell patients in Cameroon. *Plos One* **9**(2014).

59. Cato, L.D. *et al.* Genetic regulation of fetal hemoglobin across global populations. *medRxiv* (2023). https://doi.org/10.1101/2023.03.24.23287659
60. Hara, Y. *et al.* Genomic discovery and functional validation of MRP1 as a novel fetal hemoglobin modulator and potential therapeutic target in sickle cell disease. *medRxiv* (2023). https://doi.org/10.1101/2023.03.14.23287244
61. Wonkam, A. *et al.* FLT1 and other candidate fetal haemoglobin modifying loci in sickle cell disease in African ancestries. *Nat Commun* **16**, 2092 (2025).
62. Garner, C. *et al.* Genetic influences on F cells and other hematologic variables: A twin heritability study. *Blood* **95**, 342–6 (2000).
63. Galarneau, G. *et al.* Fine-mapping at three loci known to affect fetal hemoglobin levels explains additional genetic variation. *Nat Genet* **42**, 1049–51 (2010).
64. Gardner, K. *et al.* g(HbF): A genetic model of fetal hemoglobin in sickle cell disease. *Blood Adv* **2**, 235–39 (2018).
65. Bauer, D.E. *et al.* An erythroid enhancer of BCL11A subject to genetic variation determines fetal hemoglobin level. *Science* **342**, 253–7 (2013).
66. Frangoul, H. *et al.* CRISPR-Cas9 gene editing for sickle cell disease and beta-thalassemia. *N Engl J Med* **384**, 252–60 (2021).
67. Adekunle, M.O., Ojewunmi, O., Animasahun, A.B., Lawani, F.O., Ubuane, P.O. Prevalence, determinants and impact of haemoglobin phenotype misdiagnosis among parents of children living with sickle cell disease in Nigeria. *Journal of Pediatric Research* **8**, 239–45.
68. Aderotoye-Oni, S., Diaku-Akinwumi, I.N., Adeniran, A. & Falase, B. Unprepared and misinformed parents of children with sickle cell disease: Time to rethink awareness campaigns. *Cureus* **10**, e3806 (2018).
69. Nkya, S. *et al.* Building research capacity for sickle cell disease in Africa: Lessons and challenges from establishing a birth cohort in Tanzania. *Front Pediatr* **10**, 826199 (2022).
70. Nnodu, O. *et al.* HemoTypeSC, a low-cost point-of-care testing device for sickle cell disease: Promises and challenges. *Blood Cells Mol Dis* **78**, 22–8 (2019).
71. Oluwole, E.O. *et al.* Feasibility and acceptability of early infant screening for sickle cell disease in Lagos, Nigeria: A pilot study. *PLoS One* **15**, e0242861 (2020).
72. Rahimy, M.C., Gangbo, A., Ahouignan, G. & Alihonou, E. Newborn screening for sickle cell disease in the Republic of Benin. *J Clin Pathol* **62**, 46–8 (2009).
73. Tubman, V.N. *et al.* Newborn screening for sickle cell disease in Liberia: A pilot study. *Pediatr Blood Cancer* **63**, 671–6 (2016).
74. Ndeezi, G. *et al.* Burden of sickle cell trait and disease in the Uganda sickle surveillance study (US3): A cross-sectional study. *Lancet Glob Health* **4**, e195–200 (2016).
75. Ojewunmi, O.O., Adeyemo, T.A., Ayinde, O.C., Iwalokun, B. & Adekile, A. Current perspectives of sickle cell disease in Nigeria: Changing the narratives. *Expert Rev Hematol* **12**, 609–20 (2019).
76. Galadanci, A.A. *et al.* A novel newborn screening program for sickle cell disease in Nigeria. *Int J Neonatal Screen* **10** (2024).
77. Ataga, K.I. *et al.* Crizanlizumab for the prevention of pain crises in sickle cell disease. *N Engl J Med* **376**, 429–39 (2017).
78. Niihara, Y. *et al.* A phase 3 trial of l-glutamine in sickle cell disease. *N Engl J Med* **379**, 226–35 (2018).

79. Vichinsky, E. *et al.* A phase 3 randomized trial of voxelotor in sickle cell disease. *N Engl J Med* **381**, 509–19 (2019).
80. Lobo, C.L. *et al.* The effect of hydroxcarbamide therapy on survival of children with sickle cell disease. *Br J Haematol* **161**, 852–60 (2013).
81. McGann, P.T. & Ware, R.E. Hydroxyurea therapy for sickle cell anemia. *Expert Opin Drug Saf* **14**, 1749–58 (2015).
82. Wang, W.C. *et al.* Hydroxycarbamide in very young children with sickle-cell anaemia: A multicentre, randomised, controlled trial (BABY HUG). *Lancet* **377**, 1663–72 (2011).
83. Tshilolo, L. *et al.* Hydroxyurea for children with sickle cell anemia in sub-Saharan Africa. *N Engl J Med* **380**, 121–31 (2019).
84. Abdullahi, S.U. *et al.* Hydroxyurea for primary stroke prevention in children with sickle cell anaemia in Nigeria (SPRING): A double-blind, multicentre, randomised, phase 3 trial. *Lancet Haematol* **9**, e26–e37 (2022).
85. Galadanci, N.A. *et al.* Moderate fixed-dose hydroxyurea for primary prevention of strokes in Nigerian children with sickle cell disease: Final results of the SPIN trial. *Am J Hematol* **95**, E247–50 (2020).
86. Lagunju, I., Brown, B.J. & Sodeinde, O. Hydroxyurea lowers transcranial Doppler flow velocities in children with sickle cell anaemia in a Nigerian cohort. *Pediatr Blood Cancer* **62**, 1587–91 (2015).
87. Ofakunrin, A.O.D. *et al.* Effectiveness and safety of hydroxyurea in the treatment of sickle cell anaemia children in Jos, North Central Nigeria. *J Trop Pediatr* **66**, 290–8 (2020).
88. Gluckman, E. *et al.* Sickle cell disease: An international survey of results of HLA-identical sibling hematopoietic stem cell transplantation. *Blood* **129**, 1548–56 (2017).
89. Bazuaye, N. *et al.* First successful allogeneic hematopoietic stem cell transplantation for a sickle cell disease patient in a low resource country (Nigeria): A case report. *Ann Transplant* **19**, 210–3 (2014).
90. Rwezaula, S. *et al.* Challenges and outcomes of the first stem cell transplant program in Tanzania, East Africa. *Adv Hematol* **2024**, 1937419 (2024).
91. Brito, M., Ginete, C., Ofakunrin, A., Diaku-Akinwumi, I. & Inusa, B.P.D. Treating sickle cell disease in resource-limited sub-Saharan Africa: Recent strategies and recommendations in addressing the gaps for the provision of evidence-based management. *Expert Rev Hematol* **18**, 447–62 (2025).
92. Ashley-Koch, A.E. *et al.* MYH9 and APOL1 are both associated with sickle cell disease nephropathy. *Br J Haematol* **155**, 386–94 (2011).
93. Genovese, G. *et al.* Association of trypanolytic ApoL1 variants with kidney disease in African Americans. *Science* **329**, 841–5 (2010).
94. Gbadegesin, R.A. *et al.* APOL1 Bi- and Monoallelic Variants and Chronic Kidney Disease in West Africans. *N Engl J Med* (2024).
95. Adebayo, O.C. *et al.* Clinical and genetic factors are associated with kidney complications in African children with sickle cell anaemia. *Br J Haematol* **196**, 204–14 (2022).
96. Geard, A. *et al.* Clinical and genetic predictors of renal dysfunctions in sickle cell anaemia in Cameroon. *Br J Haematol* **178**, 629–39 (2017).

97. Wonkam, A. *et al.* Genetic modifiers of long-term survival in sickle cell anemia. *Clin Transl Med* **10**, e152 (2020).
98. Arbuthnot, P., Maepa, M.B., Ely, A. & Pepper, M.S. The state of gene therapy research in Africa, its significance and implications for the future. *Gene Ther* **24**, 581–9 (2017).
99. Bukini, D. *et al.* Strengthening advanced therapy for sickle cell disease in Africa: Experience from sickle cell disease centre in Dar es Salaam, Tanzania. *BMJ Glob Health* **10** (2025).
100. Ofori-Acquah, S.F. & SickleGenAfrica, N. The SickleGenAfrica network. *Lancet Glob Health* **8**, e12556 (2020).

Chapter 7

Genomics and Cardiometabolic Diseases in Africa

Amy R. Bentley[1*], *Opeyemi Soremekun*[2*], *and Adebowale A. Adeyemo*[1]

[1]Department of Clinical and Biomedical Sciences, Faculty of Health and Life Sciences, University of Exeter, Exeter EX4 4QJ, UK

[2]School of Health Sciences, University of KwaZulu-Natal, Westville Campus, Durban, South Africa

7.1 INTRODUCTION

Some of the diseases responsible for the largest burden of mortality and morbidity in the world are cardiovascular disease (CVD) and type 2 diabetes (T2D). Importantly, however, considerable interplay exists between these diseases. For instance, T2D is a major risk factor for CVD, and more than 30% of T2D patients worldwide have CVD (1). CVD is the major cause of mortality for T2D patients (1,2). T2D and CVD share common risk factors (such as obesity) and preventative strategies. In appreciation of this overlap, it is often more appropriate to discuss risk of cardiometabolic diseases (CMD) together and, indeed, the metabolic syndrome, defined by thresholds related to obesity, hypertension, and dyslipidemia, is designed to identify elevated risk of both CVD and T2D. Lifestyle factors related to CVD and T2D risk have long been appreciated and lifestyle changes are key to prevention strategies. However, genomic factors are also important in one's consideration of disease risk. For example, variants in the *TCF7L2* locus are strongly associated with an increased risk of T2D (3) and *PCSK9* variants with low-density lipoprotein cholesterol concentration (LDLC) (4). Genome-wide studies have been used to create polygenic risk scores (PRS) that are associated with levels of increased risk that are comparable to those of rare monogenic disorders (5). With the maturation of these types of strategies, it becomes increasingly possible to identify an individual's genetically inherited risk, even if that individual does not carry a high-risk, low-frequency variant but instead has multiple more moderate, more common variants. These developments promise future clinical utility for most individuals, not just those with rare Mendelian forms of CMD.

*co-first authors

Corresponding author: amy.bentley@nih.gov (Amy R. Bentley)

DOI: 10.1201/9781003436584-7

Importantly, these advances depend on underlying large-scale genomics research, and utility of these tools will favor those who are better represented in that research (6). This issue has garnered substantial focus in recent years, leading to calls for increased inclusion of diverse populations in genomic research. To address the gap in representation, major initiatives have been launched, including the Human Heredity & Health in Africa (H3Africa) (7) consortium, the All of Us Research Program (8), and Trans-Omics for Precision Medicine (TOPMed) program (9). The continent of Africa has figured prominently in discussions of representation for several reasons. First, as the birthplace of humanity and the site where humanity evolved for many millennia prior to out-of-Africa migration, the continent of Africa is home to a high degree of genetic diversity (10), which cannot be captured in studies that do not include individuals with African ancestry. Second, while diasporan populations, such as African Americans, are generally better represented in research, they are not a proxy for studying the genomics of Africans, as they have genetic admixture from other ancestry groups, they have quite a distinct environmental background from Africans, and their African ancestry is predominantly from West Africa and does not represent the entire continent (11). Finally, there remain substantial gaps in the genomic sampling across Africa, meaning that the full degree of genetic diversity is currently unknown.

In this chapter, we will review genomic research of CVD, T2D, and related traits as well as key cardiometabolic risk factors, such as obesity, serum lipids, and the metabolic syndrome (MetS). We have limited our focus to sub-Saharan Africa (SSA) and to genome-wide approaches, with emphasis on recent work (2018 and later). Background on the epidemiological landscape of cardiometabolic diseases/traits in SSA is provided for context.

7.2 EPIDEMIOLOGICAL LANDSCAPE OF CARDIOMETABOLIC DISEASES IN AFRICA

Although Africa has the lowest T2D prevalence among International Diabetes Federation regions (4.5%), the prevalence is expected to increase dramatically in coming years. Africa is anticipated to have the highest increase in prevalence at 129%. Importantly, there is considerable variability in T2D prevalence among SSA countries, with some countries experiencing more than twice the average African prevalence: Tanzania (12.3%), Zambia (11.9%), Comoros (11.7%), South Africa (10.8%), and Seychelles (8.5%). Unfortunately, 54% of T2D in Africa is undiagnosed, the highest proportion in the world, and diabetes-related expenditures are low (12). The anticipated increases in T2D prevalence, particularly with underdiagnosis, suggest an impending public health crisis that will put even more strain on the overcrowded healthcare system by increasing the already substantial prevalence of T2D complications and comorbidities in SSA (13). Studies have shown sex-specific differences in T2D within Africa as well as variations between rural and urban areas. For example, research has revealed disparities in insulin sensitivity among African populations even within rural settings. A study conducted in Kenya discovered differences in insulin resistance among ethnic groups demonstrating the impact of genetic factors and the environment (14).

CVD risk across Africa has recently been evaluated in the Africa Wits-INDEPTH Partnership for Genomic Studies (AWI-GEN) data. The study compared results from various ten-year CVD risk calculation methods in 10,349 individuals in six sites across four countries: South Africa, Kenya, Burkina Faso, and Ghana. The included sites were at different stages of epidemiological transition and with differing degrees of CVD risk factors. Three different CVD risk algorithms (Framingham, Globorisk, and WHO-CVD) were used, and there was poor correlation between the algorithms within sites. The authors emphasize the importance of assessing population-level risk using validated algorithms or context-specific approaches to guide policy making and individualized CVD treatment in Africa (15). Among AWI-GEN study participants in a low-income urban area in Nairobi, 29% had a high or medium atherogenic risk, with male sex, HIV, uncontrolled T2D, and obesity associated with higher risk (16). A study conducted among Ghanaians living in Europe and Africa highlights how one's environment shapes their CVD risk. Among men, the lowest CVD risk was found in rural Ghana, and highest in the European cities, with those living in urban Ghana intermediate. The differences among Ghanaian women by location was less pronounced. Modifiable CVD risk factors contributed to these differences across population groups. Additionally, each year of residence in a European city significantly increased the predicted ten-year CVD risk among Ghanaian migrants (17). The measurement of common carotid intima media thickness (cIMT) is considered an indicator of atherosclerosis. This condition is characterized by mild inflammation that starts early in life, progresses with aging, and can be influenced by various CVD risk factors. Recent studies have shown that both cIMT and microalbuminuria are closely linked to an increased risk of hypertension (18) and atherosclerotic cardiovascular disease (19) in SSA. Left ventricular hypertrophy (LVH) is a type of cardiac remodeling associated with markedly increased risk of heart failure and stroke, and a study of women in South Africa found that LVH was significantly associated with obesity, hypertension, and tobacco use (20). The annual heart failure mortality rate in Africa is twice the world average (21). The mechanisms underlying these high rates are likely multi-factorial, with contributors including a high prevalence of heart failure with reduced ejection fraction, high burden of co-morbidities, infections, socioeconomic factors, late presentation of heart failure patients, sub-optimal treatment use and availability, and genetic factors (21).

Hypertension poses a significant health challenge in Africa (22). It is considered the leading preventable killer on the continent, making detection, diagnosis, management, and control of hypertension of high priority (23). A cross-sectional study covering seven communities in East and West Africa revealed that 25% of participants had hypertension with 40% of those unaware of their condition. Only half of those aware of their condition were treated, and half of those treated were controlled (24). Data from the Cardiovascular H3Africa Innovation Resource of over 30,000 individuals across 13 countries found that the age-adjusted proportion of hypertension was highest among men, in western Africa, and in individuals with obesity. Obesity was associated with an odds ratio of 2.0 for hypertension among population-based samples of a younger age and 8.8 for those at an older age (25).

Although there is considerable geographical variability across the continent, the prevalence of stroke in many SSA countries is among the highest in the world (26,27). The Stroke Investigative Research and Educational Network (SIREN) enrolled 2,118 case–control pairs for stroke in 15 sites in Nigeria and Ghana (68% ischemic stroke and 32% hemorrhagic stroke). Nearly all of the population attributable risk (98.2%) of stroke in this population was associated with the following modifiable risk factors (in descending order): Hypertension, dyslipidemia, regular meat consumption, elevated waist-to-hip ratio, T2D, low leafy green consumption, stress, added salt at the table, cardiac disease, physical inactivity, and current cigarette smoking (28). A separate investigation in the SIREN study confirmed the relationship of MetS with stroke in Africa. Over 75% of stroke cases had MetS, and MetS was associated with higher stroke severity and higher odds of fatality within 1 month among those who survived the stroke (29). Importantly, stroke outcomes in Africa are poor, and recent research from inpatient stroke fatality in 16 hospitals in Ghana and Nigeria found that nearly 22% of stroke patients admitted died as inpatients. In addition to patient-level factors associated with fatality, such as low vegetable consumption, systolic blood pressure, stroke lesion volume, National Institutes of Health Stroke Score, elevated intracranial pressure, and aspiration pneumonia, over 7% of variability in the odds of stroke fatality could be attributed to system-level factors. Notably, only 5 of 16 centers had stroke units (30).

Triglyceride concentration (TG) has been established as a biomarker of cardiometabolic diseases/traits, yet TG are generally lower among West Africans and those with a high proportion of West African ancestry, such as African Americans, compared to other populations (31). To see what effect this differing distribution has on the associations between TG and cardiometabolic outcomes, Meeks et al. evaluated the association of TG with a range of traits in those with predominantly West African ancestry living in SSA, the US, or Europe, East Africans, and those of European ancestry in Europe and the US. The associations between TG and hypertension, BMI, and waist circumference were stronger among individuals of European ancestry compared to those of African ancestry. In contrast, there was substantial variation in the association between TG and T2D across environments, as opposed to across ancestries (32).

The prevalence of obesity is increasing in SSA (33) and is variable across the continent. In a study of over 10,000 individuals in six sites, the highest prevalence was observed in South African sites (42.3–66.6% [women] and 2.8–17.5% [men]) and the lowest in the West African sites (1.3–4.2% [women] and 1.2–2.2% [men]). There was significantly higher prevalence of obesity among women compared to men in nearly all sites evaluated. Higher socio-economic status and educational level were associated with increased BMI across all sites, while some factors were associated only in a subset of sites tested: being married and higher dietary intake (positive association) and smoking, alcohol, and HIV infection (inverse association) (34). In rural Burkina Faso, where overweight and obesity prevalence are relatively low, higher BMI was observed among men (median BMI 21.1 [men] and 19.8 [women]), with higher prevalence of underweight among women. Across SSA sites, socioeconomic status (men) and education (women) were the factors most strongly associated with higher BMI (35). Urbanization has also been associated with increased obesity rates in Burkina Faso. In urban families, both undernutrition and overnutrition were observed concurrently. In contrast, in rural families some children experienced wasting and stunted growth while adults tended to be underweight (36). Physical activity is associated with lower BMI, and the proportion of those meeting physical activity guidelines is high, more than 80% of AWI-GEN study participants, with most work- or transport-related physical activity (37). To identify cardiometabolic risk, researchers have explored obesity cut points tailored to Africans. Waist-to-hip ratio was found to outperform BMI and waist circumference as an index of T2D risk in Ghanaians, and the cut-offs for BMI and waist circumference in Ghanaian men was lower than what has been set by the WHO (based mainly on European ancestry individuals) (38). Similarly for CVD risk, optimal cut-offs for waist circumference in Ghanaian men were lower than those set by the WHO, while the WHO recommended cut-off performed well in Ghanaian women (39). These results suggest that use of indices not optimized for African individuals may result in under-identification of individuals, particularly men, at increased risk for CMD.

7.3 MENDELIAN RANDOMIZATION STRATEGIES TO CONFIRM RISK FACTORS FOR CARDIOMETABOLIC DISEASES

Mendelian randomization (MR) strategies are employed to validate risk factors for CMD by leveraging genetic variants that are associated with the risk factor of interest to infer causality. By utilizing genetic variants as instrumental variables, MR addresses limitations like confounding and reverse causation encountered in traditional observational studies. MR has been used to assess the relationship between lipid traits and T2D in African ancestry individuals. LDLC and total cholesterol were found to be linked to an increased risk of T2D in individuals of African ancestry while having higher levels of high-density lipoprotein cholesterol concentration (HDLC) is associated with a reduced risk of T2D (40). MR has also been used to investigate the association between metabolic traits and stroke risk, where it was discovered that the link between HDLC and stroke risk was 40% stronger in individuals with African ancestry than in those with European ancestry. The authors propose that this stronger association could be attributed to genetic and environmental factors (41). HDLC levels were found to be associated with cIMT in Europeans, while this association is not observed among those of African ancestry. Neither ancestry showed an association between triglycerides and cIMT. The presence of T2D is significantly associated with cIMT in Europeans but not in Africans. This study emphasizes the influence of genetic/environmental factors on lipid traits and T2D and their potential impact on cIMT, which serves as a marker for cardiovascular disease risk (42). In a genome-wide association study (GWAS) of serum bilirubin in West Africans, Chen et al. identified a strong association of *UGT1A1* variant rs887829. Utilizing this variant as an instrumental variable in an MR of African Americans provided evidence for a causal role of serum bilirubin levels in hypertension among African Americans, with lower levels of serum bilirubin associated with increased risk (43).

7.3.1 Genomic Factors and Cardiometabolic Diseases in Africa

One of the key motivations for studying the genomic factors underlying CMD is the evidence of familial aggregation for these traits. For instance, a meta-analysis of available literature on sub-Saharan Africans found that a positive family history was significantly associated with T2D in SSA, with an odds ratio of 3.29 [95% CI 2.40–4.52] (44). In a study of three generations of urban South African families, significant heritability was observed for cIMT (23–44%) and left ventricular mass index (21–39%) (45), estimates that are similar to those from studies outside of Africa (46–50). In contrast, significant heritability of blood pressure was not observed in this cohort of 195 individuals in 65 families, which the authors suggest may be a result of the relatively small sample size or the high prevalence of elevated blood pressure among these individuals; 88% of grandparents, 46% of parents, and 27% of children in this cohort had hypertension or prehypertension (45). Narrow-sense heritability of blood pressure was also evaluated in 4,778 Ugandans from the Uganda Genome Resource, where the heritability of systolic and diastolic blood pressure were 21% and 18%, respectively (Figure 7.1) (51). Heritability of a variety of other cardiometabolic traits was assessed in the Uganda Genome Resource, including HbA1c (41%), BMI (29%), and waist circumference (24%). For serum lipids, heritability for HDLC, LDLC, and

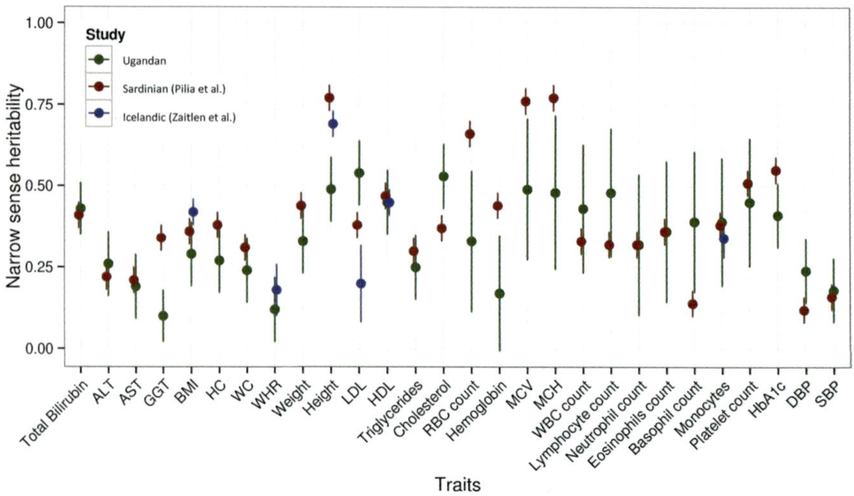

Figure 7.1 Heritabilities for 34 complex traits within the Ugandan GWAS cohort (UGWAS, *n* = 4,778) (green markers) measured using FAST-LMM, compared with those estimated in a Sardinian (red markers) and Icelandic population (blue markers) (51).

total cholesterol were relatively high: 45%, 54% and 53%, respectively. Heritability for LDLC and total cholesterol were significantly higher than had been observed in previous works among European ancestry individuals (e.g. 20–43% for LDLC). The authors speculate that a larger proportion of phenotypic variance due to environmental factors (e.g. dietary consumption and an obesogenic environment) among European ancestry populations may reduce the impact of genetic factors in those populations. In contrast, heritability of triglycerides was much more modest (25%) and similar to observations in European ancestry individuals (30%) (51).

While the increases in cardiometabolic disorders in Africa are driven by changes in lifestyle factors, there is evidence that evolutionary forces, such as natural selection, have also played a role in mechanisms that underlie these disorders (52). For instance, adaptive effects in Africans have been reported for energy metabolism, adipose tissue biology, insulin action, and salt regulation. Some of these adaptive effects appear to have been driven by climate, infectious disease, and nutritional challenges. As an example, among the Wolaita in southern Ethiopia, selection was observed for *PPARA*, which encodes proteins involved in glucose homeostasis and lipid metabolism. This signal is thought to reflect high consumption of a particular regional food crop. Unfortunately, the relative lack of genomic studies representing Africans in the diversity of environments across the continent is an important limitation in understanding selection and its role in shaping cardiometabolic risk among Africans (52).

Previous work in over one million individuals worldwide identified a significant effect of inbreeding depression in the genetic architecture of some cardiometabolic traits, such as LDLC, total cholesterol, and waist-to-hip ratio, yet this study included only a relatively small number of sub-Saharan Africans (53). Ceballos et al. were able to use data from the AWI-GEN study, with more than 10,000 sub-Saharan Africans from multiple regions of the continent (Ghana, Burkina Faso, Kenya, and South Africa), to investigate more comprehensively the potential effect of autozygosity on the genetic architecture of cardiometabolic traits in sub-Saharan Africans (54). The mean genomic inbreeding coefficient across AWI-GEN was 0.0093, with no differences between men and women or in urban versus rural sites. A significant negative inbreeding depression was observed for weight, BMI, and subcutaneous adipose tissue, with an average reduction of 0.51 kg, 0.18 kg/m^2, and 0.049 cm, respectively. Stronger effects of inbreeding depression on these traits were observed in sites with higher night light intensity (a proxy for urban development), and significantly different effects were observed by urban development for LDLC, TG, and subcutaneous adipose tissue. Differential effects of inbreeding were also observed by sex. For instance, while inbreeding in men was associated with less body weight, lower BMI, and subcutaneous fat, it was associated in women with higher values of each of these. Further analyses suggest that the inbreeding depression on these traits is caused by autozygous rare recessive alleles, and not homozygous common variants in strong linkage disequilibrium. These results highlight a complex role of inbreeding depression for some cardiometabolic traits, with interactions by sex and environmental factors.

7.3.1.1 Genetic Association Studies

Due to Africa's role as the birthplace of humanity and the location of nearly all of humanity's evolutionary history (55), there is greater genetic diversity among populations with African ancestry than among any other ancestral groups (56). As such, the importance of conducting genetic association studies in Africans is clear: some of the genetic variation present in Africans can only be investigated in Africans, as it is not present in other populations. Unfortunately, the number of large epidemiological studies that are well characterized for cardiometabolic outcomes and suitable for large-scale genomics research has been limited, curtailing the type of association studies that are possible. For instance, a 2018 systematic review of genetic studies of dyslipidemia identified only five studies in sub-Saharan African populations (57), with similar observations made in systematic reviews of T2D (44) and blood pressure (58). Additionally, many studies used candidate gene approaches and were limited to participants with specific conditions, rather than the general population (57). Another limitation is that, as noted in a 2019 review of the genetic factors associated with obesity and T2D in

Africans (59), there is an emphasis in the literature on assessing transferability of loci discovered in other ancestry groups in those of African ancestry rather than on discovery studies in Africans. This emphasis represents a failure to capitalize on the potential for novel discovery in these individuals, given well-known genetic diversity within Africa (56). Importantly, however, as discovery-focused association studies have emerged, the promise of novel findings anticipated with genomics research in African ancestry populations is being realized (60). Even modestly sized GWAS have yielded novel discoveries, such as the identification in West Africans of an association between an African-specific variant in *SEMA4D* with about 6 pounds higher weight and African-specific variants in *CTNNA3* for metabolic syndrome (59). Here, we review recent (since 2018) GWAS for cardiometabolic traits that include sub-Saharan African populations.

7.3.1.2 Genome-Wide Association Studies in African Populations

The Africa America Diabetes Mellitus (AADM) is a long-running epidemiological study of T2D, recruiting cases and controls from urban centers in Nigeria, Ghana, and Kenya. GWAS analysis of 5,231 individuals (2,342 T2D cases) yielded three genome-wide significant loci: *TCF7L2* (rs7903146), *HMGA2* (rs138066904), and *ZRANB3* (chr2:136064024) (56) *TCF7L2* is a well-established T2D loci, and, indeed, an association with *TCF7L2* has been consistently observed in T2D studies across populations, including among Africans (61–66) An association between variants in the *HMGA2* locus and T2D had been previously observed; however, the variant associated in this analysis was a deletion that was independent from those previous reports and is in much higher frequency among those with African ancestry compared to those without. The association in the *ZRANB3* locus was not previously reported and the two variants representing this locus appear to be African-ancestry specific. In a meta-analysis with 2,578 South African Zulu cases and controls, both associations at this locus remained genome-wide statistically significant. Knockdown of the zebrafish ortholog of *ZRANB3* led to a reduced number of pancreatic β-cells due to increased apoptosis in islets. Evidence from siRNA transfection of murine *Zranb3* in *MIN6* β-cells established a role of *Zranb3* in β-cell response to high glucose. Thus, this study was able to identify a novel association of African ancestry-specific *ZRANB3* variants with T2D and establish a role for this locus in β-cell mass and in the response of the pancreas to insulinogenic stimuli (67).

Chen et al. identified a significant locus near *AGMO* (rs73284431) in addition to *TCF7L2* in their T2D GWAS of 4,347 Africans (South Africa, Nigeria, Ghana, and Kenya) (68). This association had not been previously reported, and the variant was common in the population studied (MAF = 0.095), but monomorphic in most non-African populations. The authors also conducted fine-mapping of the *TCF7L2* locus, which suggested a distinct African-specific signal (indexed by rs17746147) as well as an association signal that is shared by both European and African ancestry individuals (indexed by rs7903146).

Udosen et al. conducted a GWAS meta-analysis using both a univariate approach (SPB and DBP separately) and a multivariate approach (blood pressure traits together), including nearly 78,000 African ancestry individuals drawn from the Million Veteran Program (USA), the African Partnership for Chronic Disease Research (Uganda, South Africa, Nigeria, Ghana, and Kenya), and the UK Biobank (UK) (69). Univariate meta-analyses identified 19 statistically significant loci, including 2 associations that were independent of known BP loci: rs77534700 (intergenic)-DBP and rs562545 (*MOBP*)-DBP. Bayesian fine-mapping pinpointed 11 of the lead SNPs in the 19 associated loci as having >50% posterior probability of being causal, including rs562545 from the novel association at *MOBP* (posterior probability 77%). To increase statistical power for discovery, the authors also used cross-phenotype association analysis, which considers the correlation of summary statistics among traits. This multivariate approach identified 21 independent loci for BP traits, 3 of which had not been previously observed, near *DNAJC17P1/GLULP6*, *RRM2*, and *LOC105377644*.

As part of a landmark paper describing the Uganda Genome Resource, the authors conducted GWAS of multiple cardiometabolic traits in up to 14,126 individuals from Ghana, Kenya, Nigeria, South Africa, and Uganda (51). Meta-analysis revealed two significant HbA1c associations. Importantly, one of these associations was determined to be a result of the $\alpha^{-3.7}$ thalassemia deletion. This variant explained 3% of the variation in HbA1c and this association was largely driven by the inclusion of samples from Uganda, where the allele frequency of the deletion is higher, highlighting the importance of inclusion of a variety of populations, even among those of African ancestry. The other variant associated with HbA1c, rs6724428 (*GULP1*), had not previously been reported, even though it is common across populations (for instance, effect allele frequency of 0.44 and 0.55 in AFR and EUR 1000 Genomes populations; https://www.internationalgenome.org/). GWAS meta-analysis of these same datasets for obesity traits identified two significant associations with BMI and one with hip circumference. Of these, an association with rs7798566 and BMI had not been previously reported; this variant is in the intergenic region within the *TAS2R* gene family, genes that are involved in taste sensitivity to bitter-tasting compounds. An association between rs12105634 near *SDCCAG8* and BMI had not been reported, though *SDCCAG8* has been associated with extreme childhood obesity in Europeans. Significant heterogeneity among contributing African studies was noted for both of these findings (70). GWAS of serum lipids in these pan-African samples identified 13 significant loci for serum lipids. Of these, a novel locus on chromosome 19 (chr19:21749298_A_G) was associated with HDLC; this variant was common among the African populations but absent in European ancestry populations. Distinct associations at known loci were also discovered for a deletion on chromosome 5 (5:156378584_CGGAA_C) with both LDLC and total cholesterol, and for two variants (rs12721054 [*APOC1*] and rs569795903 [*RP11–230B22.1*]) with triglycerides. Each of these associations was for variants that are not observed in EUR populations (51).

A GWAS of serum lipids carried out in 4,317 Africans from Ghana, Nigeria, and Kenya identified 28 novel associations $p < 5 \times 10^{-8}$ [34196372]. Of these, 18 were able to be tested in 9,542 African Americans and 3 replicated: rs706014 (*GPNMB*)-TG, rs9375831 (*ENPP1*)-TG, and rs115749422 (*SMARC4*)-LDLC. The *SMARC4* association was independent of nearby known *LDLR* associations. Notably, rs115749422 is rare or absent in non-African ancestry populations. Meta-analysis of Africans and African Americans identified five additional novel loci: rs138282551-TG, rs1468291761 (*PGBD5*)-HDLC, rs148194085 (*CD80*)-TG and HDLC, rs79922971 (*SLC44A1*)-total cholesterol, rs147706369 (*TLL2*)-total cholesterol, some of which were more common or only found among African ancestry populations. This study further investigated associations with serum lipids by intracontinental ancestry. Analysis of participants with predominantly West African ancestry (i.e., Nigerians, Ghanaians, and African Americans) led to the identification of associations of rs7797481 (*ORC5*)-LDLC and 20:60973327-TG. While there was a consistent association between West Africans and African Americans at the *ORC5* locus, the association among East Africans was in the opposite direction, obscuring the result when all participants were grouped together. In addition to providing more support for the potential of identifying novel loci for serum lipids traits by focusing on African ancestry populations, the results from this study also support consideration of subgroups within ancestry, where differences in genetic and environmental factors may underlie heterogeneity among associations.

Importantly, serum lipids are correlated traits, with overlap in underlying genetic and lifestyle factors influencing their distribution. In appreciation of this fact and the increased statistical power that can result from multi-trait analyses, Kamiza et al. undertook a GWAS meta-analysis and multi-trait analysis of serum lipids in individuals of African ancestry (71). Analyses were conducted in ~125,000 individuals, including ~14,000 individuals from the African Partnership for Chronic Disease Research (Nigeria, Ghana, Kenya, Uganda, and South Africa), ~11,000 individuals from the Africa Wits-INDEPTH Partnership for Genomic Studies (AWI-GEN: Burkina Faso, Ghana, Kenya, and South Africa), and ~99,000 mostly admixed African ancestry individuals from the Global Lipids Genetics Consortium. Meta-analysis identified four novel associations: Intergenic variants rs2451303 and rs6704760 with LDLC and variants in *CHRM3* and *MGAT2* with TG. Multi-trait analysis identified 14 novel associations for serum lipids, including associations for variants in *TMEM64*, *ZNF782*, and *LOC100507346*. The authors also took advantage of the generally reduced linkage disequilibrium in genomes of African ancestry to conduct fine-mapping, resulting in considerably reduced credible set size reduction. Gene expression trait colocalization of the multi-trait analysis results identified six loci with posterior probability of ≥ 80%.

Choudhury et al. conducted a GWAS of serum lipids in the 10,603 individuals of AWI-GEN (72) (Burkina Faso, Ghana, South Africa, and Kenya). In addition to significant signals in known lipids loci, this meta-analysis identified a novel LDLC locus in *GATB*. This variant is common among the population studied but is absent or rare in non-African populations. Further support for a novel locus in this region was seen upon meta-analysis with four other African cohorts (n = 23,718). Meta-analysis also identified an additional novel TG locus close to *FHIT*. Novel associated variants were also reported for known lipids loci *LPA* and *TDRD15*. Stratified GWAS also identified some differences by region within Africa. The authors conducted fine-mapping and were able to identify rs28362286 as a potential causal variant in *PCSK9* for LDLC and total cholesterol. They were also able to reduce the size of the credible set for multiple lipids loci, including *LDLR*, *LPAL2*, *LIPC*, *CELSR2*, and *APOC1*. Importantly, heterogeneity of effect sizes and allele frequencies were observed among the African cohorts included in these analyses; however, clear geographic differences were not apparent and intra-regional differences were also observed. The high level of heterogeneity has implications for the degree of agreement to be expected among African cohorts and requires further investigation.

cIMT is a key surrogate marker for cardiovascular risk. Boua et al. evaluated 7,894 individuals from Burkina Faso, Ghana, Kenya, and South Africa from the AWI-GEN study (73). Novel associations with variants in *SIRPA* and *FBXL17* were identified; notably, these variants are African ancestry specific. Sex-stratified analyses revealed additional cIMT associations: with a variant in *SNX29* in men, and variants near *LARP6* and near *CYMP* in women. Interestingly, gene-set analyses of findings from the analyses in women revealed significant enrichment of gene sets related to estrogen response. For one of the female-specific loci (rs115473055, near *CYMP*), the association of genotype with cIMT differed by menopausal status.

7.3.1.3 Polygenic Risk Scores for Cardiometabolic Disease

An exciting advance with tantalizing clinical implications is the development of polygenic risk scores (PRS), which have the potential to compute an individual's risks based on allele frequencies across the genome. As these risk scores derive from GWAS, which have been conducted unevenly across populations of different ancestries, the performance of PRS in African ancestry individuals (as well as other groups underrepresented in genomics research) has been of considerable interest. Researchers have expressed concern that PRS, which translate genomics research into information that is potentially meaningful at an individual level, will be less accurate and useful in some ancestry groups, such as Africans (59,67).

Ekoru et al. investigated the performance of PRS across ancestries by comparing the predictive utility for 12 cardiometabolic traits in sub-Saharan Africans (n = 5,200 from Nigeria, Ghana, and Kenya), African Americans (n = 9,139), and European Americans (n = 9,594) (74). PRS were constructed based on weighted sums of the number of risk alleles over all SNP-trait associations $p < 1 \times 10^{-5}$ in the GWAS catalog. Effect sizes for weighting were obtained from either UK Biobank results or the largest study in the GWAS catalog (all of which were conducted in European or East Asian individuals). Out of the 12 PRS constructed, 10 were significantly associated with the trait in European Americans, compared to only 8 and 6 in the African Americans and Africans, respectively. Similarly, the correlation coefficients were stronger between PRS and traits among the European Americans. Predictive utility of a PRS was evaluated by looking at how much a PRS improved phenotypic variance explained over a model with traditional risk factors and ancestry-derived principal components. Across all traits, PRS showed up to a 5-fold greater predictive utility for European Americans compared to African Americans and a 20-fold greater predictive utility compared to Africans. Notably, there was some variation in these patterns across traits. For HDLC, for instance, the PRS increased predictive utility more in African Americans (3.9%) than in European Americans (2.9%) or Africans (1.0%). The authors conclude that PRS for cardiometabolic traits remain suboptimal for clinical translation among sub-Saharan Africans and African Americans and emphasize the importance of inclusion of underrepresented populations in genomics research to allow for equitable clinical translation of findings.

Chikowore et al. assessed the performance of PRS for T2D derived from multi-ancestry, African American, or European American samples from the Million Veteran Program in sub-Saharan African (75). Considering a number of different parameters for PRS construction, the best predictive PRS for each ancestry source data was developed in South African Zulu (n = 1,602 cases and 981 controls); the optimized multi-ancestry, African American, and European American PRS were all significant in this target population, with the African American PRS explaining the highest phenotypic variance (1.11%). The PRS were then further validated and assessed in the Nigerians, Ghanaians, and Kenyans of the AADM study (n = 2,148 cases and 2,161 controls). In AADM, the African American and multi-ancestry PRS performed markedly better than the European American PRS (2.9% and 2.6%, respectively, vs. 0.1%). Importantly, those in the top decile of the African American PRS in AADM had a 3.6-fold higher risk of developing T2D compared to those in the lowest decile, and their average age at diagnosis was 2.6 years lower than participants in the lowest decile. While the African American and multi-ancestry PRS performed comparably in the full AADM study, when stratified by country, predictive ability varied. The African American PRS was transferable to Nigerians, Ghanaians, and Kenyans, though the predictive ability was higher in the West African countries compared to Kenya. The predictive ability of the multi-ancestry PRS was similar in the West African countries, but it was not a significant predictor of T2D in Kenyans. The authors cite differences in allele frequencies, linkage disequilibrium patterns, and heterogeneity of effect sizes as potential explanations for the limited transferability of the multi-ancestry PRS across African populations and call for larger GWAS of T2D of continental Africans.

While the poor performance of PRS derived from studies of European-ancestry individuals in sub-Saharan Africans has been established by studies such as the aforementioned, Kamiza et al. (76) found that scores derived from GWAS of African Americans or multi-ancestry samples performed better than one of European Americans in sub-Saharan Africans. Specifically, they considered PRS for serum lipids derived from African Americans, European Americans, or multi-ancestry samples (African, European, and Hispanic Americans) from the Million Veteran Program in South African Zulu (n = 2,598) and Ugandan (n = 6,407) cohorts. In the South African Zulu, the PRS for African Americans outperformed the PRS from multi-ancestry or European populations for HDLC, LDLC, and total cholesterol, while the multi-ancestry PRS performed slightly better than the African American PRS for TG. The European ancestry PRS performed significantly worse for each trait. Interestingly, the performance of these PRS was not consistent across the African populations considered. While the African American PRS for total cholesterol explained 6.345% of phenotypic variance in South African Zulu, it explained only 0.045% in the Ugandan individuals. Similarly, the correlation between the PRS and the lipids traits was lower in Ugandans versus South African Zulu. The authors noted marked differences in age, body mass index, and allele frequencies of variants included in the PRS between the two cohorts, which may contribute to the poor transferability across populations. These results are consistent with expectations that a PRS derived from data using African ancestry outperforms that derived from European-ancestry individuals among sub-Saharan African populations. However, due to genetic and environmental differences within Africa, PRS may not perform equally well across African cohorts. Thus, more research is needed to optimize PRS within sub-Saharan Africa.

The Global Lipids Genetics Consortium evaluated the question of transferability of PRS across ancestries as part of their study of serum lipids in approximately 1.65 million individuals, with emphasis on genetic diversity (77). Their GWAS meta-analyses included nearly 100,000 individuals of African ancestry (though no Africans were included in discovery analyses).

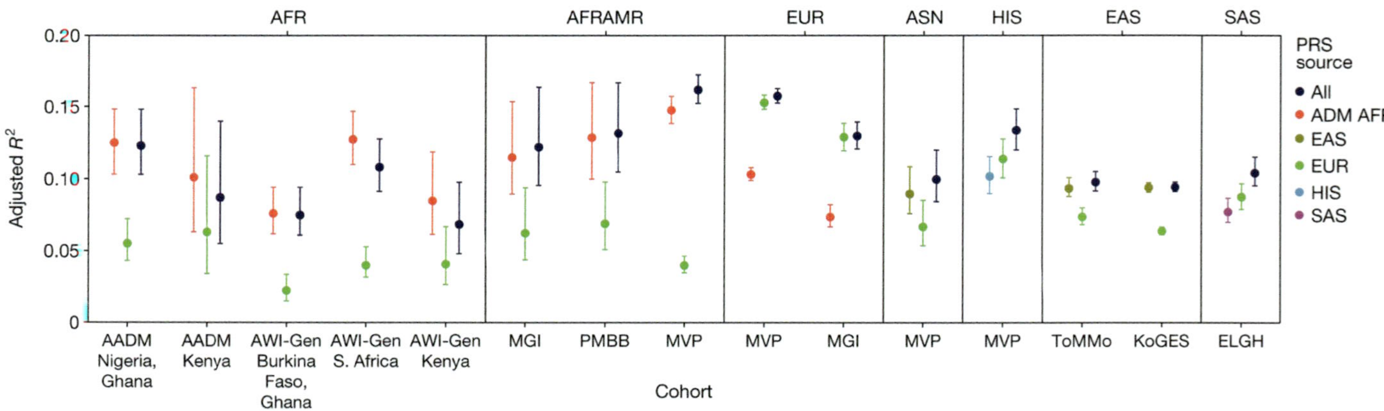

Figure 7.2 Polygenic scores generated from trans-ancestry meta-analysis show equivalent or better performance across most ancestry groups relative to ancestry-specific PRS within each cohort, whereas European ancestry-specific scores show less transferability. Adjusted R^2 is calculated with the risk score as a predictor of LDLC in a linear model with covariates. AFR: African, AFRAMR: African American, ASN: Asian American (77).

Using ancestry-specific and multi-ancestry meta-analyses, separate PRS for LDLC were created, tested in cohorts of differing ancestries that were not included in the original meta-analyses, and the performance of these PRS were compared (Figure 7.2). Notably for the purposes of this review, PRS were tested in multiple African cohorts, including individuals from Nigeria, Ghana, Kenya, Burkina Faso, and South Africa. The relative performance of the PRS was similar across African populations and in the admixed African populations evaluated. As expected, a PRS derived from European ancestry meta-analyses performed relatively poorly in Africans and admixed Africans. PRS based on GWAS in admixed Africans or multi-ancestry samples had the best performance and performed similarly across African and admixed African samples tested, though the authors noted more variation in prediction performance and mean LDLC among African compared to admixed African groups. By altering the proportion of samples of different ancestries included in the PRS source data and comparing the performance of the resulting scores, the authors concluded that LDLC PRS prediction improves most across ancestries with the addition of samples of diverse ancestry in the PRS source data. A key observation from this work is that across all ancestry groups tested, a PRS derived from multi-ancestry source data performed better or similarly to a PRS based on ancestry-matched data, providing support for the validity of using PRS derived from multi-ancestry data (with well-represented ancestral diversity) in lieu of ancestry-matched data.

While the transferability of PRS derived using different source data to sub-Saharan Africans has been generally motivated by the lack of sub-Saharan African source data for constructing a PRS, progress in the genomics of serum lipids in Africa has now made possible comparisons with the predictive ability of a PRS based on source data from sub-Saharan Africa (n = 13,115 individuals from Uganda, Ghana, Nigeria, Kenya, and South Africa) (72). The transferability of PRS for serum lipids derived from sub-Saharan African, European-ancestry, and multi-ancestry source data to the AWI-Gen dataset, with individuals from Burkina Faso, Ghana, Kenya, and South Africa were tested. The PRS from SSA had the highest predictive ability for LDLC and TG, while a multi-ancestry PRS performed the best for HDLC and total cholesterol. The European-ancestry PRS performed the worst for all traits. Interestingly, the sub-Saharan African PRS outperformed the European ancestry PRS for all traits despite having at least an order of magnitude fewer SNPs included.

7.4 FUTURE DIRECTIONS AND IMPLICATIONS

Recent years have seen marked advances in the genomics of cardiometabolic disorders in Africa. Reviews of the literature as recently as in 2020 (44) found largely candidate gene approaches to investigating the genetic factors underlying cardiometabolic traits, an approach that is optimized to confirm transferability of findings discovered in other populations to Africans. Importantly, this lack of large-scale genome-wide investigation meant that the field was missing the incredible potential for discovery in Africa, given the characteristics in the genomes of Africans as a result of human history. This issue has gained broad attention and motivated considerable investment in genomics on the continent. Chief among these is the H3Africa initiative, an investment from the US National Institutes of Health, and the Wellcome Trust into developing a pan-continental network of laboratories undertaking leading-edge genomics research (7). There have been 51 H3Africa projects supported and over 700 publications generated from the H3Africa consortium (h3africa.org accessed September 28, 2023). As a result of these and other efforts, genomics research in Africa has experienced no less than a sea change, with work from Africa spotlighted in major forums in the field and highlighted in top journals. Consequently, novel discoveries relevant for CMD are rapidly emerging from the continent, as described in this chapter. While these advances are considerable and encouraging, several themes emerge from recent work on the genomics of CMD in Africa that should be taken into consideration for future work in this field.

Although the number of SSA research participants included in genomic studies of CMD has grown considerably, the sample sizes for continental Africans are still dwarfed by those available in other world populations and even by those in populations of admixed African ancestry (59,69); Importantly, the relative paucity of Africans in research is even more critical as larger sample sizes may be necessary to capture the greater genetic variation present in the genomes of individuals from Africa. Only a comparatively small number of the around 2,000 African ethnolinguistic groups have been characterized genetically (10), and participants in genomics research on cardiometabolic traits predominantly hail from relatively few countries: Nigeria, Ghana, Kenya, Uganda, Burkina Faso, and South Africa. Importantly, the expected genomic diversity even within Nigeria is great enough to motivate the development of the Non-Communicable Diseases Genetic Heritage Study, a consortium designed to catalog the human genetic variation and to characterize non-communicable diseases with an ambitious target of 100,000 Nigerian adults (78). It is impossible to characterize how much genomic diversity is still being missed across the continent. As a result, the full impact of the genetic diversity within Africa on the study of genomics of cardiometabolic traits is largely unknown, and systematic efforts to increase representation on the continent, and particularly representation from more sites across Africa, will be necessary to address this question.

Heterogeneity is evident in the distribution of cardiometabolic traits across African populations, and genomic studies highlight heterogeneity in association analyses of these traits. This heterogeneity is unsurprising and, in fact, is even expected given the incredible diversity in terms of not only genetic background, but also those related to lifestyle, climate, and diet. In the push toward achieving needed larger sample sizes in African ancestry populations, it is imperative not to assume the homogeneity of samples from the continent. Work is still needed to determine the most appropriate subgroupings among African samples in terms of cardiometabolic traits, but evidence from recent studies suggest that a monolithic "African ancestry" grouping that includes all continental Africans as well as admixed Africans and Africans in diaspora is likely to be inappropriate. Additionally, cardiometabolic traits are the result of influences from both genetics and the environment, and not all differences observed across African populations may result from differences in genetic factors. While heterogeneity in association effects across African studies may very well proceed from regional differences in allele frequencies, for instance, they are also likely to be influenced by differences in diet, physical activity, and health practices across regions. To disentangle these effects, genomics research studies should also be well characterized regarding environmental exposures. Notably, as reviewed here, substantial research on lifestyle factors and cardiometabolic traits has been published for large-scale cohorts in Africa (15,25,30,79).

Newer technologies and approaches are allowing for efficient and novel discoveries in genomics, yet a delay in applying these advances in African populations could exacerbate disparities in genomic-driven healthcare strategies (59). Interest is growing in the use of precision medicine, and underrepresentation of Africans in genomics research will hamper its potential benefit in these individuals where new approaches are sorely needed to combat high and increasing burdens of CMD (59). For instance, researchers reviewing genomic factors and T2D in SSA determined that evidence was currently too limited and largely insufficient to inform the genetic architecture of T2D across SSA regions and to inform clinical management of T2D (44). Additionally, achieving the highest quality genomics data to evaluate research questions is still a non-trivial issue for African studies. Genotyping of some studies has been carried out using genotyping arrays that are not optimized to African populations (10). As a result, the distribution of novel, rare, and medically relevant variants across the breadth of African diversity remains to be characterized, and this gap is consequential in terms of our understanding of the contribution of genetic variation to disease (10). Also, genotype imputation is used to generate genome-wide data for association analyses, yet a recent comparison of existing reference panels highlighted systematic differences in performance of these panels across sub-Saharan geographical regions, with complexity also noted by specific ancestral groups (80). Thus, decisions regarding the most appropriate strategy for imputation for African studies may not be straightforward. As with the field of genomics in general, there is a need in research on the genomics of cardiometabolic traits in Africa to move beyond association results into understanding of the underlying biological mechanisms. Specifically, the development of omics data and resources for SSA would represent a major advance for taking research findings closer to the realm of translation.

In conclusion, there have been marked advances in the genomics of CMD using samples from SSA in recent years. While sample sizes of included individuals are still much lower than those achieved in those of European ancestry, for instance, these studies are identifying novel discoveries and helping to describe the genetic architecture of these traits on the continent. There is a need for genomic studies with larger sample sizes in Africa and particularly with representation across different parts of the continent. Further work is needed to characterize the genomic diversity across Africa and the effect of this diversity on cardiometabolic traits as well as to disentangle the genetic and environmental influences on the heterogeneity in association results across African studies.

REFERENCES

1. Einarson TR, Acs A, Ludwig C, Panton UH. Prevalence of cardiovascular disease in type 2 diabetes: A systematic literature review of scientific evidence from across the world in 2007–2017. Cardiovasc Diabetol. 2018 Jun;17(1):83.
2. Almourani R, Chinnakotla B, Patel R, Kurukulasuriya LR, Sowers J. Diabetes and cardiovascular disease: An update. Curr Diab Rep. 2019 Dec;19(12):161.
3. Del Bosque-Plata L, Martínez-Martínez E, Espinoza-Camacho MÁ, Gragnoli C. The role of TCF7L2 in type 2 diabetes. Diabetes. 2021 Jun;70(6):1220–8.
4. Barale C, Melchionda E, Morotti A, Russo I. PCSK9 biology and its role in atherothrombosis. Int J Mol Sci. 2021 May;22(11).
5. Khera AV, Chaffin M, Aragam KG, Haas ME, Roselli C, Choi SH, et al. Genome-wide polygenic scores for common diseases identify individuals with risk equivalent to monogenic mutations. Nat Genet [Internet]. 2018;50(9):1219–24. https://doi.org/10.1038/s41588-018-0183-z
6. Doumatey AP, Bentley AR, Akinyemi R, Olanrewaju TO, Adeyemo A, Rotimi C. Genes, environment, and African ancestry in cardiometabolic disorders. Trends Endocrinol Metable 2023 Oct;34(10):601–21.
7. Rotimi C, Abayomi A, Abimiku A, Adabayeri VM, Adebamowo C, Adebiyi E, et al. Research capacity: Enabling the genomic revolution in Africa. Science. 2014 Jun;344(6190):1346–8.
8. Denny JC, Rutter JL, Goldstein DB, Philippakis A, Smoller JW, Jenkins G, et al. The "all of us" research program. N Engl J Med. 2019 Aug;381(7):668–76.
9. Taliun D, Harris DN, Kessler MD, Carlson J, Szpiech ZA, Torres R, et al. Sequencing of 53,831 diverse genomes from the NHLBI TOPMed program. Nature. 2021 Feb;590(7845):290–9.
10. Choudhury A, Aron S, Botigué LR, Sengupta D, Botha G, Bensellak T, et al. High-depth African genomes inform human migration and health. Nature. 2020 Oct;586(7831):741–8.
11. Fatumo S, Choudhury A. African American genomes don't capture Africa's genetic diversity. Vol. 617, Nature. England; 2023. p. 35.

12. Sun H, Saeedi P, Karuranga S, Pinkepank M, Ogurtsova K, Duncan BB, et al. IDF Diabetes Atlas: Global, regional and country-level diabetes prevalence estimates for 2021 and projections for 2045. Diabetes Res Clin Pract [Internet]. 2022;183:109119. Available from: https://www.sciencedirect.com/science/article/pii/S0168822721004782
13. Ekoru K, Doumatey A, Bentley AR, Chen G, Zhou J, Shriner D, et al. Type 2 diabetes complications and comorbidity in Sub-Saharan Africans. EClinicalMedicine. 2019 Nov;16:30–41.
14. Hilawe EH, Yatsuya H, Kawaguchi L, Aoyama A. Differences by sex in the prevalence of diabetes mellitus, impaired fasting glycaemia and impaired glucose tolerance in sub-Saharan Africa: A systematic review and meta-analysis. Bull World Health Organ. 2013 Sep;91(9):671–82D.
15. Wagner RG, Crowther NJ, Micklesfield LK, Boua PR, Nonterah EA, Mashinya F, et al. Estimating the burden of cardiovascular risk in community dwellers over 40 years old in South Africa, Kenya, Burkina Faso and Ghana. BMJ Glob Heal. 2021 Jan;6(1).
16. Wambui D, Mohamed S, Asiki G. Prevalence of and factors associated with high atherogenic index among adults in Nairobi urban informal settlements: The AWI-Gen study. PLOS Glob Public Heal. 2022;2(7):e0000224.
17. Boateng D, Agyemang C, Beune E, Meeks K, Smeeth L, Schulze M, et al. Migration and cardiovascular disease risk among Ghanaian populations in Europe. Circ Cardiovasc Qual Outcomes [Internet]. 2017 Nov 1;10(11):e004013. https://doi.org/10.1161/CIRCOUTCOMES.117.004013
18. Nonterah EA, Boua PR, Klipstein-Grobusch K, Asiki G, Micklesfield LK, Agongo G, et al. Classical cardiovascular risk factors and HIV are associated with carotid intima-media thickness in adults from sub-Saharan Africa: Findings from H3Africa AWI-gen study. J Am Heart Assoc. 2019;8(14).
19. Nonterah EA, Boateng D, Crowther NJ, Klipstein-Grobusch K, Oduro AR, Agongo G, et al. Carotid atherosclerosis, microalbuminuria, and estimated 10-year atherosclerotic cardiovascular disease risk in sub-Saharan Africa. JAMA Netw open. 2022 Apr;5(4):e227559.
20. Kolkenbeck-Ruh A, Soepnel LM, Crouch SH, Naidoo S, Smith W, Norris SA, et al. Obesity, hypertension, and tobacco use associated with left ventricular remodeling and hypertrophy in South African women: Birth to Twenty Plus Cohort. BMC Cardiovasc Disord [Internet]. 2022;22(1):1–11. https://doi.org/10.1186/s12872-022-02837-w
21. Ajayi AA, Sofowora GG, Ladipo GO. Explaining heart failure hyper-mortality in sub Saharan Africa: Global genomic and environmental contribution review. J Natl Med Assoc. 2020 Apr;112(2):141–57.
22. Bayaraa N, Azahar NM, Kitaoka K, Kobayashi Y, Yano Y. African control of hypertension through innovative epidemiology and a vibrant ecosystem (ACHIEVE): A holistic approach for hypertension control in Africa. J Hum Hypertens [Internet]. 2023; https://doi.org/10.1038/s41371-023-00845-7
23. Parati G, Lackland DT, Campbell NRC, Ojo Owolabi M, Bavuma C, Mamoun Beheiry H, et al. How to improve awareness, treatment, and control of hypertension in Africa, and how to reduce its consequences: A call to action from the world hypertension league. Hypertension. 2022;79(9):1949–61.

24. Okello S, Muhihi A, Mohamed SF, Ameh S, Ochimana C, Oluwasanu AO, et al. Hypertension prevalence, awareness, treatment, and control and predicted 10-year CVD risk: A cross-sectional study of seven communities in East and West Africa (SevenCEWA). BMC Public Health [Internet]. 2020;20(1):1706. https://doi.org/10.1186/s12889-020-09829-5
25. Akpa OM, Made F, Ojo A, Ovbiagele B, Adu D, Motala AA, et al. Regional patterns and association between obesity and hypertension in Africa: Evidence from the H3Africa CHAIR study. Hypertens (Dallas, Tex 1979). 2020 May;75(5):1167–78.
26. Akinyemi RO, Owolabi MO, Ihara M, Damasceno A, Ogunniyi A, Dotchin C, et al. Stroke, cerebrovascular diseases, and vascular cognitive impairment in Africa. Brain Res Bull. 2019 Feb; 145:97-108.
27. Akinyemi RO, Ovbiagele B, Adeniji O, Sarfo FS, Abd-Allah F, Adoukonou T, et al. Stroke in Africa: profile, progress, prospects and priorities. Nat Rev Neruol. 2021 Oct; 17(10):364-656.
28. Owolabi MO, Sarfo F, Akinyemi R, Gebregziabher M, Akpa O, Akpalu A, et al. Dominant modifiable risk factors for stroke in Ghana and Nigeria (SIREN): A case-control study. Lancet Glob Heal. 2018;6(4):e436–46.
29. Adeoye AM, Akintunde AA, Akinyemi J, Fakunle AG, Sarfo FS, Akpalu A, et al. Determinants of metabolic syndrome and its prognostic implications among stroke patients in Africa: Findings from the Stroke Investigative Research and Educational Network (SIREN) study. J Neurol Sci. 2022 Jul;441:1–10.
30. Sarfo FS, Akpa OM, Ovbiagele B, Akpalu A, Wahab K, Obiako R, et al. Patient-level and system-level determinants of stroke fatality across 16 large hospitals in Ghana and Nigeria: A prospective cohort study. Lancet Glob Heal. 2023 Apr;11(4):e575–85.
31. Bentley AR, Rotimi CN. Interethnic differences in serum lipids and implications for cardiometabolic disease risk in African ancestry populations. Glob Heart. 2017 Jun;12(2):141–50.
32. Meeks KAC, Bentley AR, Agyemang C, Galenkamp H, van den Born B-JH, Hanssen NMJ, et al. Ancestral and environmental patterns in the association between triglycerides and other cardiometabolic risk factors. eBioMedicine [Internet]. 2023 May 1;91. https://doi.org/10.1016/j.ebiom.2023.104548
33. World obesity federation. Obesity Atlas. 2023 Mar:1–232. https://Data WorldobesityOrg/Publications/?Cat=19
34. Ramsay M, Crowther NJ, Agongo G, Ali SA, Asiki G, Boua RP, et al. Regional and sex-specific variation in BMI distribution in four sub-Saharan African countries: The H3Africa AWI-Gen study. Glob Health Action. 2018;11(sup2):1556561.
35. Boua RP, Sorgho H, Rouamba T, Nakanabo Diallo S, Bognini JD, Konkobo SZ, et al. Gender differences in sociodemographic and behavioural factors associated with BMI in an adult population in rural Burkina Faso: An AWI-Gen sub-study. Glob Health Action. 2018;11(sup2):1527557.
36. Casari S, Di Paola M, Banci E, Diallo S, Scarallo L, Renzo S, et al. Changing dietary habits: The impact of urbanization and rising socio-economic status in families fom Burkina Faso in sub-Saharan Africa. Nutrients. 2022;14.
37. Muti M, Ware LJ, Micklesfield LK, Ramsay M, Agongo G, Boua PR, et al. Physical activity and its association with body mass index: A cross-sectional analysis in middle-aged adults from 4 sub-Saharan African countries. J Phys Act Health. 2023 Mar;20(3):217–25.

38. Darko SN, Meeks KAC, Owiredu WKBA, Laing EF, Boateng D, Beune E, et al. Anthropometric indices and their cut-off points in relation to type 2 diabetes among Ghanaian migrants and non-migrants: The RODAM study. Diabetes Res Clin Pract. 2021 Mar;173:108687.
39. Commodore-Mensah Y, Agyemang C, Aboagye JA, Echouffo-Tcheugui JB, Beune E, Smeeth L, et al. Obesity and cardiovascular disease risk among Africans residing in Europe and Africa: The RODAM study. Obes Res Clin Pract. 2020;14(2):151–7.
40. Soremekun O, Karhunen V, He Y, Rajasundaram S, Liu B, Gkatzionis A, et al. Lipid traits and type 2 diabetes risk in African ancestry individuals: A Mendelian Randomization study. eBioMedicine [Internet]. 2022;78:103953. Available from: https://www.sciencedirect.com/science/article/pii/S2352396422001372
41. Fatumo S, Karhunen V, Chikowore T, Sounkou T, Udosen B, Ezenwa C, et al. Metabolic traits and stroke risk in individuals of African ancestry: Mendelian randomization analysis. Stroke [Internet]. 2021 Jul 17;0(0):STROKEAHA.121.034747. https://doi.org/10.1161/STROKEAHA.121.034747
42. Soremekun O, Slob EAW, Burgess S, Romuald Boua P, Fatumo S, Gill D. Genetically predicted lipid traits, diabetes liability, and carotid intima-media thickness in African ancestry individuals: A mendelian randomization study. Circ Genomic Precis Med [Internet]. 2022 Dec 1;15(6):e003910. https://doi.org/10.1161/CIRCGEN.122.003910
43. Chen G, Adeyemo A, Zhou J, Doumatey AP, Bentley AR, Ekoru K, et al. A UGT1A1 variant is associated with serum total bilirubin levels, which are causal for hypertension in African-ancestry individuals. NPJ Genomic Med. 2021 Jun;6(1):1–6.
44. Asamoah EA, Obirikorang C, Acheampong E, Annani-Akollor ME, Laing EF, Owiredu E-W, et al. Heritability and genetics of type 2 diabetes mellitus in sub-Saharan Africa: A systematic review and meta-analysis. J Diabetes Res. 2020;2020:3198671.
45. Ware LJ, Maposa I, Kolkenbeck-Ruh A, Norris SA, Soepnel L, Crouch S, et al. Are cardiovascular health measures heritable across three generations of families in Soweto, South Africa? A cross-sectional analysis using the random family method. BMJ Open. 2022 Sep;12(9):e059910.
46. Post WS, Larson MG, Myers RH, Galderisi M, Levy D. Heritability of left ventricular mass: The Framingham heart study. Hypertens (Dallas, Tex 1979). 1997 Nov;30(5):1025–8.
47. Jin Y, Kuznetsova T, Bochud M, Richart T, Thijs L, Cusi D, et al. Heritability of left ventricular structure and function in Caucasian families. Eur J Echocardiogr J Work Gr Echocardiogr Eur Soc Cardiol. 2011 Apr;12(4):326–32.
48. Paternoster L, Martínez González NA, Lewis S, Sudlow C. Association between apolipoprotein E genotype and carotid intima-media thickness may suggest a specific effect on large artery atherothrombotic stroke. Stroke. 2008 Jan;39(1):48–54.
49. Xiang AH, Azen SP, Buchanan TA, Raffel LJ, Tan S, Cheng LS-C, et al. Heritability of subclinical atherosclerosis in Latino families ascertained through a hypertensive parent. Arterioscler Thromb Vasc Biol. 2002 May;22(5):843–8.
50. Fox CS, Polak JF, Chazaro I, Cupples A, Wolf PA, D'Agostino RA, et al. Genetic and environmental contributions to atherosclerosis phenotypes in men and women: Heritability of carotid intima-media thickness in the Framingham heart study. Stroke. 2003 Feb;34(2):397–401.

51. Gurdasani D, Carstensen T, Fatumo S, Chen G, Franklin CS, Prado-Martinez J, et al. Uganda genome resource enables insights into population history and genomic discovery in Africa. Cell. 2019 Oct;179(4):984–1002.e36.
52. Meeks KAC, Bentley AR, Adeyemo AA, Rotimi CN. Evolutionary forces in diabetes and hypertension pathogenesis in Africans. Hum Mol Genet. 2021 Apr;30(R1):R110–8.
53. Clark DW, Okada Y, Moore KHS, Mason D, Pirastu N, Gandin I, et al. Associations of autozygosity with a broad range of human phenotypes. Nat Commun. 2019 Oct;10(1):4957.
54. Ceballos FC, Hazelhurst S, Clark DW, Agongo G, Asiki G, Boua PR, et al. Autozygosity influences cardiometabolic disease-associated traits in the AWI-Gen sub-Saharan African study. Nat Commun. 2020 Nov;11(1):5754.
55. McClellan JM, Lehner T, King M-C. Gene discovery for complex traits: Lessons from Africa. Cell. 2017 Oct;171(2):261–4.
56. Auton A, Brooks LD, Durbin RM, Garrison EP, Kang HM, Korbel JO, et al. A global reference for human genetic variation. Nature. 2015 Oct;526 (7571):68–74.
57. Noubiap JJ, Mato EPM, Guewo-Fokeng M, Kaze AD, Boulenouar H, Wonkam A. Genetic determinants of dyslipidemia in African-based populations: A systematic review. OMICS. 2018 Dec;22(12):749–58.
58. Singh S, Brandenburg J-T, Choudhury A, Gómez-Olivé FX, Ramsay M. Systematic review of genomic associations with blood pressure and hypertension in populations with African-ancestry. Front Genet. 2021;12:699445.
59. Doumatey AP, Ekoru K, Adeyemo A, Rotimi CN. Genetic basis of obesity and type 2 diabetes in Africans: Impact on precision medicine. Curr Diab Rep. 2019 Sep;19(10):105.
60. Bentley AR, Callier SL, Rotimi CN. Evaluating the promise of inclusion of African ancestry populations in genomics. NPJ Genomic Med. 2020;5:5.
61. Bouhaha R, Choquet H, Meyre D, Abid Kamoun H, Ennafaa H, Baroudi T, et al. TCF7L2 is associated with type 2 diabetes in nonobese individuals from Tunisia. Pathol Biol (Paris). 2010 Dec;58(6):426–9.
62. Cauchi S, Ezzidi I, El Achhab Y, Mtiraoui N, Chaieb L, Salah D, et al. European genetic variants associated with type 2 diabetes in North African Arabs. Diabetes Metable 2012 Oct;38(4):316–23.
63. Danquah I, Othmer T, Frank LK, Bedu-Addo G, Schulze MB, Mockenhaupt FP. The TCF7L2 rs7903146 (T) allele is associated with type 2 diabetes in urban Ghana: A hospital-based case-control study. BMC Med Genet. 2013 Sep;14:96.
64. Guewo-Fokeng M, Sobngwi E, Atogho-Tiedeu B, Donfack OS, Noubiap JJN, Ngwa EN, et al. Contribution of the TCF7L2 rs7903146 (C/T) gene polymorphism to the susceptibility to type 2 diabetes mellitus in Cameroon. J Diabetes Metab Disord. 2015;14:26.
65. Nanfa D, Sobngwi E, Atogho-Tiedeu B, Noubiap JJN, Donfack OS, Mofo EPM, et al. Association between the TCF7L2 rs12255372 (G/T) gene polymorphism and type 2 diabetes mellitus in a Cameroonian population: A pilot study. Clin Transl Med. 2015;4:17.
66. Turki A, Al-Zaben GS, Khirallah M, Marmouch H, Mahjoub T, Almawi WY. Gender-dependent associations of CDKN2A/2B, KCNJ11, POLI, SLC30A8,

and TCF7L2 variants with type 2 diabetes in (North African) Tunisian Arabs. Diabetes Res Clin Pract. 2014 Mar;103(3):e40–3.
67. Martin AR, Kanai M, Kamatani Y, Okada Y, Neale BM, Daly MJ. Clinical use of current polygenic risk scores may exacerbate health disparities. Nat Genet [Internet]. 2019;51(4):584–91. https://doi.org/10.1038/s41588-019-0379-x
68. Chen J, Sun M, Adeyemo A, Pirie F, Carstensen T, Pomilla C, et al. Genome-wide association study of type 2 diabetes in Africa. Diabetologia. 2019 Jul;62(7):1204–11.
69. Udosen B, Soremekun O, Kamiza A, Machipisa T, Cheickna C, Omotuyi O, et al. Meta-analysis and multivariate GWAS analyses in 77,850 individuals of African ancestry identify novel variants associated with blood pressure traits. Int J Mol Sci. 2023 Jan;24(3).
70. Gurdasani D, Carstensen T, Fatumo S, Chen G, Franklin CS, Prado-Martinez J, et al. Uganda genome resource enables insights into population history and genomic discovery in Africa. Cell. 2019;179(4):984–1002.e36.
71. Kamiza AB, Touré SM, Zhou F, Soremekun O, Cissé C, Wélé M, et al. Multi-trait discovery and fine-mapping of lipid loci in 125,000 individuals of African ancestry. Nat Commun. 2023 Sep;14(1):5403.
72. Choudhury A, Brandenburg J-T, Chikowore T, Sengupta D, Boua PR, Crowther NJ, et al. Meta-analysis of sub-Saharan African studies provides insights into genetic architecture of lipid traits. Nat Commun [Internet]. 2022;13(1):2578. https://doi.org/10.1038/s41467-022-30098-w
73. Boua PR, Brandenburg J-T, Choudhury A, Sorgho H, Nonterah EA, Agongo G, et al. Genetic associations with carotid intima-media thickness link to atherosclerosis with sex-specific effects in sub-Saharan Africans. Nat Commun [Internet]. 2022;13(1):855. https://doi.org/10.1038/s41467-022-28276-x
74. Ekoru K, Adeyemo AA, Chen G, Doumatey AP, Zhou J, Bentley AR, et al. Genetic risk scores for cardiometabolic traits in sub-Saharan African populations. Int J Epidemiol. 2021 Aug;50(4):1283–96.
75. Chikowore T, Ekoru K, Vujkovi M, Gill D, Pirie F, Young E, et al. Polygenic prediction of type 2 diabetes in Africa. Diabetes Care. 2022 Mar;45(3):717–23.
76. Kamiza AB, Toure SM, Vujkovic M, Machipisa T, Soremekun OS, Kintu C, et al. Transferability of genetic risk scores in African populations. Nat Med [Internet]. 2022;28(6):1163–6. https://doi.org/10.1038/s41591-022-01835-x
77. Graham SE, Clarke SL, Wu K-HH, Kanoni S, Zajac GJM, Ramdas S, et al. The power of genetic diversity in genome-wide association studies of lipids. Nature [Internet]. 2021;600(7890):675–9. https://doi.org/10.1038/s41586-021-04064-3
78. Fatumo S, Yakubu A, Oyedele O, Popoola J, Attipoe DA, Eze-Echesi G, et al. Promoting the genomic revolution in Africa through the Nigerian 100K Genome Project. Nat Genet. 2022 May;54(5):531–6.
79. Akpa OM, Okekunle AP, Asowata OJ, Chikowore T, Mohamed SF, Sarfo F, et al. Frequent vegetable consumption is inversely associated with hypertension among indigenous Africans. Eur J Prev Cardiol. 2022 Dec;29(18):2359–71.
80. Sengupta D, Botha G, Meintjes A, Mbiyavanga M, Hazelhurst S, Mulder N, et al. Performance and accuracy evaluation of reference panels for genotype imputation in sub-Saharan African populations. Cell Genomics. 2023 Jun;3(6):100332.

Chapter 8

The Genomics of Cancer in Africa

Abram B. Kamiza[1,2*] *and Solomon O. Rotimi*[3,4*]

[1]Medical Research Council, Uganda Virus Research Institute and London School of Hygiene and Tropical Medicine (MRC/UVRI &LSHTM), Entebbe, Uganda
[2]Precision Healthcare University Research Institute, Queen Mary University of London
[3]Department of Biochemistry, Covenant University, Ota, Nigeria.
[4]Cancer Genomics Laboratory, Covenant University, Ota, Nigeria.

8.1 INTRODUCTION

Cancer is an emerging public health crisis in Africa, with the World Health Organization (WHO) projecting a nearly 70% increase in new cancer cases by 2030 [1]. In 2020, Africa recorded over 1.1 million new cancer cases and 711,000 cancer-related deaths, with mortality rates far exceeding those in high-income regions owing to late diagnosis and limited access to care [1]. Breast, cervical, esophageal, and prostate cancers are the most frequent in Africa, with age-standardized incidence rates estimated to be more than 37.9, 27.6, and 26.6 per 100, 000 individuals, respectively [2,3]. Moreover, these cancer sites contribute to almost 30% of cancer-related deaths in Africa [3]. Worryingly, the many cancers in Africa are diagnosed at a relatively younger ages and in advanced stages [4–6], which worsens the prognosis and makes cancer one of the leading causes of death in Africa.

The cancer landscape in Africa is shaped by a unique and complex interplay of risk factors, including genetic susceptibility, environmental exposure, and a high burden of oncogenic infections, such as human papillomavirus (HPV), hepatitis B virus (HBV), hepatitis C virus (HCV), and human immunodeficiency virus (HIV). These overlapping drivers create distinct cancer patterns, such as the high incidence of oesophageal and cervical cancers, which remain poorly understood. Germline and somatic mutations play crucial roles in cancer development. Germline mutations occur in the DNA of reproductive cells and can be passed from one generation to the next, whereas somatic mutations occur in the DNA of non-reproductive cells and cannot be inherited. Germline and somatic mutations affect certain genes and increase the risk of cancer development. In Africa, the prevalence of

***Corresponding author:** abramkamiza@gmail.com (Abram B. Kamiza) and ola.rotimi@covenantuniversity.edu.ng (Solomon O. Rotimi)

 DOI: 10.1201/9781003436584-8

certain cancers and the distribution of germline and somatic mutations may vary among ethnic groups. A well-known example of a germline mutation is the BRCA gene mutation, which is associated with an increased risk of breast and ovarian cancers in women. Studying the prevalence of such mutations in African populations is crucial because of their unique genetic characteristics. Additionally, understanding germline and somatic mutations can contribute to the development of targeted therapies and personalized medicine for cancer treatment.

Germline and somatic mutations in cancer in Africa present a fascinating and complex landscape shaped by the continent's rich genetic diversity, varying environmental factors, and interplay of genetic and socio-economic influences. Africa is home to a diverse array of ethnic groups with distinct genetic backgrounds. This diversity extends to germline and somatic mutations associated with cancer susceptibility. Certain populations may harbour specific mutations that increase the risk of developing certain cancers. For example, oesophageal cancer is very common in eastern and southern Africa, but is very rare in West Africa, which may highlight differences in the prevalence and types of germline and somatic mutations in West and East Africa. Moreover, studies have revealed that some African populations exhibit unique *BRCA* mutations that are not commonly found in other continents. Infectious agents, such as certain viruses, play a significant role in cancer development in Africa. For example, HPV, HBV and HCV are more prevalent in some African regions and are associated with an increased risk of developing cancer, including cervical and liver cancers. This emphasizes the importance of region-specific research for tailoring preventive strategies and treatment. The interaction between germline mutations and environmental factors is important for cancer development. Africa's diverse climates, lifestyles, and exposure to various environmental agents contribute to this complex interplay. For instance, the prevalence of certain cancers in regions with higher exposure to specific carcinogens may be influenced by germline mutations or environmental factors.

As research progresses, the identification of germline and somatic mutations in African populations holds potential for advancing precision medicine. Tailoring cancer treatments based on an individual's genetic profile can enhance their efficacy and reduce the side effects. However, this requires a deeper understanding of the genetic variations specific to different African communities. The study of germline and somatic mutations in cancer in Africa is a dynamic field that requires a multifaceted approach. By considering the diverse genetic backgrounds, environmental influences, and impacts of infectious agents, researchers can unravel the complexities of germline and somatic mutations, paving the way for targeted and effective cancer prevention, diagnosis, and treatment strategies tailored to the unique needs of the African population. However, research in this area is still evolving, and efforts are underway to collect genetic data from diverse African populations to gain a comprehensive understanding of germline mutations and their implications for cancer. This knowledge can lead to more effective screening programs, early detection, and personalized treatment strategies tailored to the genetic makeup of Africans.

8.2 ROLE OF CANDIDATE GENE STUDIES IN CANCER IN AFRICA

Candidate gene studies are fundamental to genetic and genomics research. This approach often focuses on identifying specific germline and somatic variations that underpin disease risk, development, or progression. In the context of cancer in Africa, these studies have been essential for understanding the specific genes and mutations contributing to cancer susceptibility and prognosis among different populations. In Africa, candidate gene studies have primarily been conducted in patients with breast, cervical, oesophageal, and prostate cancers. These case–control studies identified *TP53*, *GSTM*, *TGFB1*, *MMP2*, *HLA-G*, *VDR*, *RANKL*, *OPG*, *SPARC*, *ERCC2*, *MTHFR*, *MICA*, *ABCB1*, *IL-6*, *SIRT1*, *RUNX1*, *TBX3*, *TTN*, *XRCC1*, *VEGF*, *XPD*, *IL-8*, *CXCR2*, and mtDNA-*HVR II* as key genetic factors associated with breast cancer susceptibility in Africa. *IL-10*, *HLA-DRB1*, *HLA-DQB1*, *TNF*, *TLR*, *VEGF*, *IL*, *CCR2*, *IL-1*, and *TP53* were identified as genetic factors associated with cervical cancer in Africa. In addition, a review article by Kuguyo et al. and Kamiza et al. summarized some important genetic factors and pathways associated with cervical cancer in Africa [7,8]. Interestingly, *TRAK2*, *CASP8*, *CHEK2*, *NAT1*, *NAT2*, *RUNX1*, *PLCE1*, *MSH3*, *MLH3*, *PMS1*, *GSTT2B*, *ALDH2*, *ADH1B*, *COX-2*, *MGMT*, *GSTT1*, *GSTP1*, *XPC*, *GSTM1*, *COMT*, *COX-2*, *NSMCE2*, *MSMB*, *CYP17*, *HSP70*, and *VEGF* were associated with oesophageal and prostate cancer, respectively.

8.3 SPECIFIC GENES STUDIED IN AFRICAN CANCERS

More than 100 years ago, Randle et al. in 1910 observed important familial hereditary male breast cancers in the Ijebu tribe in Nigeria, yet early worldwide distribution research on *BRCA1* and *BRCA2* germline mutations did not include African populations. The earliest *BRCA* studies were conducted on the North African population from Tunisia, Egypt, and Algeria [9–13], with the first published in 2002 by Monastiri et al. (2002) using protein truncation assay. The variations reported in these North African populations were similar to those previously described in European and Middle Eastern populations. In sub-Saharan Africa, South African and Nigerian populations were those with pioneering reports of *BRCA* variation unique to the continental Africans. Hence, only data from South Africa and Nigeria were included in a global distribution analysis. It was unclear whether the study participants accurately reflected the genetic and ethnic variety found in these nations. Notwithstanding this restriction, this global study noted that the specific variations found in African American families were unique to those seen in non-African ethnic groups and were of African origin [14–16]. A study by Sluiter et al. (2009) on 52 hereditary breast and/or ovarian South African families (36 were Afrikaners) using the newly developed sequencing technology that allowed for the rapid detection of large genomic rearrangements identified novel deletions of both exons 23 and 24 [17]. However, these variations were of Greek ancestry. Hence, there are gaps in the knowledge of novel mutations of indigenous African origin.

8.4 RESEARCH ON *BRCA* MUTATIONS IN SUB-SAHARA AFRICAN POPULATIONS

Candidate gene studies have investigated the prevalence of pathogenic variants in *BRCA1* and *BRCA2*, which are associated with a higher risk of breast and ovarian cancers. These studies have identified specific mutations that are more common in certain sub-Saharan African populations, contributing to a better understanding of hereditary breast and ovarian cancers in Africa. The earliest study reported an ancient *BRCA1* mutation (Y101X) in breast cancer patients in Yoruba in Nigeria [18]. Researchers discovered a non-pathogenic *BRCA1* exon 21 deletion (c.5277+480_5332+672del) in Nigeria, as well as a harmful *BRCA1* mutation (c. 1949_1950delTA) in a woman from Senegal ([19]. BRCA2 mutation was discovered in the Bantu-speaking Xhosa community in South Africa [20].

A more thorough investigation by Zheng et al. (2018), which focused on Nigerian women, produced important new information. The identified hereditary mutations in *BRCA1*, *BRCA2*, *PALB2*, and *TP53* have been linked to up to 20% of cases of inherited invasive breast cancer in Nigeria. These results are consistent with previous data by Fackenthal et al. (2012), who showed that individuals with breast cancer from Nigeria had a remarkably high prevalence of *BRCA1* and *BRCA2* mutations. In another study on Cameroonian and Ugandan women, Adedokun et al. (2020) reported that 15.8% of breast cancer cases carried pathogenic or likely pathogenic mutations in a breast cancer susceptibility gene, with *BRCA1* and *BRCA2* being the most common. They observed recurrent mutations in the *BRCA1* (c.4484G>T) and novel mutations in *BRCA1* in our sample were c.2966_2967del and c.4323_4329del, and in *BRCA2* were c.1053del, c.1964del, c.2937del, c.4693_4694dup, c.5633dup, c600dup, c.6987_6993del, and *BRCA1* (c.1796_1800delCTTAT), which were previously reported in Nigerian women [21]. Of note is the high proportion of variants of uncertain clinical significance (VUS) which were found in 96 individuals (25.2%).

The high level of VUS among women of African ancestry presently constitutes a setback in the utilization of genetic screening for risk stratification of breast, ovarian, prostate, and pancreatic cancers. The occurrence of VUS has also been reported in Burkina Faso [22], where two VUS mutations in addition to six pathogenic mutations (three in *BRCA1* and three in *BRCA2*) were observed. Two of the six pathogenic mutations are recurrent in unrelated women [22].

Studying candidate genes in familial cancer cases allows the identification of founder mutations that can be utilized for population-based screening and risk stratification, as demonstrated in Burkina Faso by Kiendrebeogo et al. (2022). To identify such founder mutations in Afrikaner families, Reeves et al. (2004) studied *BRCA1* mutations in breast and/or ovarian cancers in South Africa. However, this community also has European ancestry, and the mutations found were comparable to those found in Ashkenazi Jewish and the Netherlands. Importantly, Reeves et al. (2004) described the association of variations in *PALB2* with early onset breast cancer among Afrikaners.

Although new *BRCA* variants have been found in several African countries, such as the Democratic Republic of the Congo, Senegal, Mauritius, South Africa, and Maghrebian countries (Egypt, Tunisia, Algeria, Morocco, and Sudan), as reviewed by Abbad et al. (2018), the high level of genetic diversity in Africa limits the application of these data in the control of familial cancers. However, some studies have linked the variation in candidate genes, such as *BRCA*, to the tumour phenotype. According to Pitt et al. (2018), specific *BRCA* mutations are associated with increased structural variation and aggressive tumour biology in Nigerian women, especially in cases of HR+/HER2– breast cancers [23]. Mahfoudh et al. (2019) found that triple-negative breast cancer is partly caused by the 5382 insC mutation in *BRCA1* among Tunisian women [24]. Among all breast cancer subtypes, triple-negative breast cancer has the worst prognosis and is more common in women of West African heritage [25]. Identifying *BRCA* mutations in African populations may make it possible to treat triple-negative breast cancer patients with PARP inhibitors, which would enhance clinical outcomes and overall survival.

8.5 GENE AND ENVIRONMENT INTERACTIONS IN CANCER SUSCEPTIBILITY IN AFRICA

The major drivers of DNA damage and cancer development are exogenous agents, such as viruses and xenobiotics. Viruses that cause cancer include hepatitis viruses for liver cancer, human papillomavirus (HPV) for cervical and head and neck cancers, and Epstein-Barr virus (EBV) for Burkitt's lymphoma. It is important to highlight here that the role of EBV in breast cancer has been the subject of much research, including reports from Algeria [26], Eritrea [27], Egypt [28], and Sudan [26]. EBV is a mosquito-transmitted virus that was discovered as a human oncogenic virus in 1964 In Uganda

[29]. Other viruses, such as the mouse mammary tumour virus, have been linked to 74% of breast cancers in Tunisia [30,31].

The importance of infectious agents as drivers of the high cancer burden in Africa cannot be overemphasized. Approximately 24.5% of cancers in Africa and 28.7% in sub-Saharan Africa are linked to infectious agents [32]. Understanding germline genetics that increases the predisposition of individuals to oncogenesis has been the subject of interest for many studies. This is because oncogenic viruses affect pathways that are dysregulated in human cancers through mechanisms that include interfering with the cell cycle, targeting pRB and p53, preventing apoptosis, and perturbing the immune response [33].

Studies have shown that the *TP53* R72P mutation increases susceptibility to HPV-associated cervical and hepatitis virus-linked liver cancers [34,35]. This exon 4 single nucleotide polymorphism (SNP) of *TP53* results in an arginine (R) to proline (P) substitution at codon 72 of the p53 protein. The arginine variant is more common in Caucasian populations than in African populations and this ethnic/geographical variation has been linked to an increased risk of cervical cancer development associated with the arginine p53 variant [34,36]. This P variant and its concomitant clinical implication have been reported in Gabon, Senegal, Sudan, Morocco, and South African populations [37–39].

Beyond TP53, the reported polymorphisms that increase the risk of HPV-associated cervical cancer in Africa include immune responses and inflammation. These include the *TLR* 2/3/4/9 and IL 1/10/15 genes in Tunisia and –308 promoter polymorphism of *TNF-α* in South Africa [40–42]. The prognosis and susceptibility of breast cancer in Tunisia were reported to be influenced by polymorphisms in genes such as *TNF-**q***, *TNF* receptor II, *FASL*, *HLA-DQB1*, and *HLA-G* [43–45]. Other inflammatory genes that have been examined in Tunisian and Egyptian breast cancer patients include *NRF2*, *IL1-alpha*, *IL1-beta*, *IL6*, *IL8*, and *CXCR2* [46–50]. Chronic inflammation stimulates the development and progression of malignant neoplasms [51], which is a hallmark of cancer. Emerging evidence lends credence to the hypothesis that the aggression of cancers that disproportionately affect people of African ancestry, such as triple-negative breast cancer and prostate cancer, is underpinned by inflammation [52]. Hence, emerging therapies are now focused on altering the inflammatory microenvironment of such tumours and increasing their sensitivity to treatment [53,54].

The ability of Africans to adapt to infectious diseases is one of the key evolutionary processes that has sharpened their genomes. This adaption is believed to be responsible for the selection of variants, such as the silencing of the Duffy Antigen Receptor for Chemokines (ACKR1/DARC) gene, which increases resistance to malaria infection, particularly *Plasmodium vivax* and *Plasmodium knowlesi* [55,56]. This is because the invasion of erythrocytes by malarial parasites is facilitated by an interaction between the parasite's Duffy binding protein (DBP) and erythrocyte DARC protein. However, DARC also functions as a chemokine-scavenging or decoy receptor that can regulate chemokine bioavailability and leukocyte recruitment, a process required for inhibiting tumour progression and metastasis [56,57]. Hence, its silencing in people of African Ancestry has been linked to increased pancreatic, colorectal, breast, and prostate cancer aggression, as well as increased susceptibility to HIV and its associated cancers [57–61].

Beyond biotic cancer-causing agents, the human body also has to deal with many carcinogenic xenobiotics. One such is aflatoxins, which are products of fungi and are endemic in the tropical African climate [62]. In addition, industrialization and modernization increase exposure to a myriad of manmade chemicals in cosmetics, lifestyle, and food products. Consequently, genetic variations in genes associated with detoxification of these chemicals have been studied across African populations as candidate genes for cancer risk.

Phase I xenobiotic-metabolizing enzymes of the cytochrome P450 (CYP) family are the main enzymes involved in the detoxification of carcinogenic substances. These gene polymorphisms can affect enzyme activity and predict the effects of harmful substances on the body. Polymorphisms of *CYP1A1* and *CYP1B1* in Nigeria and Egypt [63,64], *CYP2D6* in South Africa [20], and *CYP1A2* in Tunisia have been documented in cancer patients. These CYP genes are also important in hormone-responsive tumours because they are involved in steroid metabolism and influence responses to endocrine therapies. In African males, prostate cancer is the leading cause of cancer death and is significantly influenced by germline genetics. Hence, studies from Morocco, Tunisia, Nigeria, South Africa, and Senegal have linked polymorphisms in *CYP3A4*, *CYP3A5*, *CYP1A1*, and *CYP17* to the prostate cancer risk [65–68]. In North Africa and the Ivory Coast, changes in GGN and CAG repeats in androgen receptor genes have been linked to an increased risk of prostate cancer [69,70]. CYP are mono-oxygenases whose function is to improve the polarity of xenobiotics by attaching an oxygen moiety to exogenous chemicals, thereby preventing the generation of reactive electrophilic species that interact with nucleophilic macromolecules such as DNA and proteins, which is the first step in carcinogenesis.

The phase I reaction is a series of phase II enzymatic reactions that increase the polarity of xenobiotics and facilitate their excretion. Reports have shown that polymorphisms in phase II enzymes, including glutathione-*S*-transferases (GSTT1 and GSTM1), could influence the early development and prognosis of breast cancer in women from Tunisia [71]. In premenopausal Nigerian women, the number of TA repeats in the low-activity UGT1A1 gene promoter was shown to be protective against breast cancer [72,73]. Furthermore, in Egypt and Rwanda, polymorphisms in the genes for paraoxonase, cyclooxygenase, glyoxalase, and glutathione peroxidase have also been implicated in breast cancer [74–77]. In Tunisia and Algeria, genetic variants of GSTM1, GSTT1, UDP-glucuronosyltransferase, and sulfotransferase have been associated with an increased risk of prostate cancer [65,68,78,79].

Most lung cancer cases are associated with cigarette exposure or air pollution. Hence, studies in North Africa have reported polymorphisms in *CYP1A1*, *CYP1A2*, *CYP2F1*, *CYP2A6*2*, and *CYP2A6*9* [76,80–84] as risk markers for lung cancer. Hussein et al. (2014) concluded that Egyptian smokers with *CYP1A1* rs4646903 m1 and *CYP1A1* m2 rs1048943 were more likely to develop squamous cell carcinoma.

Also, other studies from the North African countries of Egypt and Tunisia [85–87] have linked polymorphisms in the *CYP2D6* and *CYP2D*1A* genes to bladder cancer. The mechanism of bladder cancer involves the induction of carcinogenic N-nitrosamine by *Schistosoma haematobium* in the bladder [88], particularly in individuals with *TP53* mutations [85,89]. Other investigators have reported that polymorphisms in the phase enzyme genes *GST* null genotypes and *NAT*5* (341T>C) increase bladder cancer risk in populations of the North African countries [90–92].

Although liver cancer is caused by aflatoxin, a few studies have examined the role of candidate xenobiotic-metabolizing genes in the development of this disease. Tiemersma et al. (2001) and Abd El-Moneim et al. (2008) showed that individuals with the deleted *GSTM1* and T1 haplotypes are more susceptible to developing hepatocellular carcinoma when exposed to aflatoxin. This discovery stresses the relevance of genetic factors in determining an individual's sensitivity to environmental toxins as well as the necessity for tailored interventions and preventive measures among high-risk African populations. Further research is needed to completely understand the mechanisms underlying the link between these genetic deletions and the development of hepatocellular carcinoma.

Developing precision preventive and treatment strategies for cancer in Africa requires an understanding of the interactions between external agents and hereditary variables. The intricate genetic and environmental variables impacting cancer on the continent are being further researched in this area.

8.6 GENES RELATED TO VITAMIN D

The vitamin D signal transduction system also involves a series of CYP-containing sterol hydroxylases, which generate and degrade the active hormone, 1α 25-dihydroxycholecalciferol. Active hormones serve as ligands for vitamin D receptor (VDR)-mediated transcriptional gene expression. Similar to oestrogen and androgen receptors, VDR is activated by 1α 25-dihydroxycholecalciferol and it regulates many genes that are important in cancer [93]. Its action has been reported to be particularly important in tumours of African ancestry [94] and demonstrated that even non-malignant prostate cells of African ancestry have the greatest dynamic protein content in the VDR complex, as well as greater 1α 25-dihydroxycholecalciferol-regulated chromatin accessibility and a significant enhancer-enriched *VDR* cistrome. Through this mechanism, vitamin D regulates inflammation and influences cancer aggression in individuals of African ancestry. The risk of breast cancer in Egyptian women has been reported to be elevated by polymorphisms in *VDR* genes, such as ApaI and TaqI [95–97], and the *VDR* FokI rs2228570 GG genotype was associated with an increased risk of breast cancer among Ethiopian women [98]. However, the hypothesis linking vitamin D to breast cancer risk was weakened by a genome-wide association study (GWAS) that examined SNPs in the vitamin D–related pathway genes using data from a GWAS of breast cancer in the African diaspora [99], which observed no significant association between vitamin D or the SNPs within the *VDR* binding regions and breast cancer risk; however, breast cancer risk was associated with the pigment synthesis/metabolism pathway as driven by rs41302073 in *TYRP1*.

8.7 GENOMIC STUDIES UNVEILING CANCER INSIGHTS IN AFRICA

The drive to understand the genomic landscape of cancer in Africa spans a spectrum from candidate gene studies scrutinizing individual genes to comprehensive genomic investigations that encompass the entire genome, which allows for an integrated understanding of the interplay between germline and somatic mutations. Although candidate gene studies have allowed us to discern the roles of specific genes in cancer risk, progression, and aggression, they are limited in addressing the complexity of cancer [62]. Cancer is a group of complex diseases that present with diverse phenotypes and multiple heterogeneous genetic drivers. Researchers have leveraged next-generation sequencing (NGS), a transformative technology that has revolutionized the field of genomics. NGS has enabled researchers to cast a wider net by exploring the intricate interplay of multiple genes within the broader genetic landscape. This transition from candidate gene studies to genomics research has unveiled a holistic understanding of the genetic facets that underlie cancer in Africa [100].

The dearth of genomic studies on cancer in Africa coupled with the projected rising burden of the disease has made it increasingly important to comprehend the genetic foundations of cancer in Africa. Hence, this section provides an overview of the translational impact of important findings from whole-exome sequencing and genome-wide research, shedding light on the genetic variations, susceptibility SNPs, and genomic alterations that add to the complexity of cancer in Africa, ranging from breast cancer to prostate cancer and colorectal cancer. These discoveries are leading the way in opening the door to precision medicine interventions that cater to the distinct genetic profiles of African patients with cancer.

A series of genome-wide studies have provided valuable and pioneering insights into unravelling the genetic intricacies of breast cancer among African populations. These included GWAS conducted in Tunisia and South Africa [101–103], as well as whole-exome sequencing of breast tumours in Tunisia and Egypt populations [104–106]. In Tunisian studies, Shan et al. (2012) and Hamdi et al.(2018) identified several breast cancer susceptibility SNPs, including rs1219648, rs2981582, rs8051542, rs889312, and rs889312. Notably, rs9911630 emerged as a SNP with a profound impact on the expression of *BRCA1*, along with two long non-coding RNAs (*NBR2* and *LINC008854*).

In South Africa, Lupicki et al. (2018) employed genome-wide copy number alteration analysis to uncover notable amplifications on Xp22.3 and 6p21-p25, as well as other regions impacting well-known cancer-related genes such as *CCND1*, *CDKN1A*, *MDM2*, *TP53*, and *SMAD2* breast cancer. Meanwhile, the whole-exome sequencing efforts in Tunisia and Egypt linked breast cancer in Tunisian women to alterations in genes like *MMS19*, *DNAH3*, *POLK*, *KAT*β6, and *RCC1*. These genetic variants, which are distinct from those associated with DNA damage repair genes, such as *BRCA1* and *BRCA2*, are linked to functional genes, such as *NBPF10*, *ZNF750*, *CHTI5*, *NPIPB11*, and *PHIP*. These genes are involved in critical processes such as RNA binding, transcriptional regulation, extracellular matrix, structural proteins, and signal transduction.

Pitt et al. (2018) examined the molecular features of breast cancer in Nigerian patients using whole-exome sequencing and RNA-seq approaches [23], and compared them to patients from The Cancer Genome Atlas (TCGA) dataset, providing insights into the genomic characteristics of their breast cancer. Nigerian tumours with HR+/HER2– subtype showed increased homologous recombination deficiency (HRD) signature, *TP53* mutations, and greater structural variation, indicating aggressive biology. *GATA3* mutations are also more frequent in Nigerians, regardless of subtype.

Studies on the genetic risk and genomic drivers of prostate cancer among African men include GWAS studies in Tunisia, Ghana, and Uganda [107–109], as well as whole-genome sequencing in South Africa [110] and whole-exome sequencing in Nigeria [111]. Intriguingly, these GWAS did not identify common high-risk variants for prostate cancer, underscoring the genetic heterogeneity among African populations. A Tunisian study revealed three distinct regions containing 14 significant SNPs; these SNPs showed some overlap with Caucasian populations [109] in Ghana, with identification of 30 highly significant SNPs distributed across various chromosomes. Similarly, a Ugandan study identified risk alleles on multiple chromosomes [108]. Remarkably, the Ugandan and Ghanaian populations shared cytoband 6p21.32 in common, highlighting potential genetic adaptations to the African rainforest environment. Furthermore, whole-exome sequencing of Nigerian prostate cancer cases revealed a high prevalence of germline alterations in HRD genes, such as *BRCA1*, *BARD1*, *BRCA2*, *APC*, and *ATM* [111]. Remarkably, some variants identified by White et al. (2022) are in accordance with the report on breast cancer by Pitt et al. (2018), and were also found in the urinary DNA of South African men with prostate cancer [112]. These findings reinforce the need to consider genetic lineages within the Black population when investigating genomics contributions to disease disparities.

Molecular and phenotypic profiling of colorectal tumours in Nigerian patients has provided intriguing insights. The work of Alatise et al. (2021) on colorectal cancer in Nigerian patients revealed a significantly higher rate of microsatellite instability compared with cohorts from TCGA and Memorial Sloan Kettering Cancer Center. Notably, these microsatellite-stable colorectal tumours displayed distinct genetic alterations, as exemplified by a higher propensity for RAS pathway alterations, but a reduced likelihood of *APC* mutations and *WNT* pathway alterations.

These genomic studies provide crucial building blocks for understanding the genomic underpinnings of cancer in Africa, offering potential avenues for precision medicine and tailored therapeutic interventions within diverse African populations. Poly ADP-ribose polymerase (PARP) inhibitors, such as olaparib and talazoparib, can be useful in breast and prostate cancer patients with *BRCA1* or *BRCA2* mutations [113]. Patients with prostate cancer harbouring a DNA mismatch repair deficiency, as well as patients with colorectal cancer who have high microsatellite instability (MSI-H) or DNA mismatch repair (dMMR) deficiency, are appropriate candidates for immune checkpoint inhibitors such as pembrolizumab and nivolumab [114]. These tissue-agnostic therapies are changing the face of cancer clinical care, and the identification of these mutations in the African population is a pointer to the benefits that the population can derive from precision oncology care.

The discovery of African-specific genomic alterations, such as somatic mutations in key driver genes like *NCOA2* and *STK19*, as reported by Jaratlerdsiri et al. (2022), as well as a pronounced burden of structural variations (SVs) defining a hyper-SV tumour subtype [115], underscores the significance of African genomics in advancing cancer control globally. Although these features were identified in African-derived tumours, their biological mechanisms are applicable to other populations. For instance, *NCOA2* mutations linked to aberrant hormone receptor signalling and STK19 alterations affecting MAPK pathways offer promising targets for hormone-based therapies and kinase inhibitors, respectively. Similarly, the hyper-SV subtype, associated with genomic instability and poor outcomes, may reveal broader therapeutic vulnerabilities to PARP inhibitors, DNA damage response blockers, and immune checkpoint therapies. These insights demonstrate that beyond filling the diversity gaps, African genomics data, indeed, is pioneering and improving the robustness of the toolkit for precision oncology worldwide.

In addition to the aforementioned studies, cancer genomic studies in people of African descent have laid the foundation for potentially clinically relevant findings that could be validated in African populations with potential to strengthen the global cancer understanding and control. For example, over 125,000 novel DNA sequences specific to African genomes (APG contigs) were reported by Martini et al. (2024), some of which are with predicted CpG islands and transcriptional activity in triple-negative breast cancer. These regions are absent from standard reference genomes (GRCh37/38), emphasizing the superiority of African genomics data in understanding genomics diseases like cancer [116].

8.8 GENOME-WIDE ASSOCIATION STUDIES

GWAS on cancer in the African population have been limited. The only GWAS to investigate the association between genetic variants and cancer in an African population was conducted in Ghanaian men with prostate cancer [107]. This GWAS included 474 prostate cancer cases and 458 controls, and no significant association was found between genetic variants and prostate cancer [107]. Although the results of this GWAS were not genome-wide significant, the most suggestive significant SNP was located near *GATA3*, which has been associated with prostate cancer in other ethnic populations [117–119]. Moreover, ten SNPs previously reported to be associated with prostate cancer in the European population were replicated by this GWAS, which provided insights into the role of host genetic factors in the pathogenesis of cancer.

Although the number of GWAS on cancer have been limited in Africa, currently, there are ongoing GWAS across Africa aiming to identify genetic variants associated with cancer through various research group and consortia, including Evolving Risk Factors for Cancers in African Populations (ERICA), African Female Breast Cancer Epidemiology (AFBRECANE) study, and Africa Esophageal Cancer Consortium (AfrECC). Currently, ERICA is investigating the contribution of host genetic variants to the development of breast, cervical, and oesophageal cancers in the Black sub-Saharan African population using the Johannesburg Cancer Study in South Africa [120]. The AFBRECANE study through the Human Heredity & Health for Africa (H3Africa) consortium is conducting a GWAS of breast cancer and its molecular subtypes in Nigeria women, whereas AfrECC, a consortium of oesophageal cancer researchers in Africa extending from Ethiopia to South Africa (oesophageal cancer corridor), will be conducting a GWAS of oesophageal cancer in Africa [121]. The results from these GWAS will unravel the key genetic factors and biological pathways involved in the pathogenesis of different cancer sites in Africa and provide a platform for genetic risk prediction and precision medicine. Moreover, breast and ovarian cancer GWAS resulting from the genomic and oncogenic groups in Tunisia will shed more light on the role of host genetic factors in the aetiology of breast and ovarian cancers in North Africa. In addition, plans are underway to form African oncogenetic groups under the African Organization for Research and Training in Cancer, which will focus on the molecular and genetic aspects of cancer in the continent.

However, Africa has a long way to go before precision medicine can become a reality. The main reasons for this are inadequate genetic studies, a lack of replication among studies, and stark differences in the incidence and burden of cancer in Africa. Prostate cancer is more frequent in West Africa, whereas cervical and oesophageal cancers are more common in eastern and southern Africa [1,121]. These differences may be attributed to the admixture and differences in environmental exposure among African regions [122]. Hence, more regional GWAS are required to identify the genetic factors and biological pathways involved in cancer pathogenesis in Africa. Interestingly, H3Africa is fostering capacity building in genetic studies in Africa [123]; with this, together with the low cost of genotyping arrays, the number of GWAS in Africa is expected to increase over the years, and having precision medicine may become a reality in Africa.

8.9 EPIGENETICS STUDIES IN AFRICAN CANCERS

The functional impact of genetic and genomic variation occurs through gene expression, which is influenced by the environment. However, this effect is tightly controlled by epigenetics, which is a heritable change in gene expression that occurs without alterations to the DNA sequence. This phenomenon has emerged as a powerful lens through which to understand the hallmarks of cancer [124], including mechanisms such as DNA methylation, non-coding RNAs, and histone modification. Consequently, the interplay between the rich genetic diversity, diverse climatic environments, and environmental exposures and epigenetic factors have appeared to play a pivotal role in shaping cancer risk, prognosis, and overall outcomes on the continent.

Epigenetic factors have been implicated in the risk and prognosis of breast cancer in Africa. These factors encompass various roles, including the influence of tissue microRNA, circulating free mRNA, circulating long non-coding RNA [10,125–128], and the DNA methylation status of breast cancer susceptibility genes such as *APC*, *ERα*, *RASSF1A*, *UCHL1*, *COX-2*, and *FHIT* [125,129–131]. These epigenetic alterations have been identified as significant contributors to breast tumour research development across continents. In the realm of liver cancer, researchers have attempted to develop biomarkers using epigenetic factors, notably miRNAs. Prominent examples include serum Mir-224, Mir-215, Mir-143, Mir-122, Mir-199a, and Mir-16 [74,132]. Among these, Mir-122 and Mir-222 levels have been identified as discriminating biomarkers for distinguishing liver injury from liver cancer [133]. Moreover, studies have reported associations between LncRNA *HULC* rs7763881, *MALAT* rs619586, and decreased susceptibility to liver cancer in Egyptian hepatitis virus-persistent carriers [134].

Epigenetic changes have also been investigated in the context of colorectal cancer carcinogenesis in Africa [135–139]. For instance, methylation of genes such as *UCH1* and p14/*ARF* has been identified as a driving force for colorectal cancer, particularly in the presence of *TP53* mutations in Tunisia [135,136,140,141]. Studies on North African populations have delved into the influence of polymorphisms in the telomere and mitochondrial D-loop regions on the clinicopathological characteristics of colorectal cancers among their patients [142,143].

Aberrant methylation patterns in specific genes have shown promise for screening purposes in Senegal for cervical cancer, either as stand-alone markers or in combination with cytology [144]. Building on this background, Gagliardi et al. (2020) conducted a comprehensive profiling of cancer genomes, transcriptomes, and epigenomes in 118 Ugandan women with cervical cancer, including HIV-positive individuals [145]. Their study revealed HPV clade-specific differences in cervical tumour DNA methylation, promoter- and enhancer-associated histone marks, gene expression, and pathway dysregulation. Furthermore, this study unravelled the correlations between changes in histone modification at HPV integration sites and the upregulation of nearby genes.

Overall, the few epigenetic investigations on African cancers have provided valuable insights into the molecular mechanisms underlying cancer development and progression and have opened an avenue for the development of novel biomarkers for improved diagnosis. These findings lay the foundation for the development of novel diagnostic and therapeutic strategies tailored to the unique genetic and epigenetic profiles observed in African populations.

8.10 TRANSLATIONAL IMPACT AND CALL TO ACTION

The escalating incidence and burden of cancer across Africa necessitate a paradigm shift in prevention, detection, and treatment strategies that leverage African genomics. Therefore, understanding founder mutations in familial cancers across African populations [22,146] can enable population-based screening and targeted prevention. Such data are critical to developing Afrocentric population–specific variant panels to identify individuals at high risk of developing breast, ovarian, prostate, and pancreatic cancers. Such panels are also essential for accurate identification of eligible patients and ensuring equitable access to life-saving targeted therapies like PARP inhibitors. In order to improve the ease of cancer diagnosis and treatment, it imperative to accelerate the research and translation of less evasive genomic diagnostic technologies like circulating tumour DNA into clinical care. With advances in artificial intelligence (AI) technologies, we hope that African genomics data would be integrated with AI-powered mobile applications that could facilitate public awareness and self-screening for early detection.

Overall, despite significant progress through consortia initiatives like H3Africa, ERICA, AFBRECANE, and AfrECC, genomics research in Africa faces challenges such as limited resources, infrastructure, and a scarcity of comprehensive studies (Rotimi et al., 2021). However, the unique genetic diversity of African populations offers an unparalleled opportunity to advance global cancer control. African genomics data, including novel DNA sequences, provide critical insights into cancer aetiology and progression not found elsewhere; therefore, accelerating drug discovery and therapeutic development will benefit all populations. A bold vision for the future necessitates intentional sustained investment in local capacity building, state-of-the-art infrastructure, and the training of a new generation of African genomic scientists. This commitment will not only transform cancer care on the continent but also ensure Africa's unique and invaluable contribution to the global fight against cancer, across the whole cancer control continuum, fostering health equity and innovation worldwide.

REFERENCES

1. Parkin DM, Bray F, Ferlay J, Jemal A. Cancer in Africa 2012. Cancer Epidemiol Biomarkers Prev. 2014;23: 953–966. https://doi.org/10.1158/1055-9965.EPI-14-0281
2. Ferlay J, Soerjomataram I, Dikshit R, Eser S, Mathers C, Rebelo M, et al. Cancer incidence and mortality worldwide: Sources, methods and major patterns in GLOBOCAN 2012. Int J Cancer. 2015;136: E359–E386. https://doi.org/10.1002/ijc.29210
3. Bray F, Ferlay J, Soerjomataram I, Siegel RL, Torre LA, Jemal A. Global cancer statistics 2018: GLOBOCAN estimates of incidence and mortality worldwide for 36 cancers in 185 countries. CA Cancer J Clin. 2018;68: 394–424. https://doi.org/10.3322/caac.21492
4. Irabor D, Adedeji OA. Colorectal cancer in Nigeria: 40 years on. A review. Eur J Cancer Care (Engl). 2009;18: 110–115. https://doi.org/10.1111/j.1365-2354.2008.00982.x
5. Brewster AM, Chavez-MacGregor M, Brown P. Epidemiology, biology, and treatment of triple-negative breast cancer in women of African ancestry. Lancet Oncol. 2014;15: e625–e634. https://doi.org/10.1016/S1470-2045(14)70364-X
6. Plummer M, de Martel C, Vignat J, Ferlay J, Bray F, Franceschi S. Global burden of cancers attributable to infections in 2012: A synthetic analysis. Lancet Glob Health. 2016;4: e609–e616. https://doi.org/10.1016/S2214-109X(16)30143-7
7. Kuguyo O, Tsikai N, Thomford NE, Magwali T, Madziyire MG, Nhachi CFB, et al. Genetic susceptibility for cervical cancer in African populations: What are

the host genetic drivers? OMICS. 2018;22: 468–483. https://doi.org/10.1089/omi.2018.0075

8. Kamiza AB, Kamiza S, Mathew CG. HLA-DRB1 alleles and cervical cancer: A meta-analysis of 36 case-control studies. Cancer Epidemiol. 2020;67: 101748. https://doi.org/10.1016/j.canep.2020.101748
9. Ibrahim SS, Hafez EE, Hashishe MM. Presymptomatic breast cancer in Egypt: Role of BRCA1 and BRCA2 tumor suppressor genes mutations detection. J Exp Clin Cancer Res. 2010;29: 82. https://doi.org/10.1186/1756-9966-29-82
10. Hafez MM, Hassan ZK, Zekri ARN, Gaber AA, Al Rejaie SS, Sayed-Ahmed MM, et al. MicroRNAs and metastasis-related gene expression in Egyptian breast cancer patients. Asian Pac J Cancer Prev. 2012;13: 591–598. https://doi.org/10.7314/apjcp.2012.13.2.591
11. Monastiri K, Ben Ahmed S, Presneau N, Bignon JY, Chouchane L. [Rapid detection of BRCA-1 germline mutations by the protein truncation test in Tunisian families]. Tunis Med. 2002;80: 515–518.
12. Troudi W, Uhrhammer N, Sibille C, Dahan C, Mahfoudh W, Bouchlaka Souissi C, et al. Contribution of the BRCA1 and BRCA2 mutations to breast cancer in Tunisia. J Hum Genet. 2007;52: 915–920. https://doi.org/10.1007/s10038-007-0195-5
13. Uhrhammer N, Abdelouahab A, Lafarge L, Feillel V, Ben Dib A, Bignon Y-J. BRCA1 mutations in Algerian breast cancer patients: High frequency in young, sporadic cases. Int J Med Sci. 2008;5: 197–202. https://doi.org/10.7150/ijms.5.197
14. Rebbeck TR, Friebel TM, Friedman E, Hamann U, Huo D, Kwong A, et al. Mutational spectrum in a worldwide study of 29,700 families with BRCA1 or BRCA2 mutations. Hum Mutat. 2018;39: 593–620. https://doi.org/10.1002/humu.23406
15. Zhang J, Fackenthal JD, Huo D, Zheng Y, Olopade OI. Searching for large genomic rearrangements of the BRCA1 gene in a Nigerian population. Breast Cancer Res Treat. 2010;124: 573–577. https://doi.org/10.1007/s10549-010-1006-9
16. Zhang J, Fackenthal JD, Zheng Y, Huo D, Hou N, Niu Q, et al. Recurrent BRCA1 and BRCA2 mutations in breast cancer patients of African ancestry. Breast Cancer Res Treat. 2012;134: 889–894. https://doi.org/10.1007/s10549-012-2136-z
17. Sluiter M, Mew S, van Rensburg EJ. PALB2 sequence variants in young South African breast cancer patients. Fam Cancer. 2009;8: 347–353. https://doi.org/10.1007/s10689-009-9241-0
18. Zhang B, Fackenthal JD, Niu Q, Huo D, Sveen WE, DeMarco T, et al. Evidence for an ancient BRCA1 mutation in breast cancer patients of Yoruban ancestry. Fam Cancer. 2009;8: 15–22. https://doi.org/10.1007/s10689-008-9205-9
19. Diez O, Pelegrí A, Gadea N, Gutiérrez-Enríquez S, Masas M, Tenés A, et al. Novel BRCA1 deleterious mutation (c.1949_1950delTA) in a woman of Senegalese descent with triple-negative early-onset breast cancer. Oncol Lett. 2011;2: 1287–1289. https://doi.org/10.3892/ol.2011.390
20. van der Merwe NC, Hamel N, Schneider S-R, Apffelstaedt JP, Wijnen JT, Foulkes WD. A founder BRCA2 mutation in non-Afrikaner breast cancer patients of the Western Cape of South Africa. Clin Genet. 2012;81: 179–184. https://doi.org/10.1111/j.1399-0004.2010.01617.x

21. Zheng Y, Walsh T, Gulsuner S, Casadei S, Lee MK, Ogundiran TO, et al. Inherited breast cancer in Nigerian women. J Clin Oncol. 2018;36: 2820–2825. https://doi.org/10.1200/JCO.2018.78.3977
22. Biancolella M, Ouédraogo NLM, Zongo N, Zohoncon TM, Testa B, Rizzacasa B, et al. Breast cancer in West Africa: Molecular analysis of BRCA genes in early-onset breast cancer patients in Burkina Faso. Hum Genomics. 2021;15: 65. https://doi.org/10.1186/s40246-021-00365-w
23. Pitt JJ, Riester M, Zheng Y, Yoshimatsu TF, Sanni A, Oluwasola O, et al. Characterization of Nigerian breast cancer reveals prevalent homologous recombination deficiency and aggressive molecular features. Nat Commun. 2018;9: 4181. https://doi.org/10.1038/s41467-018-06616-0
24. Mahfoudh W, Bettaieb I, Ghedira R, Snoussi K, Bouzid N, Klayech Z, et al. Contribution of BRCA1 5382insC mutation in triple negative breast cancer in Tunisia. J Transl Med. 2019;17: 123. https://doi.org/10.1186/s12967-019-1873-8
25. Onyia AF, Nana TA, Adewale EA, Adebesin AO, Adegboye BE, Paimo OK, et al. Breast cancer phenotypes in Africa: A scoping review and meta-analysis. JCO Glob Oncol. 2023;9: e2300135. https://doi.org/10.1200/GO.23.00135
26. Yahia ZA, Adam AA, Elgizouli M, Hussein A, Masri MA, Kamal M, et al. Epstein Barr virus: A prime candidate of breast cancer aetiology in Sudanese patients. Infect Agent Cancer. 2014;9: 9. https://doi.org/10.1186/1750-9378-9-9
27. Fessahaye G, Elhassan AM, Elamin EM, Adam AAM, Ghebremedhin A, Ibrahim ME. Association of Epstein—Barr virus and breast cancer in Eritrea. Infect Agent Cancer. 2017;12: 62. https://doi.org/10.1186/s13027-017-0173-2
28. El-Naby NEH, Hassan Mohamed H, Mohamed Goda A, El Sayed Mohamed A. Epstein-Barr virus infection and breast invasive ductal carcinoma in Egyptian women: A single center experience. J Egypt Natl Canc Inst. 2017;29: 77–82. https://doi.org/10.1016/j.jnci.2017.02.002
29. Harford JB. Viral infections and human cancers: The legacy of Denis Burkitt. Br J Haematol. 2012;156: 709–718. https://doi.org/10.1111/j.1365-2141.2011.09017.x
30. Levine PH, Pogo BG-T, Klouj A, Coronel S, Woodson K, Melana SM, et al. Increasing evidence for a human breast carcinoma virus with geographic differences. Cancer. 2004;101: 721–726. https://doi.org/10.1002/cncr.20436
31. Hachana M, Trimeche M, Ziadi S, Amara K, Gaddas N, Mokni M, et al. Prevalence and characteristics of the MMTV-like associated breast carcinomas in Tunisia. Cancer Lett. 2008;271: 222–230. https://doi.org/10.1016/j.canlet.2008.06.001
32. Parkin DM, Hämmerl L, Ferlay J, Kantelhardt EJ. Cancer in Africa 2018: The role of infections. Int J Cancer. 2020;146: 2089–2103. https://doi.org/10.1002/ijc.32538
33. Moukassa D, Boumba AM, Ngatali CF, Ebatetou A, Mbon JBN, Ibara J-R. Virus-induced cancers in Africa: Epidemiology and carcinogenesis mechanisms. OJ Pathology. 2017;8: 1–14. https://doi.org/10.4236/ojpathology.2018.81001
34. Kamiza AB, Kamiza S, Singini MG, Mathew CG. Association of TP53 rs1042522 with cervical cancer in the sub-Saharan African population: A meta-analysis. Trop Med Int Health. 2020;25: 666–672. https://doi.org/10.1111/tmi.13397

35. de Araujo Souza PS, Villa LL. Genetic susceptibility to infection with human papillomavirus and development of cervical cancer in women in Brazil. Mutat Res. 2003;544: 375–383. https://doi.org/10.1016/j.mrrev.2003.06.013
36. Sousa H, Santos AM, Pinto D, Medeiros R. Is there a biological plausability for p53 codon 72 polymorphism influence on cervical cancer development? Acta Med Port. 2011;24: 127–134.
37. Eltahir HA, Elhassan AM, Ibrahim ME. Contribution of retinoblastoma LOH and the p53 Arg/Pro polymorphism to cervical cancer. Mol Med Rep. 2012;6: 473–476. https://doi.org/10.3892/mmr.2012.942
38. Lahsen AO, Baba H, Bensghir R, Fayssel N, Sodqi M, Marih L, et al. TP53 R72P polymorphism and susceptibility to human papillomavirus infection among women with human immunodeficiency virus in Morocco: A case-control study. J Cancer Prev. 2017;22: 248–253. https://doi.org/10.15430/JCP.2017.22.4.248
39. Ndiaye R, Dem A, Mbaye PM, Guèye PM, Diop G, Diop PA, et al. Study of codon 72 of p53 gene as a risk-factor in cervical cancer in Senegal. Bull Cancer. 2014;101: 789–794. https://doi.org/10.1684/bdc.2014.1911
40. Govan VA, Constant D, Hoffman M, Williamson A-L. The allelic distribution of -308 Tumor Necrosis Factor-alpha gene polymorphism in South African women with cervical cancer and control women. BMC Cancer. 2006;6: 24. https://doi.org/10.1186/1471-2407-6-24
41. Zidi S, Gazouani E, Stayoussef M, Mezlini A, Ahmed SK, Yacoubi-Loueslati B, et al. IL-10 gene promoter and intron polymorphisms as genetic biomarkers of cervical cancer susceptibility among Tunisians. Cytokine. 2015;76: 343–347. https://doi.org/10.1016/j.cyto.2015.05.028
42. Zidi S, Stayoussef M, Alsaleh BL, Gazouani E, Mezlini A, Ebrahim BH, et al. Relationships between common and novel Interleukin-6 gene polymorphisms and risk of cervical cancer: A case-control study. Pathol Oncol Res. 2017;23: 385–392. https://doi.org/10.1007/s12253-016-0127-9
43. Baccar Harrath A, Yacoubi Loueslati B, Troudi W, Hmida S, Sedkaoui S, Dridi A, et al. HLA class II polymorphism: Protective or risk factors to breast cancer in Tunisia? Pathol Oncol Res. 2006;12: 79–81. https://doi.org/10.1007/BF02893448
44. Mestiri S, Bouaouina N, Ben Ahmed S, Chouchane L. A functional polymorphism of the tumor necrosis factor receptor-II gene associated with the survival and relapse prediction of breast carcinoma. Cytokine. 2005;30: 182–187. https://doi.org/10.1016/j.cyto.2005.01.007
45. Zidi I, Dziri O, Zidi N, Sebai R, Boujelebene N, Ben Hassine A, et al. Association of HLA-G +3142 C>G polymorphism and breast cancer in Tunisian population. Immunol Res. 2016;64: 961–968. https://doi.org/10.1007/s12026-015-8782-6
46. Al Azhary NM, Kamel MM, Ismail YM, Mahmoud AA, Radwan EM. The role of genetic polymorphisms in Nrf2 and P73 in Egyptian women with breast cancer. Asian Pac J Cancer Prev. 2016;17: 4945–4949. https://doi.org/10.22034/APJCP.2016.17.11.4945
47. Snoussi K, Mahfoudh W, Bouaouina N, Fekih M, Khairi H, Helal AN, et al. Combined effects of IL-8 and CXCR2 gene polymorphisms on breast cancer

susceptibility and aggressiveness. BMC Cancer. 2010;10: 283. https://doi.org/10.1186/1471-2407-10-283

48. Snoussi K, Mahfoudh W, Bouaouina N, Ahmed SB, Helal AN, Chouchane L. Genetic variation in IL-8 associated with increased risk and poor prognosis of breast carcinoma. Hum Immunol. 2006;67: 13–21. https://doi.org/10.1016/j.humimm.2006.03.018
49. Snoussi K, Strosberg AD, Bouaouina N, Ben Ahmed S, Chouchane L. Genetic variation in pro-inflammatory cytokines (interleukin-1beta, interleukin-1alpha and interleukin-6) associated with the aggressive forms, survival, and relapse prediction of breast carcinoma. Eur Cytokine Netw. 2005;16: 253–260.
50. Soliman AS, Kleer CG, Mrad K, Karkouri M, Omar S, Khaled HM, et al. Inflammatory breast cancer in north Africa: Comparison of clinical and molecular epidemiologic characteristics of patients from Egypt, Tunisia, and Morocco. Breast Dis. 2011;33: 159–169. https://doi.org/10.3233/BD-2012-000337
51. Todoric J, Umemura A, Taniguchi K, Karin M. Inflammation and Cancer. Holland-Frei Cancer Medicine. John Wiley & Sons, Ltd; 2017. pp. 1–8. https://doi.org/10.1002/9781119000822.hfcm027
52. Zhang W. The role of inflammation in breast cancer and prostate cancer. Clin Oncol Cancer Res. 2011;8: 77–84. https://doi.org/10.1007/s11805-011-0563-y
53. Pandey MK, Suskil MV, Chitren R, Al-Odat O, Jonnalagadda SC, Aggarwal BB. Cancer on fire: Role of inflammation in prevention and treatment. Current Advances for Development of Functional Foods Modulating Inflammation and Oxidative Stress. Elsevier; 2022. pp. 605–626. https://doi.org/10.1016/B978-0-12-823482-2.00024-8
54. Seledtsov VI, Darinskas A, Von Delwig A, Seledtsova GV. Inflammation control and immunotherapeutic strategies in comprehensive cancer treatment. Metabolites. 2023;13: 123. https://doi.org/10.3390/metabo13010123
55. McManus KF, Taravella AM, Henn BM, Bustamante CD, Sikora M, Cornejo OE. Population genetic analysis of the DARC locus (Duffy) reveals adaptation from standing variation associated with malaria resistance in humans. PLoS Genet. 2017;13: e1006560. https://doi.org/10.1371/journal.pgen.1006560
56. Rappoport N, Simon AJ, Amariglio N, Rechavi G. The Duffy antigen receptor for chemokines, ACKR1,- "Jeanne DARC" of benign neutropenia. Br J Haematol. 2019;184: 497–507. https://doi.org/10.1111/bjh.15730
57. Yeyeodu ST, Kidd LR, Kimbro KS. Protective innate immune variants in racial/ethnic disparities of breast and prostate cancer. Cancer Immunol Res. 2019;7: 1384–1389. https://doi.org/10.1158/2326-6066.CIR-18-0564
58. Elson JK, Beebe-Dimmer JL, Morgenstern H, Chilkuri M, Blanchard J, Lentsch AB. The Duffy Antigen/Receptor for Chemokines (DARC) and prostate-cancer risk among Jamaican men. J Immigr Minor Health. 2011;13: 36–41. https://doi.org/10.1007/s10903-010-9330-z
59. Martini R, Founta K, Maurice S, White J, Balogun O, Chambwe N, et al. Abstract 2948: Unveiling hidden genomic diversity: Exploring epigenetic potential and gene expression in African Pan Genome contig sequences. Cancer Research (Chicago, Ill). 2024;84: 2948–2948. https://doi.org/10.1158/1538-7445.AM2024-2948
60. Vazquez ED, Fang X, Levesque LA, Huynh M, Venegas C, Lu N, et al. Chemokine receptors differentially expressed by race category and molecular

subtype in the breast cancer TCGA cohort. Sci Rep. 2022;12: 10825. https://doi.org/10.1038/s41598-022-14734-5
61. Zhou S, Liu M, Hu Y, An W, Liang X, Yu W, et al. Expression of Duffy Antigen Receptor for Chemokines (DARC) is down-regulated in colorectal cancer. J Recept Signal Transduct Res. 2015;35: 462–467. https://doi.org/10.3109/10799893.2015.1009113
62. Rotimi SO, Rotimi OA, Salhia B. A Review of cancer genetics and genomics studies in Africa. Front Oncol. 2020;10: 606400. https://doi.org/10.3389/fonc.2020.606400
63. Ibrahim MH, Rashed RA, Hassan NM, Al-Azhary NM, Salama AI, Mostafa MN. Association of Cytochrome P450–1B1 gene polymorphisms with risk of breast cancer: An Egyptian study. Asian Pac J Cancer Prev. 2016;17: 2861–2866.
64. Okobia MN, Bunker CH, Garte SJ, Zmuda JM, Ezeome ER, Anyanwu SN, et al. Cytochrome P450 1B1 Val432Leu polymorphism and breast cancer risk in Nigerian women: A case control study. Infect Agent Cancer. 2009;4 Suppl 1: S12. https://doi.org/10.1186/1750-9378-4-S1-S12
65. Fernández-Santander A, Novillo A, Gaibar M, Romero-Lorca A, Moral P, Sánchez-Cuenca D, et al. Cytochrome and sulfotransferase gene variation in north African populations. Pharmacogenomics. 2016;17: 1415–1423. https://doi.org/10.2217/pgs-2016-0016
66. Novillo A, Romero-Lorca A, Gaibar M, Bahri R, Harich N, Sánchez-Cuenca D, et al. Genetic diversity of CYP3A4 and CYP3A5 polymorphisms in North African populations from Morocco and Tunisia. Int J Biol Markers. 2015;30: e148–e151. https://doi.org/10.5301/jbm.5000118
67. Souiden Y, Mahdouani M, Chaieb K, Elkamel R, Mahdouani K. CYP17 gene polymorphism and prostate cancer susceptibility in a Tunisian population. Cancer Epidemiol. 2011;35: 480–484. https://doi.org/10.1016/j.canep.2010.11.008
68. Souiden Y, Mahdouani M, Chaieb K, Elkamel R, Mahdouani K. Polymorphisms of glutathione-S-transferase M1 and T1 and prostate cancer risk in a Tunisian population. Cancer Epidemiol. 2010;34: 598–603. https://doi.org/10.1016/j.canep.2010.06.002
69. Akinloye O, Gromoll J, Simoni M. Variation in CAG and GGN repeat lengths and CAG/GGN haplotype in androgen receptor gene polymorphism and prostate carcinoma in Nigerians. Br J Biomed Sci. 2011;68: 138–142. https://doi.org/10.1080/09674845.2011.11730341
70. Esteban E, Rodon N, Via M, Gonzalez-Perez E, Santamaria J, Dugoujon J-M, et al. Androgen receptor CAG and GGC polymorphisms in Mediterraneans: Repeat dynamics and population relationships. J Hum Genet. 2006;51: 129–136. https://doi.org/10.1007/s10038-005-0336-7
71. Khedhaier A, Remadi S, Corbex M, Ahmed SB, Bouaouina N, Mestiri S, et al. Glutathione S-transferases (GSTT1 and GSTM1) gene deletions in Tunisians: Susceptibility and prognostic implications in breast carcinoma. Br J Cancer. 2003;89: 1502–1507. https://doi.org/10.1038/sj.bjc.6601292
72. Adegoke OJ, Shu XO, Gao Y-T, Cai Q, Breyer J, Smith J, et al. Genetic polymorphisms in uridine diphospho-glucuronosyltransferase 1A1 (UGT1A1) and risk of breast cancer. Breast Cancer Res Treat. 2004;85: 239–245. https://doi.org/10.1023/B:BREA.0000025419.26423.b8

73. Huo D, Kim H-J, Adebamowo CA, Ogundiran TO, Akang EE, Campbell O, et al. Genetic polymorphisms in uridine diphospho-glucuronosyltransferase 1A1 and breast cancer risk in Africans. Breast Cancer Res Treat. 2008;110: 367–376. https://doi.org/10.1007/s10549-007-9720-7
74. Fawzy MS, Aly NMM, Shalaby SM, El-Sawy WH, Abdul-Maksoud RS. Cyclo-oxygenase-2 169C>G and 8473T>C gene polymorphisms and prostaglandin E2 level in breast cancer: A case-control study. Gene. 2013;527: 601–605. https://doi.org/10.1016/j.gene.2013.06.007
75. Habyarimana T, Bakri Y, Mugenzi P, Mazarati JB, Attaleb M, El Mzibri M. Association between glutathione peroxidase 1 codon 198 variant and the occurrence of breast cancer in Rwanda. Mol Genet Genomic Med. 2018;6: 268–275. https://doi.org/10.1002/mgg3.367
76. Hussein AG, Pasha HF, El-Shahat HM, Gad DM, Toam MM. CYP1A1 gene polymorphisms and smoking status as modifier factors for lung cancer risk. Gene. 2014;541: 26–30. https://doi.org/10.1016/j.gene.2014.03.003
77. Mohammad MA, Zeeneldin AA, Abd Elmageed ZY, Khalil EH, Mahdy SME, Sharada HM, et al. Clinical relevance of cyclooxygenase-2 and matrix metallo-proteinases (MMP-2 and MT1-MMP) in human breast cancer tissue. Mol Cell Biochem. 2012;366: 269–275. https://doi.org/10.1007/s11010-012-1305-z
78. Benabdelkrim M, Djeffal O, Berredjem H. GSTM1 and GSTT1 polymorphisms and susceptibility to prostate cancer: A case-control study of the Algerian population. Asian Pac J Cancer Prev. 2018;19: 2853–2858. https://doi.org/10.22034/APJCP.2018.19.10.2853
79. Novillo A, Gaibar M, Romero-Lorca A, Chaabani H, Amir N, Moral P, et al. UDP-glucuronosyltransferase genetic variation in North African populations: A comparison with African and European data. Ann Hum Biol. 2018;45: 516–523. https://doi.org/10.1080/03014460.2018.1559354
80. B'chir F, Pavanello S, Knani J, Boughattas S, Arnaud MJ, Saguem S. CYP1A2 genetic polymorphisms and adenocarcinoma lung cancer risk in the Tunisian population. Life Sci. 2009;84: 779–784. https://doi.org/10.1016/j.lfs.2009.03.008
81. Ezzeldin N, El-Lebedy D, Darwish A, El Bastawisy A, Abd Elaziz SH, Hassan MM, et al. Association of genetic polymorphisms CYP2A6*2 rs1801272 and CYP2A6*9 rs28399433 with tobacco-induced lung Cancer: Case-control study in an Egyptian population. BMC Cancer. 2018;18: 525. https://doi.org/10.1186/s12885-018-4342-5
82. Ezzeldin N, El-Lebedy D, Darwish A, El-Bastawisy A, Hassan M, Abd El-Aziz S, et al. Genetic polymorphisms of human cytochrome P450 CYP1A1 in an Egyptian population and tobacco-induced lung cancer. Genes Environ. 2017;39: 7. https://doi.org/10.1186/s41021-016-0066-4
83. Pavanello S, B'chir F, Pulliero A, Saguem S, Ben Fraj R, El Aziz Hayouni A, et al. Interaction between CYP1A2-T2467DELT polymorphism and smoking in adenocarcinoma and squamous cell carcinoma of the lung. Lung Cancer. 2007;57: 266–272. https://doi.org/10.1016/j.lungcan.2007.04.004
84. Tournel G, Cauffiez C, Leclerc J, Billaut-Laden I, Allorge D, Chevalier D, et al. CYP2F1 genetic polymorphism: Identification of interethnic variations. Xenobiotica. 2007;37: 1433–1438. https://doi.org/10.1080/00498250701644403

85. Abdel-Rahman SZ, Anwar WA, Abdel-Aal WE, Ghoneim MA, Au WW. The CYP2D6 extensive metabolizer genotype is associated with increased risk for bladder cancer. Cancer Lett. 1997;119: 115–122. https://doi.org/10.1016/s0304-3835(97)00265-6
86. Feki-Tounsi M, Khlifi R, Louati I, Fourati M, Mhiri M-N, Hamza-Chaffai A, et al. Polymorphisms in XRCC1, ERCC2, and ERCC3 DNA repair genes, CYP1A1 xenobiotic metabolism gene, and tobacco are associated with bladder cancer susceptibility in Tunisian population. Environ Sci Pollut Res Int. 2017;24: 22476–22484. https://doi.org/10.1007/s11356-017-9767-x
87. Ouerhani S, Marrakchi R, Bouhaha R, Ben Slama MR, Sfaxi M, Ayed M, et al. The role of CYP2D6*4 variant in bladder cancer susceptibility in Tunisian patients. Bull Cancer. 2008;95: E1–E4. https://doi.org/10.1684/bdc.2008.0583
88. Shaw ME, Elder PA, Abbas A, Knowles MA. Partial allelotype of schistosomiasis-associated bladder cancer. Int J Cancer. 1999;80: 656–661. https://doi.org/10.1002/(sici)1097-0215(19990301)80:5<656::aid-ijc4>3.0.co;2-a
89. Habuchi T, Takahashi R, Yamada H, Ogawa O, Kakehi Y, Ogura K, et al. Influence of cigarette smoking and schistosomiasis on p53 gene mutation in urothelial cancer. Cancer Res. 1993;53: 3795–3799.
90. El Nouby KA, Abd El Hameed AH, Negm OE, Hamouda HE, El Gamal OM, Ismail GM. Genetic polymorphism of glutathione-S-transferase (GST-M1 and GST-T1) in schistosomiasis -associated bladder cancer in Egyptian patients. J Egypt Soc Parasitol. 2008;38: 991–1006.
91. Goerlitz D, El Daly M, Abdel-Hamid M, Saleh DA, Goldman L, El Kafrawy S, et al. GSTM1, GSTT1 null variants, and GPX1 single nucleotide polymorphism are not associated with bladder cancer risk in Egypt. Cancer Epidemiol Biomarkers Prev. 2011;20: 1552–1554. https://doi.org/10.1158/1055-9965.EPI-10-1306
92. Hadami K, Dakka N, Bensaid M, El Ahanidi H, Ameur A, Chahdi H, et al. Evaluation of glutathione S-transferase pi 1 expression and gene promoter methylation in Moroccan patients with urothelial bladder cancer. Mol Genet Genomic Med. 2018;6: 819–827. https://doi.org/10.1002/mgg3.449
93. Adelani IB, Rotimi OA, Maduagwu EN, Rotimi SO. Vitamin D: Possible therapeutic roles in Hepatocellular Carcinoma. Front Oncol. 2021;11: 642653. https://doi.org/10.3389/fonc.2021.642653
94. Siddappa M, Hussain S, Wani SA, White J, Tang H, Gray JS, et al. African American prostate cancer displays quantitatively distinct vitamin D receptor cistrome-transcriptome relationships regulated by BAZ1A. Cancer Res Commun. 2023;3: 621–639. https://doi.org/10.1158/2767-9764.CRC-22-0389
95. Abd-Elsalam EA-E, Ismaeil NA, Abd-Alsalam HS. Vitamin D receptor gene polymorphisms and breast cancer risk among postmenopausal Egyptian women. Tumour Biol. 2015;36: 6425–6431. https://doi.org/10.1007/s13277-015-3332-3
96. El-Shorbagy HM, Mahmoud NH, Sabet S. Association of vitamin D receptor gene polymorphisms with breast cancer risk in an Egyptian population. Tumour Biol. 2017;39: 1010428317727738. https://doi.org/10.1177/1010428317727738
97. Hemida MA, AbdElmoneim NA, Hewala TI, Rashad MM, Abdaallah S. Vitamin D Receptor in breast cancer tissues and its relation to Estrogen Receptor

Alpha (ER-α) gene expression and serum 25-hydroxyvitamin D levels in Egyptian breast cancer patients: A case-control study. Clin Breast Cancer. 2019;19: e407–e414. https://doi.org/10.1016/j.clbc.2018.12.019

98. Ahmed JH, Makonnen E, Fotoohi A, Yimer G, Seifu D, Assefa M, et al. Vitamin D status and association of VDR genetic polymorphism to risk of breast cancer in Ethiopia. Nutrients. 2019;11: 289. https://doi.org/10.3390/nu11020289
99. Wang S, Huo D, Kupfer S, Alleyne D, Ogundiran TO, Ojengbede O, et al. Genetic variation in the vitamin D related pathway and breast cancer risk in women of African ancestry in the root consortium. Int J Cancer. 2018;142: 36–43. https://doi.org/10.1002/ijc.31038
100. Odedina FT, Rotimi S. Promoting cancer genomics research in Africa: A roadmap. Nat Rev Cancer. 2021;21: 409–410. https://doi.org/10.1038/s41568-021-00359-9
101. Hamdi Y, Ben Rekaya M, Jingxuan S, Nagara M, Messaoud O, Benammar Elgaaied A, et al. A genome wide SNP genotyping study in the Tunisian population: Specific reporting on a subset of common breast cancer risk loci. BMC Cancer. 2018;18: 1295. https://doi.org/10.1186/s12885-018-5133-8
102. Shan J, Mahfoudh W, Dsouza SP, Hassen E, Bouaouina N, Abdelhak S, et al. Genome-Wide Association Studies (GWAS) breast cancer susceptibility loci in Arabs: Susceptibility and prognostic implications in Tunisians. Breast Cancer Res Treat. 2012;135: 715–724. https://doi.org/10.1007/s10549-012-2202-6
103. Lupicki K, Elifio-Esposito S, Fonseca AS, Weber SH, Sugita B, Langa BC, et al. Patterns of copy number alterations in primary breast tumors of South African patients and their impact on functional cellular pathways. Int J Oncol. 2018;53: 2745–2757. https://doi.org/10.3892/ijo.2018.4589
104. Hamdi Y, Boujemaa M, Ben Rekaya M, Ben Hamda C, Mighri N, El Benna H, et al. Family specific genetic predisposition to breast cancer: Results from Tunisian whole exome sequenced breast cancer cases. J Transl Med. 2018;16: 158. https://doi.org/10.1186/s12967-018-1504-9
105. Kim YC, Soliman AS, Cui J, Ramadan M, Hablas A, Abouelhoda M, et al. Unique features of germline variation in five Egyptian familial breast cancer families revealed by exome sequencing. PLoS One. 2017;12: e0167581. https://doi.org/10.1371/journal.pone.0167581
106. Riahi A, Radmanesh H, Schürmann P, Bogdanova N, Geffers R, Meddeb R, et al. Exome sequencing and case-control analyses identify RCC1 as a candidate breast cancer susceptibility gene. Int J Cancer. 2018;142: 2512–2517. https://doi.org/10.1002/ijc.31273
107. Cook MB, Wang Z, Yeboah ED, Tettey Y, Biritwum RB, Adjei AA, et al. A genome-wide association study of prostate cancer in West African men. Hum Genet. 2014;133: 509–521. https://doi.org/10.1007/s00439-013-1387-z
108. Du Z, Lubmawa A, Gundell S, Wan P, Nalukenge C, Muwanga P, et al. Genetic risk of prostate cancer in Ugandan men. Prostate. 2018;78: 370–376. https://doi.org/10.1002/pros.23481
109. Shan J, Al-Rumaihi K, Rabah D, Al-Bozom I, Kizhakayil D, Farhat K, et al. Genome scan study of prostate cancer in Arabs: Identification of three genomic regions with multiple prostate cancer susceptibility loci in Tunisians. J Transl Med. 2013;11: 121. https://doi.org/10.1186/1479-5876-11-121

110. Jaratlerdsiri W, Chan EKF, Gong T, Petersen DC, Kalsbeek AMF, Venter PA, et al. Whole-genome sequencing reveals elevated tumor mutational burden and initiating driver mutations in African men with treatment-Naïve, high-risk prostate cancer. Cancer Res. 2018;78: 6736–6746. https://doi.org/10.1158/0008-5472.CAN-18-0254
111. White JA, Kaninjing ET, Adeniji KA, Jibrin P, Obafunwa JO, Ogo CN, et al. Whole-exome sequencing of Nigerian prostate tumors from the Prostate Cancer Transatlantic Consortium (CaPTC) reveals DNA repair genes associated with African ancestry. Cancer Res Commun. 2022;2: 1005–1016. https://doi.org/10.1158/2767-9764.CRC-22-0136
112. Temilola DO, Wium M, Paccez J, Salukazana AS, Rotimi SO, Otu HH, et al. Detection of cancer-associated gene mutations in urinary cell-free DNA among prostate cancer patients in South Africa. Genes (Basel). 2023;14: 1884. https://doi.org/10.3390/genes14101884
113. Fong PC, Boss DS, Yap TA, Tutt A, Wu P, Mergui-Roelvink M, et al. Inhibition of poly(ADP-ribose) polymerase in tumors from BRCA mutation carriers. N Engl J Med. 2009;361: 123–134. https://doi.org/10.1056/NEJMoa0900212
114. Makaremi S, Asadzadeh Z, Hemmat N, Baghbanzadeh A, Sgambato A, Ghorbaninezhad F, et al. Immune checkpoint inhibitors in colorectal cancer: Challenges and future prospects. Biomedicines. 2021;9: 1075. https://doi.org/10.3390/biomedicines9091075
115. Gong T, Jaratlerdsiri W, Jiang J, Willet C, Chew T, Patrick SM, et al. Genome-wide interrogation of structural variation reveals novel African-specific prostate cancer oncogenic drivers. Genome Med. 2022;14: 100. https://doi.org/10.1186/s13073-022-01096-w
116. Martini R, Chen Y, Jenkins BD, Elhussin IA, Cheng E, Hoda SA, et al. Investigation of triple-negative breast cancer risk alleles in an International African-enriched cohort. Sci Rep. 2021;11: 9247. https://doi.org/10.1038/s41598-021-88613-w
117. Na R, Helfand BT, Chen H, Conran CA, Crawford SE, Hayward SW, et al. A genetic variant near GATA3 implicated in inherited susceptibility and etiology of benign prostatic hyperplasia (BPH) and lower urinary tract symptoms (LUTS). Prostate. 2017;77: 1213–1220. https://doi.org/10.1002/pros.23380
118. Nguyen AHT, Tremblay M, Haigh K, Koumakpayi IH, Paquet M, Pandolfi PP, et al. Gata3 antagonizes cancer progression in Pten-deficient prostates. Hum Mol Genet. 2013;22: 2400–2410. https://doi.org/10.1093/hmg/ddt088
119. Bezerra SM, Lotan TL, Faraj SF, Karram S, Sharma R, Schoenberg M, et al. GATA3 expression in small cell carcinoma of bladder and prostate and its potential role in determining primary tumor origin. Hum Pathol. 2014;45: 1682–1687. https://doi.org/10.1016/j.humpath.2014.04.011
120. Chen WC, Singh E, Muchengeti M, Bradshaw D, Mathew CG, Babb de Villiers C, et al. Johannesburg Cancer Study (JCS): Contribution to knowledge and opportunities arising from 20 years of data collection in an African setting. Cancer Epidemiol. 2020;65: 101701. https://doi.org/10.1016/j.canep.2020.101701
121. Van Loon K, Mwachiro MM, Abnet CC, Akoko L, Assefa M, Burgert SL, et al. The African esophageal cancer consortium: A call to action. J Glob Oncol. 2018;4: 1–9. https://doi.org/10.1200/JGO.17.00163

122. Busby GB, Band G, Si Le Q, Jallow M, Bougama E, Mangano VD, et al. Admixture into and within sub-Saharan Africa. Elife. 2016;5: e15266. https://doi.org/10.7554/eLife.15266
123. Adedokun BO, Olopade CO, Olopade OI. Building local capacity for genomics research in Africa: Recommendations from analysis of publications in Sub-Saharan Africa from 2004 to 2013. Glob Health Action. 2016;9: 31026. https://doi.org/10.3402/gha.v9.31026
124. Paro R, Grossniklaus U, Santoro R, Wutz A. Introduction to Epigenetics. Cham (CH): Springer; 2021. Available: http://www.ncbi.nlm.nih.gov/books/NBK585706/
125. Debouki-Joudi S, Trifa F, Khabir A, Sellami-Boudawara T, Frikha M, Daoud J, et al. CpG methylation of APC promoter 1A in sporadic and familial breast cancer patients. Cancer Biomark. 2017;18: 133–141. https://doi.org/10.3233/CBM-160005
126. Hamdi K, Blancato J, Goerlitz D, Islam M, Neili B, Abidi A, et al. Circulating cell-free miRNA expression and its association with clinicopathologic features in inflammatory and non- inflammatory breast cancer. Asian Pac J Cancer Prev. 2016;17: 1801–1810. https://doi.org/10.7314/apjcp.2016.17.4.1801
127. Pollard J, Burns PA, Hughes TA, Ho-Yen C, Jones JL, Mukherjee G, et al. Differential expression of MicroRNAs in breast cancers from four different ethnicities. Pathobiology. 2018;85: 220–226. https://doi.org/10.1159/000488456
128. Zidan HE, Karam RA, El-Seifi OS, Abd Elrahman TM. Circulating long non-coding RNA MALAT1 expression as molecular biomarker in Egyptian patients with breast cancer. Cancer Genet. 2018;220: 32–37. https://doi.org/10.1016/j.cancergen.2017.11.005
129. Karray-Chouayekh S, Trifa F, Khabir A, Boujelbane N, Sellami-Boudawara T, Daoud J, et al. Aberrant methylation of RASSF1A is associated with poor survival in Tunisian breast cancer patients. J Cancer Res Clin Oncol. 2010;136: 203–210. https://doi.org/10.1007/s00432-009-0649-6
130. Karray-Chouayekh S, Trifa F, Khabir A, Boujelbene N, Sellami-Boudawara T, Daoud J, et al. Methylation status and overexpression of COX-2 in Tunisian patients with ductal invasive breast carcinoma. Tumour Biol. 2011;32: 461–468. https://doi.org/10.1007/s13277-010-0139-0
131. Trifa F, Karray-Chouayekh S, Jmaa ZB, Jmal E, Khabir A, Sellami-Boudawara T, et al. Frequent CpG methylation of ubiquitin carboxyl-terminal hydrolase 1 (UCHL1) in sporadic and hereditary Tunisian breast cancer patients: Clinical significance. Med Oncol. 2013;30: 418. https://doi.org/10.1007/s12032-012-0418-2
132. Mamdouh S, Khorshed F, Aboushousha T, Hamdy H, Diab A, Seleem M, et al. Evaluation of Mir-224, Mir-215 and Mir-143 as serum biomarkers for HCV associated hepatocellular carcinoma. Asian Pac J Cancer Prev. 2017;18: 3167–3171. https://doi.org/10.22034/APJCP.2017.18.11.3167
133. Motawi TMK, Sadik NAH, Shaker OG, Ghaleb MH. Elevated serum microRNA-122/222 levels are potential diagnostic biomarkers in Egyptian patients with chronic hepatitis C but not hepatic cancer. Tumour Biol. 2016;37: 9865–9874. https://doi.org/10.1007/s13277-016-4884-6

134. Motawi TMK, El-Maraghy SA, Sabry D, Mehana NA. The expression of long non coding RNA genes is associated with expression with polymorphisms of HULC rs7763881 and MALAT1 rs619586 in hepatocellular carcinoma and HBV Egyptian patients. J Cell Biochem. 2019;120: 14645–14656. https://doi.org/10.1002/jcb.28726
135. Abdelmaksoud-Dammak R, Saadallah-Kallel A, Miladi-Abdennadher I, Ayedi L, Khabir A, Sallemi-Boudawara T, et al. CpG methylation of ubiquitin carboxyl-terminal hydrolase 1 (UCHL1) and P53 mutation pattern in sporadic colorectal cancer. Tumour Biol. 2016;37: 1707–1714. https://doi.org/10.1007/s13277-015-3902-4
136. Chaar I, Amara S, Elamine OE, Khiari M, Ounissi D, Khalfallah T, et al. Biological significance of promoter hypermethylation of p14/ARF gene: Relationships to p53 mutational status in Tunisian population with colorectal carcinoma. Tumour Biol. 2014;35: 1439–1449. https://doi.org/10.1007/s13277-013-1198-9
137. Chan AO, Soliman AS, Zhang Q, Rashid A, Bedeir A, Houlihan PS, et al. Differing DNA methylation patterns and gene mutation frequencies in colorectal carcinomas from Middle Eastern countries. Clin Cancer Res. 2005;11: 8281–8287. https://doi.org/10.1158/1078-0432.CCR-05-1000
138. Nieminen TT, Shoman S, Eissa S, Peltomäki P, Abdel-Rahman WM. Distinct genetic and epigenetic signatures of colorectal cancers according to ethnic origin. Cancer Epidemiol Biomarkers Prev. 2012;21: 202–211. https://doi.org/10.1158/1055-9965.EPI-11-0662
139. Sekal M, Ameurtesse H, Chbani L, Ouldim K, Bennis S, Abkari M, et al. Epigenetics could explain some Moroccan population colorectal cancers peculiarities: Microsatellite instability pathway exploration. Diagn Pathol. 2015;10: 77. https://doi.org/10.1186/s13000-015-0326-9
140. Arfaoui AT, Kriaa LBM, El Hadj OEA, Ben Hmida MA, Khiari M, Khalfallah T, et al. Association of a p73 exon 2 GC/AT polymorphism with colorectal cancer risk and survival in Tunisian patients. Virchows Arch. 2010;457: 359–368. https://doi.org/10.1007/s00428-010-0942-4
141. Chaar I, Amara S, Khiari M, Ounissi D, Dhraif M, Ben Hamida AEM, et al. Relationship between MDM2 and p53 alterations in colorectal cancer and their involvement and prognostic value in the Tunisian population. Appl Immunohistochem Mol Morphol. 2013;21: 228–236. https://doi.org/10.1097/PAI.0b013e31825f1c20
142. Kassem AM, El-Guendy N, Tantawy M, Abdelhady H, El-Ghor A, Abdel Wahab AH. Mutational hotspots in the mitochondrial D-loop region of cancerous and precancerous colorectal lesions in Egyptian patients. DNA Cell Biol. 2011;30: 899–906. https://doi.org/10.1089/dna.2010.1186
143. Mzahma R, Kharrat M, Fetiriche F, Bouasker null, Ben Moussa M, Ben Safta Z, et al. The relationship between telomere length and clinicopathologic characteristics in colorectal cancers among Tunisian patients. Tumour Biol. 2015;36: 8703–8713. https://doi.org/10.1007/s13277-015-3545-5
144. Feng Q, Balasubramanian A, Hawes SE, Toure P, Sow PS, Dem A, et al. Detection of hypermethylated genes in women with and without cervical neoplasia. J Natl Cancer Inst. 2005;97: 273–282. https://doi.org/10.1093/jnci/dji041

145. Gagliardi A, Porter VL, Zong Z, Bowlby R, Titmuss E, Namirembe C, et al. Analysis of Ugandan cervical carcinomas identifies human papillomavirus clade-specific epigenome and transcriptome landscapes. Nat Genet. 2020;52: 800–810. https://doi.org/10.1038/s41588-020-0673-7
146. Ouedraogo SY, Zoure AA, Zeye MMJ, Kiendrebeogo TI, Zhou X, Sawadogo AY, et al. BRCA1, BRCA2, TP53, PIK3CA, PTEN and AKT1 genes mutations in Burkina Faso breast cancer patients: Prevalence, spectrum and novel variant. Mol Genet Genomics. 2022;297: 1257–1268. https://doi.org/10.1007/s00438-022-01914-1

Chapter 9

Psychiatric Genetic Research in Africa

Allan Kalungi[1,2,3*], *Jared G. Maina*[4], *Sylvanus Toikumo*[5,6], *and Lukoye Atwoli*[7,8]

[1]Department of Non-communicable Disease Epidemiology (NCDE), London School of Hygiene and Tropical Medicine, London, UK.

[2]Medical Research Council/Uganda Virus Research Institute and London School of Hygiene and Tropical Medicine (MRC/UVRI and LSHTM) Uganda Research Unit, Entebbe, Uganda.

[3]Department of Medical Biochemistry, College of Health Sciences, Makerere University, Kampala, Uganda.

[4]Social, Genetic and Developmental Psychiatry Centre, Institute of Psychiatry, Psychology & Neuroscience, King's College London, de Crespigny Park, London, SE5 8AF UK.

[5]Mental Illness Research, Education and Clinical Center, Crescenz VAMC, Philadelphia, PA, USA.

[6]Department of Psychiatry, University of Pennsylvania Perelman School of Medicine, Philadelphia, PA, USA.

[7]Department of Mental Health, School of Medicine, Moi University College of Health Sciences, Eldoret, Kenya.

[8]Brain and Mind Institute and Department of Internal Medicine, Medical College East Africa, the Aga Khan University, Nairobi, Kenya.

9.1 INTRODUCTION

Psychiatric disorders are collections of complex traits comprising a variety of combinations of thoughts, emotions, and behaviors.[1] The fifth edition of the *Diagnostic and Statistical Manual of Mental Disorders* (*DSM-5*) defines psychiatric disorders as syndromes characterized by clinically significant disturbance in an individual's cognition, emotion regulation, or behavior that reflects dysfunction in the psychological, biological, or developmental processes underlying mental functioning. Psychiatric disorders exert clinically significant disturbances, unveiling profound dysfunctions that degrade mental functioning.[2] They encompass a wide range of mental conditions (see Table 9.1).

Globally, psychiatric disorders are among the top ten contributors to disease burden,[3] affecting an estimated 1 in every 8 people.[4] There is a huge

*Corresponding author: allan.kalungi@mrcuganda.org (Allan Kalungi)

DOI: 10.1201/9781003436584-9

Table 9.1 Common Psychiatric Disorders and Their Key Symptoms

Psychiatric Disorder	*Key Symptoms*
Depression	Persistent sadness, loss of interest in activities, feelings of hopelessness, fatigue, difficulty concentrating
Generalized anxiety disorder	Excessive worry, restlessness, difficulty concentrating, irritability, muscle tension, sleep disturbances
Bipolar disorder	Mood swings ranging from depressive lows to manic highs, increased activity levels during mania, feelings of worthlessness during depressive episode
Schizophrenia	Hallucinations, delusions, disorganized speech, severe emotional flatness or inappropriate reactions
Obsessive-compulsive disorder	Recurrent, unwanted thoughts (obsessions) and repetitive behaviors (compulsions), distress over the obsessions, urge to perform certain rituals
Post-traumatic stress disorder	Flashbacks, nightmares, severe anxiety, uncontrollable thoughts of the previously experienced traumatic event, emotional numbness
Eating disorders	Extreme emotions, attitudes, and behaviors surrounding weight and food issues (e.g., severe restriction of food, binge eating, purging)
Attention-deficit hyperactivity disorder	Inattention, hyperactivity, impulsive behavior, difficulty in maintaining focus, disorganization
Personality disorders	Varying moods, unstable self-image, behavioral challenges, difficulties in personal and social relationships (e.g., intense fear of abandonment, frequent mood swings)
Substance use disorders	Compulsive use of substances despite harmful consequences, withdrawal symptoms, tolerance, neglect of normal activities in favor of substance use

burden of psychiatric disorders in Africa.[3] The age-standardized prevalence for any psychiatric disorder has been estimated at 13.4% in sub-Saharan Africa, compared to the global estimate of 12.3%.[3] Table 9.2 compares lifetime prevalence estimates of several disorders in Africa to global estimates.

The burden of psychiatric disorders in Africa is not only substantial and growing but also often underestimated. Mental and substance use disorders account for a significant proportion of disability-adjusted life years (DALYs) in sub-Saharan Africa, contributing to profound personal, societal, and economic consequences.[3,5] Despite this burden, mental health research remains underfunded and under-prioritized, with many individuals going undiagnosed or untreated.[6] This public health crisis necessitates urgent investment in understanding the underlying determinants, including genetic factors, which may interact with Africa's unique sociocultural and environmental landscape.[7]

There are notable discrepancies in estimated lifetime prevalence rates for various mental disorders in Africa as compared with global estimates

Table 9.2 Estimated Prevalence of Various Mental Disorders in Africa and the World

Psychiatric Disorder	*Estimated Lifetime Prevalence in Africa (%)*	*Estimated Lifetime Global Prevalence (%)*
Depression	4.59[5]	3.76[5]
Generalized anxiety disorder	3.59[5]	4.05[5]
Bipolar disorder	0.59[5]	0.53[5]
Schizophrenia	0.22[5]	0.32[5]
Obsessive-compulsive disorder	**	2.30[8]
Post-traumatic stress disorder	2.30[9]	3.90[9]
Eating disorders	0.11[5]	0.18[5]
Attention-deficit hyperactivity disorder	0.61[5]	1.14[5]
Personality disorders	**	7.80[10]
Substance use disorder	1.48[5]	2.17[5]

Note: **, data not available or limited.

(Table 9.2). These discrepancies may stem from a variety of underlying factors. For instance, the slightly higher prevalence of depressive disorders in Africa compared to the global average could suggest regional variations in exposure to stressors or differing methodologies in data collection. Conversely, the generally lower rates of disorders such as generalized anxiety disorder and PTSD in Africa, despite significant socio-economic challenges and conflict exposure, might indicate substantial underreporting and underdiagnosis. This underrepresentation could be due to limited mental health infrastructure, cultural stigmas surrounding mental health issues, or a lack of trained professionals to diagnose and treat these conditions. Additionally, the absence of comprehensive data for disorders like obsessive-compulsive disorder and personality disorders in Africa highlights a critical gap in research and awareness, emphasizing the need for increased mental health resources and targeted studies to better understand and address these conditions within the continent.

The assessment of psychiatric disorders is most commonly informed by the *DSM-IV* or *DSM-5* or the *International Statistical Classification of Diseases and Related Health Problems*, 10th Revision (*ICD-10*) or the combination of *ICD-10* and *DSM-IV* or *DSM-5*.[11] The *ICD-10* published by the World Health Organization (WHO) categorizes psychiatric disorders into the following: (1) Organic disorders, which stem from identifiable causes such as cerebral diseases, brain injury, or other insults leading to cerebral dysfunction. This dysfunction can be primary, directly affecting the brain, or secondary, affecting the brain only as part of the multiple organs or systems of the body that are involved. Notable examples include dementia; (2) substance use disorders arising from the consumption of psychoactive substances, whether medically prescribed or not; (3) psychotic disorders, which encompass conditions like schizophrenia characterized by profound distortions in thinking, perception, and affect, often resulting in a significant departure from reality; (4) mood affective disorders, characterized by fundamental disruptions in mood with or without associated anxiety. Examples include bipolar affective disorder and major depressive disorder (MDD); (5) anxiety disorders, primarily marked by excessive and persistent feelings of worry that dominate the individual's experience; (6) eating disorders, characterized by disturbances in eating behaviors, often intertwined with physiological and physical factors, exemplified by conditions like anorexia nervosa; (7) personality disorders, which manifest as deviations from societal norms in behavior and personality traits, often leading to significant impairment in social and occupational functioning; (8) disorders of psychological development, such as autism, are characterized by impairment or delay in various psychological functions crucially tied to the maturation of the central nervous system, typically evident from the early stages of development.[12] In essence, the *ICD-10* offers a comprehensive framework for understanding the multifaceted landscape of psychiatric disorders, encompassing a diverse array of conditions with distinct etiologies, symptoms, and treatment approaches.

Psychiatric disorders demonstrate moderate to high heritability with genetic susceptibility variation across different disorders.[13] This variability includes unique genetic factors specific to individual disorders and shared genetic predispositions among them.[13] In contrast, the lower heritability observed in specific psychiatric disorders suggests a significant impact of modifiable risk factors, including social[14] and environmental determinants.[11,15] These factors hugely drive the onset and progression of such disorders, highlighting the complex interplay between genetic predisposition and environmental influences in their development.

There exist evidenced psychological and pharmacological treatments. These have been reported to minimize symptoms to allow an individual to function and these have been briefly described elsewhere for depression, post-traumatic stress disorder, alcohol abuse, generalized anxiety disorder, and attention-deficit hyperactivity disorder. However, regardless of whether the psychological or pharmacological treatments are used individually or in a combination, the response rate among patients has been reported to be utmost 50% among a large umbrella review and meta-analytic meta-analysis which investigated the efficacy of psychotherapies and pharmacotherapies for mental disorders among adult participants.[16] Psychiatric genetic research has the potential to identify new mechanistic pathways for psychiatric disorder, which can lead to new drugs or drug targets. This chapter explores the landscape of psychiatric genetic research in Africa.

9.1.1 Global Psychiatric Genetics

Much of what we know about the genetics of psychiatric disorders has been spearheaded by the Psychiatric Genomics Consortium (PGC) (https://pgc.unc.edu/). Formed in 2007, the PGC is a collective effort consisting of nearly

1,000 researchers across ~40 countries involved in 16 domains in the psychiatric field. The PGC is organized into working groups, with each representing a specific psychiatric and neurodevelopmental disorder. By pooling studies from multiple research groups globally, the PGC is able to tackle the sample size issue in complex disease genomics. Larger sample sizes often translate to increased statistical power for identifying genetic determinants of diseases. For instance, the MDD working group of the PGC (PGC-MDD) has demonstrated the success of the PGC. In 2013, the first freeze of the PGC-MDD was released.[17] In this study, the authors analyzed 18,759 participants (9,240 cases, 9,519 controls) of European ancestry but identified no genome-wide significant associations. Fast forward to 2025, the third and latest freeze of the PGC-MDD working group was published.[18] This meta-analysis comprised of more than 5 million participants (~688,000 cases and ~4,600,000 controls) and identified 697 genome-wide associations at 635 genomic *loci*. This expanded study provided evidence for involvement of amygdala and hippocampal excitatory neurons linked to depressive behaviors. Similarly, the PGC-bipolar disorder working group has been instrumental in advancing our understanding of the genetics of bipolar disorder. The current study has nearly 3 million participants (~160,000 cases and ~2,700,000 controls) and identified 337 genome-wide associations at 298 *loci* implicating several factors in the pathophysiology of bipolar disorder, such as GABAergic interneurons and medium spiny neurons.[19] Similar success stories have been reported across the other PGC working groups (https://pgc.unc.edu/for-researchers/working-groups/). Figure 9.1 shows the number of independent hits significantly associated with various psychiatric disorders from PGC studies.

However, a key limitation of current large-scale genomic studies of psychiatric disorders has been the lack of ancestral diversity in the study participants.[7] The majority of global psychiatric genetics research has been conducted among individuals of European ancestry (Table 9.3).

In multi-ancestry studies, representation of African ancestry is predominantly by diaspora Africans living in the Americas or Europe who are part of studies, such as the Million Veteran Program (https://www.mvp.va.gov/pwa/), the All of Us research program (https://allofus.nih.gov/), and the United Kingdom Biobank (https://www.ukbiobank.ac.uk/). However, the genetic makeup of diaspora African Americans is not an accurate representation of African genetic diversity. It has been demonstrated that they largely represent the Niger-Congo sub-Saharan population group, which is only one of the five major African ethnolinguistic divisions.[25]

While African populations remain underrepresented in global psychiatric genetic research, their inclusion is not solely a matter of equity. Due to their evolutionary history, African genomes exhibit the highest levels of genetic diversity globally, including a high proportion of private alleles that are unique to Africa.[26] There is potential for discovery of unique causal variants and potentially protective alleles shaped by distinct evolutionary and

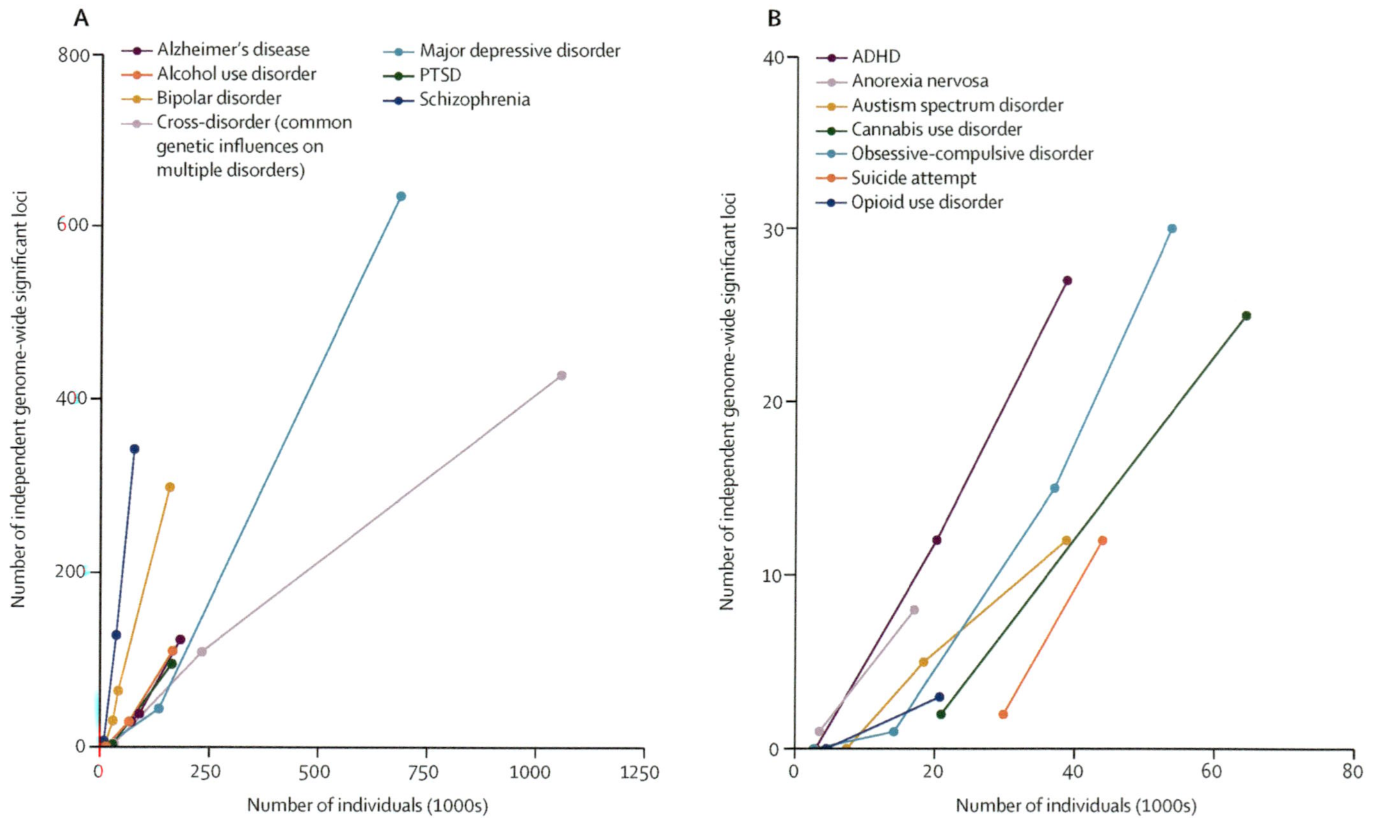

Figure 9.1 Progression of genome-wide significant loci with time and increasing number of cases for different psychiatric disorders from the PGC studies. (Adapted from: The Psychiatric Genomics Consortium: Discoveries and Direction, Agrawal et al., PsyArXiv 2025; doi.org/10.31234/osf.io/xykj2_v1).

Table 9.3 Sample Sizes of Select Psychiatric Conditions Genetics Studies: European and African Ancestry Representation

Phenotype	*Total Sample Size*	*Number of European Participants (%)*	*Number of African Participants (%)*
Major depressive disorder[18]	5,053,033	3,887,532 (76.93)	131,996 (2.61)
Bipolar disorder[19]	2,954,535	2,454,385 (83.07)	98,426 (3.33)
Schizophrenia[20]	175,799	161,405 (91.81)	10,070 (5.73)
Substance use disorders[21]	1,118,180	1,025,550 (91.72)	92,630 (8.28)
Suicide attempt[22]	958,896	816,178 (85.12)	86,486 (9.02)
Post-traumatic stress disorder[23]	1,280,990	1,222,882 (95.46)	51,034 (3.98)
Major anxiety disorders[24]	852,222	852,222 (100)	0 (0)

Note: The percentages do not add up to 100 for all rows as other ancestries in those studies are not shown.

environmental pressures. Also, although several genetic variants associated with several disorders have been identified, the most likely causal variants are yet to be discovered. The highly genetically diverse data from Africa has potential to improve fine-mapping of loci to aid the discovery of causal variants. Far from merely filling a gap, integrating African genomics data can illuminate mechanisms of etiology that remain invisible in homogeneous datasets and contribute informative insights into global psychiatric genetics.

9.2 PSYCHIATRIC GENETICS RESEARCH CONDUCTED THUS FAR IN AFRICA

Psychiatric disorders have substantial a genetic component with quantifiable heritability estimates at approximately 60% accounting for the risk.[27] Efforts to disentangle the role of genetics in psychiatric disorder etiology has led to study designs that explore genetic variation in "biologically plausible" candidate genes, across the genome, or gene–environment interactions. The past decade has witnessed increasing use of genome-wide association studies (GWAS) for identifying genetic factors associated with psychiatric disorders. Most of these studies have been conducted in individuals of European ancestry and have been deposited in publicly available databases such as the GWAS catalog (https://www.ebi.ac.uk/gwas/) or large collaborative efforts, such as the PGC.[28] Although Africa makes up 15% of the world's population, only about 2.4% of African representation is included in the GWAS catalog, which represents about 7% of the reported phenotypic associations.[29] This signals the need for inclusion of African populations in genetics research effort toward achieving better precision psychiatry and targeted therapeutics in the continent, and, by extension, the world at large.[30]

Africa's marked genetic diversity together with its unique epidemiological and sociocultural contexts presents a powerful setting to investigate the intricate interactions between genetic predispositions, environmental exposures, and cultural frameworks in shaping mental health outcomes. The continent bears a high burden of chronic infectious diseases such as HIV/AIDS, malaria, and tuberculosis, as well as frequent exposure to socio-political conflict, poverty, displacement, and climate-related stressors. These conditions serve as potent environmental stressors that can interact with underlying genetic vulnerabilities, potentially modifying the risk, onset, and course of psychiatric disorders.[31,32]

Cultural beliefs and practices also play a fundamental role in how mental illness is perceived, experienced, and addressed across African societies. In many communities, symptoms of depression and anxiety are somatically expressed or attributed to spiritual and supernatural causes, which in turn affects care-seeking behavior and the interpretation of psychiatric symptoms.[33,34] Traditional healing systems and community-based care remain dominant mental health frameworks in both rural and urban areas.[35] This underscores the importance of integrating cultural context into psychiatric assessment and genetic research.

These sociocultural and environmental dynamics offer a unique opportunity to explore gene–environment and gene–culture interactions, which are increasingly recognized as central to understanding the etiology of complex psychiatric disorders.[36] For example, early-life adversity, trauma, and chronic social stress are common in many African settings and have been linked to epigenetic modifications that alter gene expression, particularly in stress-related pathways.[37] Such mechanisms may help explain population-level variability in psychiatric disorder expression and resilience.

Understanding how community living structures, spiritual coping, stigma, and psychosocial resilience interact with genetic risk loci in African populations could illuminate novel pathways of both vulnerability and protection. This integrative approach holds potential not only for advancing psychiatric genetics but also for informing culturally grounded and effective mental health interventions tailored to African settings.

9.2.1 Gene Discovery Efforts for Psychiatric Disorders in Africa

Tremendous success has been recorded toward identifying replicable genetic loci associated with anxiety disorders, albeit in Europeans only.[38] GWAS of anxiety disorders in individuals from Africa have been hampered by limited statistical power due to small sample sizes. In a cross-ancestry meta-analysis of postpartum depression that comprised women of African ancestry (n = 1,711), no genetic variant reached genome-wide significance.[39] Only very recently, an MDD GWAS that included individuals from continental Africa identified a suggestive association for a genetic variant (rs6902879) upstream of the melanin-concentrated hormone receptor gene (*MCHR2*).[38] Similar to MDD, recent evidence has shown that larger sample sizes can aid successful detection of loci influencing PTSD in African populations.

A plethora of psychiatric genetic studies conducted on the African continent have investigated candidate genes associated with various disorders. These candidate gene studies are detailed in a curated database (www.psychgenafrica.org). In the first GWAS and polygenic risk scores (PRS) of PTSD in a South African population (n = 603), whereas not one occurrence of single nucleotide polymorphism (SNP) reached genome-wide significance, a suggestive risk loci in *PARK2* was shown to have modest effect on the development of PTSD.[40] Interestingly, the same variant had genome-wide significance in the multi-ancestry meta-analysis of PTSD from the PGC-PTSD workgroup (n = 200,000),[41] suggesting that bigger is better, and therefore large GWAS are critical for bridging the global health disparity.[42]

A recent study in a South African population found no association between a schizophrenia-derived PRS and methamphetamine associated psychosis or brain structure measures.[43] However, a whole exome analysis in a sample of ~1,800 self-identified South Africa Xhosa participants from the Human Heredity & Health in Africa (H3Africa) (http://h3africa.org/) initiative identified schizophrenia-associated rare and damaging genetic variants localized in brain-expressed genes.[11] Some of these genes, for example, *GRUA2*, *CACNA1A*, and *CNTNAP1* are involved in synaptic activity-related pathways ranging from calcium channel signaling to cell adhesion—consistent with findings previously reported in other populations.[44] These findings demonstrate that, beyond increasing sample size for GWAS—identifying subtle changes in severe ultrarare genes in African cohorts can unravel novel disease biology.

9.2.2 Integrating Multi-Omics into Psychiatric Research in Africa

In the past decade, high throughput technologies for analysis of omics data, such as epigenomics– have offered some mechanistic insight into the etiology of psychiatric disorders. While the brain remains the tissue of interest for psychiatry research, the role of peripheral tissues as biological end points for profiling disease state and mechanism has earned attention. This is because the use of post-mortem tissues, especially in the African context, requires substantial ethical considerations such as informed consent, language barriers, and limited access to healthcare.[45] Moreover, ascertaining phenotypic information for deceased individuals remains a key ethical issue for psychiatric genetics research.[42]

Despite the aforementioned hurdles, some progress has been recorded toward deciphering the mechanism of psychiatric disease development in Africa. One such mechanism commonly studied is epigenetic variation, in particular DNA methylation (DNAm) at cytosine-guanine (CpG) dinucleotide sites. DNAm regulates gene expression and provides the mechanistic framework for studying gene-environment interactions.[46] Studies have illustrated how environmental risk factors, such as trauma,[47] rape,[48] war,[49–52] prenatal maternal stress[19] are biologically embedded through altered DNAm patterns, thereby influencing psychiatric health outcomes. These studies, using an epigenome-wide association studies (EWAS) or candidate (*epi*)gene approach, have prioritized various differentially methylated genes as potential end-points for biomarker development.

PTSD symptoms among individuals exposed to rape in a South African cohort have been shown to be mediated by decreased methylation of genes involved in stress regulation (*ADCYAP1*) and neuronal polarization and synaptic development (*BRSK2*).[48] Another EWAS of PTSD also identified decreased methylation in genes involved in memory formation (*NTRK2*) among survivors of the rebel war in northern Uganda or the Rwandan genocide.[49,50] Methylation changes in another stress-regulatory gene—*NR3C1* have also been recorded in response to war-related trauma among individuals exposed to the Rwandan genocide and war-associated rape.[51] Among women from Tutsi ethnicity, in utero exposure to the genocide was associated with increased methylation in three core genes, including *BCOR*, *PRDM8*, and *VWDE*.[52] In another study, prenatal war trauma was shown to relate to decreased methylation in a long non-coding RNA (*TENM3-AS1*) just upstream of *TENM3*,[53] which is involved in neuronal development.[54]

In relation to differential candidate DNAm profiles, increased methylation of genes implicated in dopaminergic neurotransmission (*NR4A2*) and synaptic plasticity (*AUTS2*) have been associated with reduced PTSD symptom severity in participants receiving forensic offender rehabilitation.[55] Also, an intron-7 methylation profile, consistent with lower *FKBP5* expression and elevated glucocorticoid receptor levels, has been associated with higher anxiety proneness.[47]

In sum, these studies highlight genes and biological pathways that improve our understanding of the mechanisms underlying key environmental risk factors for psychiatric disorders that are applicable to Africa. These findings lay the groundwork for a more comprehensive approach to the study of psychiatric disorders, using multiple omics sources, including genome, transcriptome, and epigenome, to attain a more integrative understanding of their biological origins.

9.3 PSYCHIATRIC GENETICS RESEARCH INITIATIVES IN AFRICA

Saddled with the responsibility to close the knowledge gap and disparity between Africa and the rest of the world,[56] several ongoing psychiatric genetic studies are making significant strides in Africa, focusing on understanding the genetic underpinnings of various psychiatric disorders and improving mental health equity across the continent.

One of the major initiatives is the Neuropsychiatric Genetics of African Populations (NeuroGAP) project (https://www.broadinstitute.org/stanley-center-psychiatric-research/stanley-global/neuropsychiatric-genetics-african-populations-neurogap). This collaboration between the Stanley Center for Psychiatric Research at the Broad Institute, the Harvard T.H. Chan School of Public Health, and research institutions and universities in Ethiopia, Kenya, Uganda, and South Africa aims to bridge the diversity gap in psychiatric genetic research. The NeuroGAP project includes specific research arms like NeuroGAP-Psychosis study,[57] which focuses on genetics of psychosis (schizophrenia and bipolar disorder), and NeuroDev, which focuses on the genetics of childhood neurodevelopmental disorders such as autism spectrum disorders and intellectual disability.[58] The NeuroGAP-Psychosis study has successfully recruited 43,100 participants (21,550 psychosis cases and 21,550 psychosis-free controls) from Uganda, Kenya, Ethiopia, and South Africa, whose DNA has been sequenced using the blended-genome-exome sequencing technology. The NeuroDev study, on the other hand, seeks to recruit 5,400 participants (~1,800 cases of neurodevelopmental disorders and 1,800 controls and ~1,800 parent controls) from Kenya and South Africa, whose DNA will be whole exome sequenced. Preliminary phenotypic and genetic data have been collected from 600 participants, including 200 children with neurodevelopmental disorders.[59]

The NeuroGAP project also sought to enhance infrastructure and research capacity in African countries involved in the study. The project included the Global Initiative for Neuropsychiatric Genetics Education in Research program of Harvard and MIT (GINGER, https://ginger.sph.harvard.edu/), which provides training and career development in neuropsychiatric genetics. Since 2017, GINGER has trained 24 fellows from Africa in psychiatric genetic research methods, creating an impact on the continent. The NeuroGAP project also included the Global Initiative in Neuropsychiatric Ethics (NeuroGenE), focusing on the ethical aspects of conducting such research in diverse populations (https://neurogene.org). This initiative is conducting research as well building capacity on the ethics of diagnosis, prevention, and interventions relating to psychiatric and neurodevelopmental disorders in the world, with a focus on low- and middle-income countries, and has several collaborators across Africa.

An important outcome of the NeuroGAP-Psychosis study is that there are efforts to extend it to other sites in Kenya, with the latest effort being in Nairobi, based at the Mathari National Teaching and Referral Hospital with controls being recruited from communities in Nairobi. The goal is to recruit 5,000 cases and 5,000 controls and contribute to the existing repositories.

Another ongoing initiative is the Depression Genetics in Africa (DepGenAfrica) project, which aims to improve our understanding of MDD in continental Africa individuals.[60] DepGenAfrica is a Wellcome Trust–funded initiative that is executed through a UK–Africa collaboration from 2022 to 2027. Each of the participating institution in each of the African countries is paired with a UK institution. The project has three overarching aims:

1. To characterize MDD in Africa and identify the genetic architecture of MDD among 12,000 people (8,000 cases and 4,000 controls) living in Ethiopia, Malawi, and Nigeria.
2. To compare the symptoms and genetic architecture of MDD in African participants with other ancestral populations, including diaspora Africans.
3. To perform downstream analyses of African MDD genetic data to better understand underlying biological mechanisms.

Consenting study participants will be drawn from existing clinical populations including inpatient admissions and outpatient clinics in Ethiopia and Nigeria. Additionally, existing research participants in community studies in Malawi (Generation Malawi: https://www.meiru.info/generation-malawi/ and Healthy Lives Malawi: https://www.meiru.info/healthy-lives-malawi/) will be included in DepGenAfrica. Among consenting participants, MDD diagnosis (current and lifetime) will be made based on clinical assessment using the Schedules for Clinical Assessment in Neuropsychiatry, version 2.1. Other psychiatric phenotypes related to MDD, such as depressive and anxiety symptoms, history of antidepressant use, family history of depression and depression associated with the antenatal period, will be assessed. Data on psychoactive substance abuse, social factors, anthropometry and cognition will also be collected using relevant and context tested instruments. Samples will be genotyped and/or sequenced using the most suitable arrays and sequencing platforms at the Sydney Brenner Institute for Molecular Biosciences at the University of the Witwatersrand in South Africa. Once complete, DepGenAfrica will deliver the first large-scale genomic study of MDD among continental Africans. This project will help fill the major gap in our understanding of MDD among Africans and enhance diversity in current MDD genetics. Importantly, DepGenAfrica will also establish a pan-African depression genetics network, improve capacity in terms of personnel and infrastructure, and source for further funding to support African-led research science.

Another ongoing initiative is the Genetic Risk for Depression in Africa (GENDA) study. This study seeks to investigate the genetic risk for depression among continental Africans (https://wellcome.org/research-funding/funding-portfolio/funded-grants/genetic-risk-major-depressive-disorder-among-three). This study investigates the genetic risk for depression among a sample of about 10,000 participants from three studies/cohorts: (1) the Uganda Genome Resource, (2) the NeuroGAP-Psychosis study,[57] and (3) the Malawi Epidemiology and Interventions Research Unit.[61] To support PGC Africa's goal of advancing psychiatric genetic research on the continent using Pan-African genomic data, this study will, in the long-term, integrate data and findings from other genetic studies of depression in Africa, including the DepGenAfrica study.[62] This effort will contribute to the development of a pan-African database for depression, thereby advancing depression genetics research across the continent.

Another ongoing initiative is the Genetic Risk for PTSD in African Ancestry Populations: Enhancing Discovery by Addressing Inequality (the GPA study). This study aims to recruit about 16,000 participants from Uganda and Kenya, who will be assessed for PTSD and will have their DNA sequenced using the blended genome–exome technology.[7] In Kenya alone, a total of 2,000 participants had been recruited by December 2024 and the study is on track to meet and probably exceed its target of 8,000 participants.

Another ongoing initiative is the Genomics of Brain Disorders. This initiative by the South African Medical Research Council includes various projects, such as the epigenomic and genomic analysis of PTSD in South African populations and investigations into the molecular etiology of PTSD, schizophrenia, and Parkinson's disease using human-induced pluripotent stem cells (https://www.sun.ac.za/english/faculty/healthsciences/psychiatry/research/research-units/samrc-su-extramural-unit-on-the-genomics-of-brain-disorders).

Other research efforts include the following: (1) the Drakenstein Child Health Study, a multi-disciplinary birth cohort study in South Africa that uses genetic and epigenetic data[63] to examine the role of early life psychobiological and psychosocial factors in child health outcomes,[64] (2) the South African Xhosa study, which investigated the genetic basis of schizophrenia among 2,200 Xhosa people in South Africa (https://h3africa.org/index.php/consortium/the-genomics-of-schizophrenia-in-the-south-african-xhosa-people/), (3) the transgenerational epigenomics of trauma and PTSD in Rwanda study, which investigates the transgenerational genomic impact of genocide exposure and PTSD among women survivors of the Rwanda genocide and their offspring (https://h3africa.org/index.php/consortium/transgenerational-epigenomics-of-trauma-and-ptsd-in-rwanda/), and (4) the SHARED Roots initiative, which seeks to understand psychiatric disorder comorbidities by investigating both genomic and environmental factors contributing to PTSD, Parkinson's disease, schizophrenia and cardio-vascular disease risk. This initiative includes studies investigating the genomic, neural, cellular, and environmental signatures that are common to these comorbidities in the mixed ancestry South African population group. By investigating microbiome,[65] transcriptomic,[66] and epigenetic data,[67] this study takes an integrative approach to understanding psychiatric disease risk and development.

Overall, the studies outlined highlight a paucity of genetic research in Africa. The few psychiatric genetic studies conducted have been candidate gene studies. Candidate gene studies limit our understanding of the genetic nature of psychiatric disorders in Africa as they focus on a small number of genes. There is an urgent need for large-scale psychiatric genetic research in Africa. Psychiatric genetic research on the continent is essential for the validity and applicability of existing findings, particularly in a continent as genetically and culturally diverse as Africa. By including more genetically diverse populations, there is potential for discovery of novel genetic information that could lead to better diagnostic tools and treatments tailored to diverse genetic backgrounds. Several factors have impeded psychiatric genetic research in Africa and these have been described in a recent commentary, which also suggests ways through which these can be addressed to achieve this research in Africa.[7] These factors are briefly highlighted in the following sections with ways through which they can be mitigated.

9.3.1 Factors Impeding Psychiatric Genetic Research in Africa

Psychiatric genetic research in Africa faces significant barriers, primarily stemming from historical mistrust, cultural perceptions, and infrastructural challenges. Historically, exploitative research practices have left a legacy of African skepticism toward foreign-led studies. Additionally, cultural stigmatization of mental health issues complicates the acceptance and understanding of psychiatric genetics within communities. Infrastructure deficits, including limited access to advanced genomic technologies and a shortage of trained local researchers, further hinder progress. These issues are briefly highlighted now with proposed approaches to mitigate them:

9.3.1.1 Vast Cultural, Language, and Genetic Diversity

African genomes have the highest genetic diversity globally, including a greater percentage of private alleles, which are unique to Africa.[26] The diversity within Africa extends beyond genetics, to include ethnic and linguistic diversity, with over 3,000 ethnic groups and 2,000 languages spoken (https://www.ethnologue.com). Despite this diversity offering a unique opportunity to understand human genetic evolution and the basis of disease, it poses challenges for genetic analysis due to the complexity of genetic variation.[7] Language has been reported to contribute to population stratification/substructure in Africa.[68] Population stratification is of big concern in psychiatric genetic research in Africa. If not accounted for, differences in population stratification can easily result in spurious genetic findings. However, genetic analysis methods, which account for population stratification, are evolving and can be used in Africa. One such is principal component analysis, which calculates principal components and incorporates them as covariates in regression models to account for population stratification during analysis of GWAS. Other analytical methods which can be used to control for population stratification include admixture analysis (https://dalexander.github.io/admixture/) and structure analysis (https://web.stanford.edu/group/pritchardlab/structure.html).

There is currently no standard for conducting GWAS for different psychiatric disorders in Africa. However, various GWAS software packages have been developed, which can be optimized to handle the highly genetically diverse genetic data from Africa. These software packages include Fast-GWA, which can be optimized for large-scale data[69]; GEMMA, which uses linear mixed models to account for population stratification and relatedness and has potential to control for both observed and hidden population stratification and kinship, which are common in genetically diverse populations[70]; Regenie, which uses ridge regression to perform whole-genome regression and prediction and can handle very large genetic datasets efficiently[71]; SAIGE, which uses a mixed model approach to control for population stratification and relatedness in large-scale genetic association studies, making it highly applicable to Africa[72]; and TRACTOR, which allows inclusion of any admixed samples in GWAS by leveraging local ancestry.[73]

PRS aggregate the effects of numerous genetic variants to predict the likelihood of developing psychiatric disorders. Current research shows poor prediction accuracies for PRS in Africans, most especially when the base data come from non-African populations.[74,75] Developing PRS based on data derived from African populations could improve their predictive accuracy for these groups. This could lead to earlier interventions and tailored treatment strategies, addressing the specific genetic risks present. PRS calculation tools, which allow cross-ancestry and trans-ethnic predictions, are evolving and these will be key to advancing psychiatric genetic research in Africa, once optimized. These tools include the following: PRS-CSx, a Bayesian polygenic prediction method that improves cross-population prediction accuracy by integrating GWAS summary statistics from multiple populations[76]; polypred, a method that improves trans-ethnic polygenic prediction by incorporating causal effects in addition to tagging effects, or polypred+, in the special case where a large training sample is available in the non-European target population[77]; SbayesR, a Bayesian multi-regression model that outperforms gold standard pruning and thresholding performance, particularly across diverse ancestries[78]; and CT-SLEB, a software program that calculates PRS based on ancestry-specific GWAS summary statistics from multi ancestry samples.[79]

Rather than being viewed solely as a barrier, this multidimensional complexity, which arises from genetic diversity, environmental variation, cultural practices, and social determinants, offers a unique opportunity to build globally relevant genomic epidemiology models. African settings allow the interrogation of gene–environment and gene–culture interactions at a scale and richness unparalleled in other regions, positioning Africa as a critical testing ground for advancing the sophistication of psychiatric genomics.

Whole-genome sequencing (WGS) provides a comprehensive method for analyzing the entire genome, which is crucial in a continent like Africa, where many genetic variations remain uncharted. This method can uncover both common and rare genetic variants that contribute to psychiatric disorders, providing a richer genetic map to inform more personalized medicine approaches in African populations. WGS has in the past not been accessible to Africa due to very costs. However, current advances in WGS sequencing technology are leading to innovative and cost-friendly WGS approaches. An example is blended genome–exome technology, in which introns (non-coding regions of the genome) are sequenced at shallow depth, while exons (coding regions of the genome that directly contribute to protein synthesis) are sequenced at greater depth, thereby enabling unbiased, high-quality and cost-effective capture of genetic variation.[80] The largest genetic initiative in Africa, the NeuroGAP-Psychosis study, employed this technology to sequence the DNA of 43,100 participants.

There is also need to strengthen other insightful technologies that are currently underutilized in Africa, such as epigenomic studies to understand mechanisms through which gene expression may be regulated by environmental and lifestyle factors, without altering the DNA sequence. This research is important in Africa, where diverse environmental factors come into play. Understanding the epigenetic changes that contribute to psychiatric disorders could lead to interventions that are tailored to the environmental and cultural context of different populations within the continent. There is need to also strengthen neuroimaging genetic research in Africa. These studies have potential to offer insights into how genetic variations affect brain structure and function, leading to better-targeted therapies. Also, induced pluripotent stem cell (iPSC) technology is underdeveloped on the continent. The iPSC technology involves reprogramming adult cells to an embryonic stem cell-like state, allowing researchers to study the cellular models of psychiatric disorders. This technology can be particularly useful in African settings to investigate the cellular impacts of genetic variations specific to African genomes, offering insights that are directly applicable to the local population.

9.3.1.2 Mistrust Issues

Various mistrust issues surround genetics research in Africa generally, and these have impeded psychiatric genetics research initiatives on the continent as well. For example, there are historical issues of ethical dumping in Africa, whereby research and clinical trials have been conducted in Africa with less stringent ethical standards compared to those applied in developed countries.[7] There are also issues with cultural sensitivities as mental health issues are highly stigmatized in many African communities, alongside cultural beliefs and myths which surrounding the origin of psychiatric disorders.[81] Issues around informed consent always arise when participants do not fully understand the purpose of research or how their data will be used or the implications of participating in research. This is sometimes due to language barriers and complex scientific concepts.[82] There is also a concern that genetic research might lead to increased stigma or discrimination against individuals diagnosed with the disorders, particularly if genetic predispositions are misinterpreted as deterministic or if communities are labelled as genetically vulnerable to certain conditions. There are, as well, concerns about confidentiality and security of genetic data.[83] Participants often worry about who has access to their information and how it might be used beyond the scope of original research. There is also often a perceived lack of equitable partnerships in research Africa. African researchers and institutions may feel that they are not equally involved in the decision-making process or in the intellectual leadership of research projects. This can ultimately diminish trust in the motives and outcomes of research.

Various approaches have been suggested to address mistrust issues around psychiatric genetic research in Africa.[7] Such approaches include strengthening community engagements and ethical review mechanisms, ensuring that African researchers fully get involved in the analysis and publication of their data.[7]

9.3.1.3 Limited Funding and Institutional Capacity

Psychiatric genetics research in Africa has been limited by funding. This is mainly because it has been overshadowed by more immediate public health concerns, such as infectious disease like malaria and HIV. Without adequate funding, setting up and maintaining the necessary infrastructure to conduct psychiatric genetics research is challenging.

There is also a notable shortage of trained geneticists and bioinformaticians and other specialists needed to conduct the highly sophisticated psychiatric genetic research. This shortage is further exacerbated by a brain drain, where emerging psychiatric geneticists from Africa emigrate to developed countries for better opportunities. Also, institutions often lack the capacity to provide advanced training in statistical genetic methods, which are relevant to psychiatric genetic research—this limits the scope of research questions that can be pursued locally.

Substantial training and infrastructure has been proposed as a strategy which can address the critical need for skilled genetic researchers in Africa.[7] Training initiatives which have contributed to psychiatric genetic capacity building in Africa are worthy of note and these include the GINGER program and the Mental Health Data prize Africa (https://mentalhealthdataprizeafrica.aphrc.org). Initiated by the African Population and Health Research Center in partnership with the Wellcome Trust, this initiative over a 5-month period trains early and mid-career researchers from Africa in advanced statistical analysis and machine learning algorithms tailored for mental health applications in Africa.

Several approaches which could be employed to address limited funding for psychiatric genetic research in Africa have been proposed.[7] Funding schemes which specifically target African psychiatric genetic researchers have been proposed. Increasing public awareness and engagement about the importance of psychiatric genetic research in Africa can also help in lobbying for more funding and support from stakeholders and philanthropists. Collaborations within Africa have also been proposed as mechanisms through which African researchers can share resources; the recently inaugurated Psychiatric Genetics Consortium Africa is taking initiative to bring African psychiatric genetic researchers with similar interests to foster collaborations among themselves.

9.3.1.4 Limited Exposure and Visibility of African Scientists

Generally, researchers from Africa often have fewer opportunities to engage with the international scientific community. Important networking events, conference, and collaborations that could enhance their visibility and integrate their work into larger research dialogues are less accessible due to financial constraints or geopolitical factors. African researchers also face challenges in publishing their research in high-impact journals due to lack of access to resources, such as language editing services and the high costs associated with publishing—including article-processing charges in open-access journals. Also, there is often bias in the global scientific community, which may undervalue research coming from institutions perceived as less prestigious, which unfortunately includes many African institutions. This bias can affect peer review processes and the acceptance rate of papers from Africa. Limited funding also reduces the opportunities for presenting research at international platforms, where visibility is enhanced and collaborations are formed.

There is need for collaborations between psychiatric genetic researchers based in Africa and their counterparts in other parts of the world.[7] These collaborations can provide platforms for joint publications and sharing of resources and knowledge. Investing in the training of Africa researchers in psychiatric genetic analysis methods can also elevate the overall quality and impact of their research. Programs designed to improve research methodology, scientific writing, and grantsmanship skills would be beneficial. Supporting and promoting regional conferences can provide more accessible platforms for African researchers to showcase their work. Similarly, strengthening local scientific journal can help in providing mor publication opportunities.

Increasing the exposure and visibility of African scientists is critical not only for advancing psychiatric genetic research but also for ensuring that the global scientific community benefits from diverse perspectives and approaches. This inclusivity can lead to more comprehensive and culturally sensitive research outcomes, ultimately benefiting the global fight against psychiatric disorders. The factors that impede psychiatric genetic research in Africa and the proposed approaches to mitigate them have been summarized in Figure 9.2.

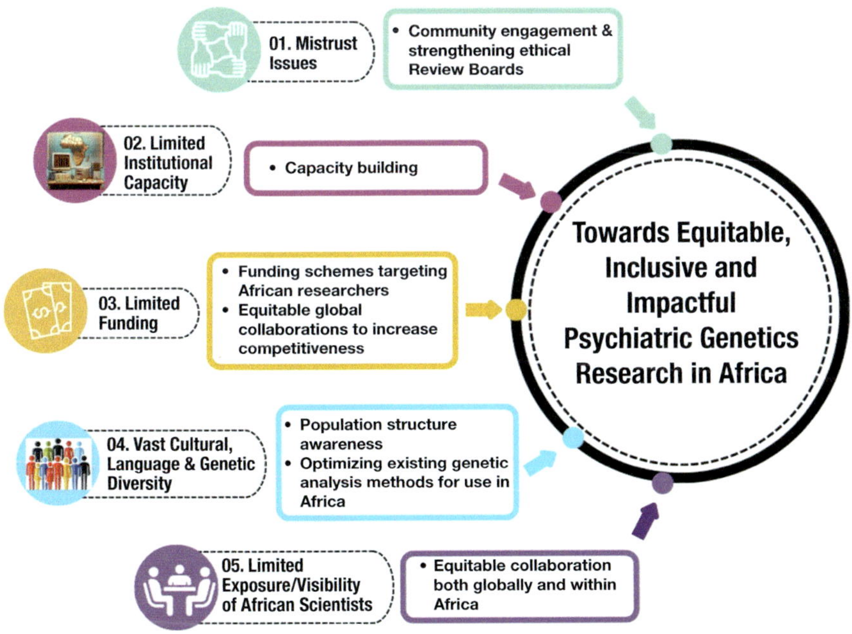

Figure 9.2 Challenges (numbered) and proposed strategies to facilitate impactful psychiatric genetic research in Africa. (Source: Kalungi et al., 2024)

9.4 POTENTIAL CLINICAL AND THERAPEUTIC IMPLICATIONS FOR PSYCHIATRIC GENETIC RESEARCH IN AFRICA

Psychiatric genetic data from Africa has the potential to significantly contribute to clinical and therapeutic advancements globally. The immense genetic diversity in African populations offers unique insights that can lead to improved understanding and treatment of psychiatric disorders. Following is a brief discussion of how genetic data from Africa can be pivotal to global genetics research.

African has the highest level of genetic diversity among human populations.[26] Studying these genetic variations can help identify unique genetic markers associated with psychiatric disorders, which might be rare or absent in other populations. Understanding how specific genetic variants contribute to psychiatric disorders in African populations can reveal new biological pathways and mechanisms that could be pivotal in disease epidemiology. The unique genetic insights can also lead to the discovery of novel targets for drug development, ultimately leading to new therapeutic options for psychiatric disorders. More importantly, leveraging African psychiatric genetic data enables the development of culturally appropriate mental health interventions by moving beyond Western-centric diagnostic approaches and treatment modalities, which may not align with African sociocultural contexts.

Psychiatric genetic data from Africa has potential to improve PRS prediction models. Current PRS prediction models are primarily based on data from European populations. Incorporating data from African populations can improve the accuracy and applicability of these models globally, especially for populations of African descent worldwide. Incorporating African genomics data into public health planning can drive precision psychiatry, allowing for more targeted ancestry-informed prevention and treatment strategies at the population level, ultimately reducing the burden of psychiatric disorders on the continent.

In conclusion, including genetic data from African populations in global psychiatric genetic research addresses a significant gap in the current understanding of psychiatric genetics, which is predominantly based on participants of European ancestry. This inclusivity not only enriches the scientific understanding of psychiatric disorders but also enhances therapeutic approaches, ensuring that they are more effective, inclusive, and equitable. This approach benefits populations in Africa and contributes valuable insights that have global implications in the fight against psychiatric disorders.

9.5 CONCLUSION AND FUTURE DIRECTIONS

In conclusion, psychiatric genetic research in Africa holds transformative potential not only for the continent but also for the global understanding of psychiatric disorders. Despite facing significant challenges, such as underrepresentation in global genetic studies, ethical concerns, and limited infrastructure, there is a clear path forward. Increased investments in local research capacities, robust ethical frameworks, and enhanced collaborative efforts are essential. Future efforts should focus on expanding GWAS that include diverse African populations to uncover unique genetic markers and pathways. Additionally, leveraging advancements in multi-omics technologies can provide deeper insights into the interplay between genetic and environmental factors influencing psychiatric conditions. Promoting equitable partnerships that empower African scientists and institutions will be crucial in ensuring that the benefits of psychiatric genetic research are shared widely. By embracing these strategies, we can anticipate not only more personalized and effective interventions for psychiatric disorders but also a greater representation of African genetic diversity in the global research landscape, enriching our collective knowledge and approach to mental health.

To realize the full potential of psychiatric genetics in Africa, there must be a deliberate and sustained commitment to equitable investment in

African-led research. This is not merely about addressing historical underrepresentation but unlocking a critical frontier for global mental health discovery. Investment in African psychiatric genetic research will accelerate innovation, uncover novel biological mechanisms, and contribute to the development of culturally relevant interventions and precision public health strategies. Prioritizing African leadership, infrastructure, and capacity building will ensure that this scientific momentum translates into tangible and equitable health benefits across the continent and beyond.

REFERENCES

1. Gratten J, Wray NR, Keller MC, Visscher PM. Large-scale genomics unveils the genetic architecture of psychiatric disorders. *Nature Neuroscience*. June 1 2014;17(6):782–790. https://doi.org/10.1038/nn.3708
2. Stein DJ, Palk AC, Kendler KS. What is a mental disorder? An exemplar-focused approach. *Psychological Medicine*. Apr 2021;51(6):894–901. https://doi.org/10.1017/s0033291721001185
3. GBD Mental Disorders Collaborators. Global, regional, and national burden of 12 mental disorders in 204 countries and territories, 1990–2019: A systematic analysis for the Global Burden of Disease Study 2019. *The Lancet Psychiatry*. 2022;9(2):137–150.
4. World Health Organization. Mental disorders key facts. Accessed 11 January 2025. https://www.who.int/news-room/fact-sheets/detail/mental-disorders
5. Castaldelli-Maia JM, Bhugra D. Analysis of global prevalence of mental and substance use disorders within countries: Focus on sociodemographic characteristics and income levels. *International Review of Psychiatry*. 2022;34(1):6–15.
6. World Health Organization. Barriers to mental health care in Africa. World Health Organization. Accessed 2 June 2025. https://www.afro.who.int/news/barriers-mental-health-care-africa
7. Kalungi A, Stein DJ, Okewole N, Fatumo S. Approaches to enable equitable psychiatric genetic research in Africa. *Communications Medicine*. Oct 25 2024;4(1):216. https://doi.org/10.1038/s43856-024-00639-z
8. Ruscio AM, Stein DJ, Chiu WT, Kessler RC. The epidemiology of obsessive-compulsive disorder in the National Comorbidity Survey Replication. *Molecular Psychiatry*. 2010;15(1):53–63.
9. Koenen KC, Ratanatharathorn A, Ng L, et al. Posttraumatic stress disorder in the world mental health surveys. *Psychological Medicine*. 2017;47(13):2260–2274.
10. Winsper C, Bilgin A, Thompson A, et al. The prevalence of personality disorders in the community: A global systematic review and meta-analysis. *The British Journal of Psychiatry*. 2020;216(2):69–78.
11. Greene MC, Yangchen T, Lehner T, et al. The epidemiology of psychiatric disorders in Africa: A scoping review. *Lancet Psychiatry*. Aug 2021;8(8):717–731. https://doi.org/10.1016/s2215-0366(21)00009-2
12. https://icd.who.int/browse10/2019/en#/V

13. Lee SH, Ripke S, Neale BM, et al. Genetic relationship between five psychiatric disorders estimated from genome-wide SNPs. *Nature Genetics*. 2013;45(9):984–994.
14. Shim RS, Compton MT. The social determinants of mental health: Psychiatrists' roles in addressing discrimination and food insecurity. *Focus (Am Psychiatr Publ)*. Jan 2020;18(1):25–30. https://doi.org/10.1176/appi.focus.20190035
15. Cataldo I, Lepri B, Neoh MJY, Esposito G. Social media usage and development of psychiatric disorders in childhood and adolescence: A review. *Front Psychiatry*. 2020;11:508595. https://doi.org/10.3389/fpsyt.2020.508595
16. Leichsenring F, Steinert C, Rabung S, Ioannidis JP. The efficacy of psychotherapies and pharmacotherapies for mental disorders in adults: An umbrella review and meta-analytic evaluation of recent meta-analyses. *World Psychiatry*. 2022;21(1):133–145.
17. Ripke S, Wray NR, Lewis CM, et al. A mega-analysis of genome-wide association studies for major depressive disorder. *Molecular Psychiatry*. Apr 1 2013;18(4):497–511. https://doi.org/10.1038/mp.2012.21
18. Adams MJ, Streit F, Meng X, et al. Trans-ancestry genome-wide study of depression identifies 697 associations implicating cell types and pharmacotherapies. *Cell*. https://doi.org/10.1016/j.cell.2024.12.002
19. O'Connell KS, Koromina M, van der Veen T, et al. Genomics yields biological and phenotypic insights into bipolar disorder. *Nature*. Jan 22 2025. https://doi.org/10.1038/s41586-024-08468-9
20. Trubetskoy V, Pardiñas AF, Qi T, et al. Mapping genomic loci implicates genes and synaptic biology in schizophrenia. *Nature*. Apr 1 2022;604(7906):502–508. https://doi.org/10.1038/s41586-022-04434-5
21. Hatoum AS, Colbert SMC, Johnson EC, et al. Multivariate genome-wide association meta-analysis of over 1 million subjects identifies loci underlying multiple substance use disorders. *Nature Mental Health*. Mar 1 2023;1(3):210–223. https://doi.org/10.1038/s44220-023-00034-y
22. Docherty AR, Mullins N, Ashley-Koch AE, et al. GWAS meta-analysis of suicide attempt: Identification of 12 genome-wide significant loci and implication of genetic risks for specific health factors. *American Journal of Psychiatry*. 2023;180(10):723–738.
23. Nievergelt CM, Maihofer AX, Atkinson EG, et al. Genome-wide association analyses identify 95 risk loci and provide insights into the neurobiology of post-traumatic stress disorder. *Nature Genetics*. May 1 2024;56(5):792–808. https://doi.org/10.1038/s41588-024-01707-9
24. Strom NI, Verhulst B, Bacanu S-A, et al. Genome-wide association study of major anxiety disorders in 122,341 European-ancestry cases identifies 58 loci and highlights GABAergic signaling. *medRxiv*. 2024:2024.07.03.24309466. https://doi.org/10.1101/2024.07.03.24309466
25. Patin E, Lopez M, Grollemund R, et al. Dispersals and genetic adaptation of Bantu-speaking populations in Africa and North America. *Science*. May 5 2017;356(6337):543–546. https://doi.org/10.1126/science.aal1988
26. Tishkoff SA, Reed FA, Friedlaender FR, et al. The genetic structure and history of Africans and African Americans. *Science*. 2009;324(5930):1035–1044.
27. Andreassen OA, Hindley GF, Frei O, Smeland OB. New insights from the last decade of research in psychiatric genetics: Discoveries, challenges and clinical implications. *World Psychiatry*. 2023;22(1):4–24.

28. Sullivan PF, Agrawal A, Bulik CM, et al. Psychiatric genomics: An update and an agenda. *American Journal of Psychiatry*. 2018;175(1):15–27.
29. Morales J, Welter D, Bowler EH, et al. A standardized framework for representation of ancestry data in genomics studies, with application to the NHGRI-EBI GWAS catalog. *Genome Biology*. 2018;19:1–10.
30. Bentley AR, Callier SL, Rotimi CN. Evaluating the promise of inclusion of African ancestry populations in genomics. *NPJ Genomic Medicine*. 2020;5(1):5.
31. Kinyanda E, Woodburn P, Tugumisirize J, Kagugube J, Ndyanabangi S, Patel V. Poverty, life events and the risk for depression in Uganda. *Social Psychiatry and Psychiatric Epidemiology*. Jan 2011;46(1):35–44. https://doi.org/10.1007/s00127-009-0164-8
32. Kalungi A, Kinyanda E, Akena DH, et al. Prevalence and correlates of common mental disorders among participants of the Uganda Genome Resource: Opportunities for psychiatric genetics research. *Molecular Psychiatry*. Jan 2025;30(1):122–130. https://doi.org/10.1038/s41380-024-02665-8
33. Okello ES, Ekblad S. Lay concepts of depression among the Baganda of Uganda: A pilot study. *Transcultural Psychiatry*. 2006;43(2):287–313.
34. Ventevogel P, Jordans M, Reis R, de Jong J. Madness or sadness? Local concepts of mental illness in four conflict-affected African communities. *Conflict and Health*. 2013;7:1–16.
35. Sorsdahl K. Traditional healers in the treatment of common mental disorders in South Africa: Erratum. *The Journal of Nervous and Mental Disease*. 2010;198(9):695.
36. Uher R. Gene–environment interactions in severe mental illness. *Frontiers in Psychiatry*. 2014;5:48.
37. Turecki G, Meaney MJ. Effects of the social environment and stress on glucocorticoid receptor gene methylation: A systematic review. *Biological Psychiatry*. 2016;79(2):87–96.
38. Meng X, Navoly G, Giannakopoulou O, et al. Multi-ancestry genome-wide association study of major depression aids locus discovery, fine mapping, gene prioritization and causal inference. *Nature Genetics*. 2024;56(2):222–233.
39. Guintivano J, Byrne EM, Kiewa J, et al. Meta-analyses of genome-wide association studies for postpartum depression. *American Journal of Psychiatry*. 2023;180(12):884–895.
40. Swart PC, Van den Heuvel LL, Lewis CM, Seedat S, Hemmings SM. A genome-wide association study and polygenic risk score analysis of posttraumatic stress disorder and metabolic syndrome in a south african population. *Frontiers in Neuroscience*. 2021;15:677800.
41. Nievergelt CM, Maihofer AX, Klengel T, et al. International meta-analysis of PTSD genome-wide association studies identifies sex-and ancestry-specific genetic risk loci. *Nature Communications*. 2019;10(1):4558.
42. Dalvie S, Koen N, Duncan L, et al. Large scale genetic research on neuropsychiatric disorders in African populations is needed. *EBioMedicine*. 2015;2(10):1259–1261.
43. Passchier RV, Stein DJ, Uhlmann A, van der Merwe C, Dalvie S. Schizophrenia polygenic risk and brain structural changes in methamphetamine-associated psychosis in a South African population. *Frontiers in Genetics*. 2020;11:1018.

44. Gulsuner S, Stein D, Susser E, et al. Genetics of schizophrenia in the South African Xhosa. *Science*. 2020;367(6477):569–573.
45. De Vries J, Bull SJ, Doumbo O, et al. Ethical issues in human genomics research in developing countries. *BMC Medical Ethics*. 2011;12:1–10.
46. Aristizabal MJ, Anreiter I, Halldorsdottir T, et al. Biological embedding of experience: A primer on epigenetics. *Proceedings of the National Academy of Sciences*. 2020;117(38):23261–23269.
47. Womersley JS, Roeh S, Martin L, et al. FKBP5 intron 7 methylation is associated with higher anxiety proneness and smaller right thalamus volume in adolescents. *Brain Structure and Function*. 2022;227(8):2809–2820.
48. Jani N, Naeemah A, Sylvanus T, et al. Genome-wide differentially methylated genes associated with posttraumatic stress disorder and longitudinal change in methylation in rape survivors. *Translational Psychiatry*. 2021;11(1).
49. Vukojevic V, Coynel D, Ghaffari NR, et al. NTRK2 methylation is related to reduced PTSD risk in two African cohorts of trauma survivors. *Proceedings of the National Academy of Sciences*. 2020;117(35):21667–21672.
50. Crombach A, Rukundo-Zeller AC, Vukojevic V, et al. Differential methylation of linoleic acid pathway genes is associated with PTSD symptoms–a longitudinal study with Burundian soldiers returning from a war zone. *Translational Psychiatry*. 2024;14(1):32.
51. Straight B, Fisher G, Needham BL, et al. Lifetime stress and war exposure timing may predict methylation changes at NR3C1 based on a pilot study in a warrior cohort in a small-scale society in Kenya. *American Journal of Human Biology*. 2021;33(4):e23515.
52. Musanabaganwa C, Wani AH, Donglasan J, et al. Leukocyte methylomic imprints of exposure to the genocide against the Tutsi in Rwanda: A pilot epigenome-wide analysis. *Epigenomics*. 2022;14(1):11–25.
53. Quinn EB, Hsiao CJ, Maisha FM, Mulligan CJ. Prenatal maternal stress is associated with site-specific and age acceleration changes in maternal and newborn DNA methylation. *Epigenetics*. 2023;18(1):2222473.
54. Zhang X, Lin P-Y, Liakath-Ali K, Südhof TC. Teneurins assemble into presynaptic nanoclusters that promote synapse formation via postsynaptic non-teneurin ligands. *Nature Communications*. 2022;13(1):2297.
55. Xulu KR, Womersley JS, Sommer J, et al. DNA methylation and psychotherapy response in trauma-exposed men with appetitive aggression. *Psychiatry Research*. 2021;295:113608.
56. Patel V. Global mental health: From science to action. *Harvard Review of Psychiatry*. 2012;20(1):6–12.
57. Stevenson A, Akena D, Stroud RE, et al. Neuropsychiatric genetics of African populations-psychosis (NeuroGAP-Psychosis): A case-control study protocol and GWAS in Ethiopia, Kenya, South Africa and Uganda. *BMJ Open*. 2019;9(2):e025469.
58. De Menil V, Hoogenhout M, Kipkemoi P, et al. The NeuroDev study: Phenotypic and genetic characterization of neurodevelopmental disorders in Kenya and South Africa. *Neuron*. 2019;101(1):15–19.
59. Kipkemoi P, Kim HA, Christ B, et al. Phenotype and genetic analysis of data collected within the first year of NeuroDev. *Neuron*. 2023;111(18):2800–2810. e5.

60. Okewole, A., Maina, J.G., Nakanga, W.P., et al. Depression Genetics in Africa (DepGenAfrica): protocol for the first large-scale case-control study of major depressive disorder among continental Africans. Welcome Open Research. 2025; 10:335
61. Crampin AC, Kayuni N, Amberbir A, et al. Hypertension and diabetes in Africa: Design and implementation of a large population-based study of burden and risk factors in rural and urban Malawi. *Emerg Themes Epidemiol.* 2016;13:3. https://doi.org/10.1186/s12982-015-0039-2
62. Maina J, Nakanga W, Mathema T, et al. F48. DEPRESSION GENETICS IN AFRICA. *European Neuropsychopharmacology.* 2024;87:231.
63. Rijlaarsdam J, Cosin-Tomas M, Schellhas L, et al. DNA methylation and general psychopathology in childhood: An epigenome-wide meta-analysis from the PACE consortium. *Molecular Psychiatry.* 2023;28(3):1128–1136.
64. Stein DJ, Koen N, Donald K, et al. Investigating the psychosocial determinants of child health in Africa: The Drakenstein child health study. *Journal of Neuroscience Methods.* 2015;252:27–35.
65. Malan-Muller S, Valles-Colomer M, Foxx CL, et al. Exploring the relationship between the gut microbiome and mental health outcomes in a posttraumatic stress disorder cohort relative to trauma-exposed controls. *European Neuropsychopharmacology.* 2022;56:24–38.
66. Hemmings SM, Swart P, Womersely JS, et al. RNA-seq analysis of gene expression profiles in posttraumatic stress disorder, Parkinson's disease and schizophrenia identifies roles for common and distinct biological pathways. *Discover Mental Health.* 2022;2(1):6.
67. Toikumo S, Malan-Muller S, Swart P, et al. Epigenome-wide association study on posttraumatic stress disorder diagnosis and metabolic syndrome in a South African population. *European Neuropsychopharmacology.* 2019.
68. Atkinson EG, Dalvie S, Pichkar Y, et al. Genetic structure correlates with ethnolinguistic diversity in eastern and southern Africa. *The American Journal of Human Genetics.* 2022;109(9):1667–1679.
69. Jiang L, Zheng Z, Qi T, et al. A resource-efficient tool for mixed model association analysis of large-scale data. *Nature Genetics.* 2019;51(12):1749–1755.
70. Zhou X, Stephens M. Genome-wide efficient mixed-model analysis for association studies. *Nature Genetics.* 2012;44(7):821–824.
71. Mbatchou J, Barnard L, Backman J, et al. Computationally efficient whole-genome regression for quantitative and binary traits. *Nature Genetics.* 2021;53(7):1097–1103.
72. Zhou W, Nielsen JB, Fritsche LG, et al. Efficiently controlling for case-control imbalance and sample relatedness in large-scale genetic association studies. *Nature Genetics.* 2018;50(9):1335–1341.
73. Atkinson EG, Maihofer AX, Kanai M, et al. Tractor uses local ancestry to enable the inclusion of admixed individuals in GWAS and to boost power. *Nature Genetics.* 2021;53(2):195–204.
74. Majara L, Kalungi A, Koen N, et al. Low and differential polygenic score generalizability among African populations due largely to genetic diversity. *Human Genetics and Genomics Advances.* 2023;4(2).
75. Kamiza AB, Toure SM, Vujkovic M, et al. Transferability of genetic risk scores in African populations. *Nature Medicine.* 2022;28(6):1163–1166.

76. Ruan Y, Lin Y-F, Feng Y-CA, et al. Improving polygenic prediction in ancestrally diverse populations. *Nature Genetics*. 2022;54(5):573–580.
77. Weissbrod O, Kanai M, Shi H, et al. Leveraging fine-mapping and non-European training data to improve trans-ethnic polygenic risk scores. *MedRxiv*. 2021.
78. Lloyd-Jones LR, Zeng J, Sidorenko J, et al. Improved polygenic prediction by Bayesian multiple regression on summary statistics. *Nature Communications*. 2019;10(1):5086.
79. Zhang H, Zhan J, Jin J, et al. A new method for multiancestry polygenic prediction improves performance across diverse populations. *Nature Genetics*. 2023;55(10):1757–1768.
80. Boltz TA, Chu BB, Liao C, et al. A blended genome and exome sequencing method captures genetic variation in an unbiased, high-quality, and cost-effective manner. *bioRxiv*. 2024.
81. Amuyunzu-Nyamongo M. The social and cultural aspects of mental health in African societies. *Commonwealth Health Partnerships*. 2013;2013:59–63.
82. Fatumo S, Chikowore T, Choudhury A, Ayub M, Martin AR, Kuchenbaecker K. A roadmap to increase diversity in genomic studies. *Nature Medicine* Feb 2022;28(2):243–250. https://doi.org/10.1038/s41591-021-01672-4
83. Martin AR, Teferra S, Möller M, Hoal EG, Daly MJ. The critical needs and challenges for genetic architecture studies in Africa. *Current Opinion in Genetics & Development*. Dec 2018;53:113–120. https://doi.org/10.1016/j.gde.2018.08.005

Chapter 10

The Past, Present, and Future of Genomics in North Africa

Challenges and Opportunities in Health

Yosr Hamdi[1], *Maroua Boujemaa*[1*], *Badr-EddineAllal*[2*], *Ichrak Benamri*[5], *Chiraz Mehemmai*[2], *Najla Kharrat*[4*], *Fouzia Radouni*[3*], *and Ahmed Rebai*[4]

[1]Laboratory of Biomedical Genomics and Oncogenetics, LR16IPT05, Institut Pasteur de Tunis, University of Tunis El Manar, Tunis, Tunisia

[2]Tamayouz_Laboratory of Cellular and Molecular Biology, University of Sciences and Technology Houari Boumediene, Algiers, Algeria.

[3]Chlamydiae and Mycoplasma Laboratory, Research Department, Institut Pasteur du Maroc, 20360, Casablanca, Morocco.

[4]Laboratory of Molecular and Cellular Screening Processes, Centre of Biotechnology of Sfax, University of Sfax, Sfax, Tunisia

[5]Laboratory of Information Technology and Modeling, Faculty of Sciences Ben M'sik, Hassan II University-Casablanca, BP 7955 Sidi Othman Casablanca, Morocco

[*]Maroua Boujemaa and Badr-EddineAllal contributed equally to this work, and Najla Kharrat and Fouzia Radouni contributed equally to this work.

10.1 INTRODUCTION

The North African population, which exceeds 220 million people, has not been represented in international genomics projects, including the 1000 Genomes Project (1000G) (1000 Genomes Project Consortium et al., 2015) and The African Genome Variation Project (Gurdasani Deepti et al., 2015). However, several local studies have been performed in the last ten years on the genetic diversity of the North African populations using different methods, from basic marker genotyping (Y-STR, mtDNA, SNPs, or single nucleotide polymorphisms) to high throughput sequencing (exome and few whole genomes). The samples used include specific ethnic groups as well as the general admixed population. Results showed that the different ethnic groups have diverse genomics patterns depending on their origin; however, the general population showed homogenous composition, indicating a highly admixed population with two major genetic components: an Amazigh component, the "natives" of North Africa whose origin traces back to 10,000 years ago, and an Arab component, brought by the Arabian

DOI: 10.1201/9781003436584-10

invasion of North Africa about 1,000 years ago, with an estimated one million individuals belonging to Arab tribes from the Najd region of the central Arabian Peninsula (Henn B. M. et al., 2012; Vilà-Valls L et al., 2024). The available data also indicated that there is likely an increasing gradient of the "Arab" component from east to west and from south to north. We can, however, find some other genetic components in specific and limited geographic regions that correspond to other waves of migration, like the Phoenician (currently Lebanon) migration about 3,000 years ago, the Andalusians coming back from southern Spain 400–500 years ago, and the Ottomans from Turkey invading and ruling North Africa from 1574 to 1705. Other data indicate two recent admixture events: One in the 13th to 14th centuries, with western sub-Saharan populations (Senegambian), and one in the 17th with an Eastern population (Sudan and Kenya), coinciding with peak of slave-trade activity (Lucas-Sánchez et al., 2023). A recent study based on whole-genome sequencing (WGS) data of 364 Africans from different regions indicated a back-to-Africa origin of North African individuals, which happened in the Epipaleolithic times, as early as 22.3 kya, and supports the theory of Arabization starting in the 11th century as the main source of Arabian ancestry in the region (Serradell et al., 2024).

The specific population structure in North Africa had a significant impact on disease prevalence. Indeed, the demographic transition and the adoption of a Western lifestyle have dramatically increased the incidence rate of non-communicable diseases, including cancer, diabetes, and cardio-vascular diseases. On the other hand, high rates of consanguinity and endogamy have resulted in high prevalence rates of autosomal recessive conditions, mainly for xeroderma pigmentosum, ataxia, and usher disorders. The impact of genetics on populations' health cannot be investigated without considering all healthcare system components, the social specificities, the environment factors, the cultural behaviors, and even historical events. Notably, the underrepresentation of African populations in global genetic databases still highlights the necessity for tailored research and interventions to address this pressing health concern (Haghighatdoost, F et al., 2024).

The genetic diversity in North Africa was particularly observed in the various responses to COVID-19 across the region. The pandemic played a crucial role in democratizing molecular biology and genomics research in North Africa. This increased focus on genomics has led to the establishment of new fields such as metagenomics and pharmacogenomics. Metagenomics provides insights into the microbiome's role in health and disease, identifying potential microbial biomarkers associated with specific conditions. Pharmacogenomics helps predict individual drug responses and adverse effects based on genetic variations, allowing for personalized treatment options.

This chapter provides an overview of the history of the North African region; the cultural, demographic, and linguistic changes that influenced the genetic landscape in the region; recent genomic studies pertaining to the region; and specific genomics profiles and their impact on health and society. The chapter also discusses the challenges and opportunities related to human genome projects and funding, data-sharing, and ethical and regulatory issues in the North African region.

10.2 NORTH AFRICAN POPULATION GENOMICS

The current North African population was formed by a mixture between an indigenous Berber majority, a relatively large Arab contribution, a less important Mediterranean component, and a very small relatively more recent sub-Saharan African contribution. The oldest anatomically modern human-like population was found in the North African region, at the Moroccan site of Jebel Irhoud, dating back to 300,000 years ago; stone artifacts found in Algeria suggest the presence of hominids in North Africa approximately 2.4 million years ago.

The ethnic mixing in North Africa resulted in a homogeneous population having a rich and unique culture (Chaabani 2017).

Results from genetic studies in Tunisia indicate that the ancestors of current Berbers had already received an ancient sub-Saharan African component (Chaabani et al., 1986; Chaâbani and Cox 1988; Bahri et al., 2008; Ben Halima et al., 2015). This component present in the entire Tunisian population, as well as other populations of North Africa, could reach 39% and dates to the first stage of the Neolithic age (El Moncer et al., 2010). In fact, this ancient sub-Saharan component distinguishes the North African populations and Berbers from those on the other side of the Mediterranean Sea, where this component is almost non-existent. However, we find it in the Middle East Arab populations (Chaabani et al., 2000; Chaabani 2002; Bahri et al., 2012; Badro et al., 2013; Triki-Fendri et al., 2016). Indeed, most of the migratory waves to North Africa originated from the Middle East, like the Neolithic expansion. The spread of the Arab culture and language as well as agriculture has had a significant impact on the population structure of North Africans, diluting the autochthonous Paleolithic component (Barrou 1982; Goodspeed 2007)

In Neolithic times, North Africa experienced gene flow from Europe as well, suggesting a complex and heterogeneous gene flow in the region, as shown by the presence of Iberian genetic ancestry in Western North Africans (Serradell et al., 2024). The Arabization waves that occurred from the 7th to the 11th centuries have also resulted in a post-Neolithization gene flow in North Africa (Fregel et al., 2018). Proximity to Mediterranean populations, contacts with south Europeans, and the trans-Saharan slave trade from Roman times to the 19th century have also significantly influenced North African demography and population (Lucas-Sánchez et al., 2023; Elkamel et al., 2017).

This complex pattern of migrations, at different temporal and geographical levels, has challenged not only the demographic reconstruction of North Africans but also their ethnolinguistic and genetic architecture. Indeed, the current North Africa populations are classified into two major linguistic groups: The Imazighen, also known as Berbers, and the North African Arabs. However, a few retreated groups maintained their customs, along with the Amazigh identity and language (i.e., Tamazight). Previous studies have shown that some Tamazight-speaking populations are genetic outliers with remarkable differences with their neighboring Arab-speaking populations. This has been attributed to isolation and drift, as well to differences in their cultural histories (Fadhlaoui-Zid et al., 2004).

Recent genetic studies have shown that Algeria and Morocco represent admixed populations with rich genetic diversity shaped by multiple ancestral lineages (Bekada et al., 2015; Lucas-Sánchez et al., 2021; Choudhury et al., 2014, Serradell et al., 2024). Both countries recognize Tamazight and Arabic as official languages. In contrast, Tunisia is characterized by different linguistical and cultural characteristics, where Arabic is the only official language, and Tamazight-speaking communities exist as small, isolated groups (Lucas-Sánchez et al., 2023). The Phoenician legacy in Tunisia is primarily embodied by the ancient city of Carthage, which was founded by Phoenician settlers from Tyre (in modern-day Lebanon) around 814 BCE. Carthage became a powerful maritime empire, dominating trade across the Mediterranean and leaving a lasting impact on the region's culture, language, and even phenotype, shaping Tunisians in ways that reflect a closer affinity with Syrian and Lebanese heritage (Zalloua et al., 2008).

Many studies using genetic markers, such as those of Y-chromosome and mitochondrial DNA (mtDNA), indicate a mild distinction between Arabs and Berbers (Arauna et al., 2017; Badache et al., 2019) and show that Algerians share a genetics background with Arab and European populations (Bekada et al., 2013; Ivanova et al., 1999). In Tunisia, Y chromosome, mitochondrial DNA, and STR markers showed the admixture of the Tunisian population and its specific genetic landscape (Ben Halim et., 2018; Fadhlaoui-Zid et al., 2011; El Moncer et al., 2010; Frigi et al.,2017). In Morocco, genetic and anthropological studies using DNA-based genetic markers have been used to assess population diversity and origin (Garrigan and Hammer 2006). A recent publication based on an analyses of 109 Moroccan whole genomes showed that 80% of the analyzed Moroccan variants consisted of four major ancestry components, namely, North African (51.2%), European (10.9%), Middle Eastern (10.7%), and West African (6.8%) (El Fahime et al., 2025). Many studies, including those from few whole genomes, revealed similar results with Algeria and Tunisia, indicating significant gene flow from Spain (K Bentayebi et al., 2014; Serradell et al., 2024).

Comparative analyses of results obtained in studies carried out on the Tunisian, Algerian, and Moroccan populations show that, if we consider representative samples of large regions, we often find a non-significant difference between the three populations. These findings reflect the genetic proximity between the two main ethnic groups (Berber and Arab) that made up the North African populations, suggesting a soft split as drivers of divergence in North Africa (Coudray et al., 2006; Serradell et al., 2024).

Genetic disorders represent a significant public health challenge, contributing to both economic and social burdens. Recent assessments indicate that over 400 genetic disorders have been reported in Tunisia (Elloumi-Zghal et al., 2018). About 60% of the reported genetic disorders are autosomal recessive, a phenomenon that can be attributed to the high prevalence of consanguineous and endogamous mating within the population. Indeed, consanguineous marriages are very frequent in North Africa, with reported frequencies ranging from 20% to over 50% among unions involving first cousins or closer relatives (Romdhane et al., 2019). Several studies have been conducted showing that the average level of consanguinity in North Africa ranges from 19.9% to 39% (Riou et al., 1989; Ben Halim et al., 2013; Talbi J et al., 2007; Tadmouri G. O et al., 2009; Guidoumi M et al., 2015; Shawky RM et al., 2011). According to available data the highest rates seem to be encountered in Libya and Mauritania, where they reach 47.2% (Romdhane L et al., 2019; Mezzi N et al., 2021).

Scientific research led by initiatives like the Human Heredity & Health in Africa (H3Africa), the AGenDA initiative, and the Consortium and the Catalog of Transmission Genetics in Arabs (CTGA) are playing a role in addressing the gaps and promoting genomics research and data sharing in North Africa. Data from the CTGA show that Egypt is the most represented country regarding disease and gene records, followed by Tunisia, Morocco, Algeria, and Libya, respectively (Figure 10.1).

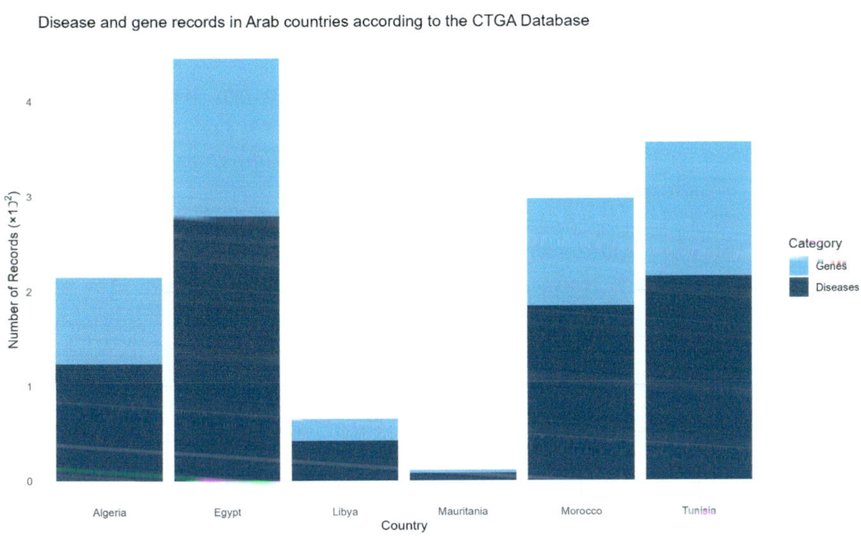

Figure 10.1 Disease and gene records in North African countries according to the CTGA database (as of September 2024).

10.3 COMMON AND COMPLEX HEREDITARY DISEASES IN NORTH AFRICA

North Africa is currently facing a demographic transition heavily impacting health and disease prevalence. Indeed, chronic non-communicable diseases, which are largely driven by lifestyle changes and genetic factors, have risen dramatically in the region over the past two decades.

10.3.1 Oncology

Cancer remains a significant health challenge in North Africa, with 336,905 new cases annually and 204,760 deaths reported. Lung, breast, and liver cancers are the most prevalent tumor types, representing 28% of all cases (Figure 10.2A). The cancer genomics landscape in North Africa is shaped by a combination of genetic predispositions, environmental factors, lifestyle, and the unique demographic structure of the population and its diverse ethnic groups. Several studies have been conducted to investigate the genetic patterns of several cancer types with lung, breast, colon, and hematological malignancies being the most studied ones. The huge progress in sequencing technologies has revolutionized our understanding of this malignant disease, allowing for more precise identification of risk factors, prognostic markers, and potential therapeutic targets.

Most cancer research in North Africa has focused on somatic mutations (Chemlal N et al., 2013), but several studies have also been performed on inherited germline genetic variations (Figure 10.2B).

10.3.1.1 Lung Cancer

Lung cancer is a significant global public health concern. Research on lung cancer in the Tunisian population has identified several key molecular markers that guide treatment strategies mainly for non-small-cell lung carcinoma (NSCLC), which is the most common molecular subtype. *EGFR* mutations were present in 19% to 44% of patients, with some studies finding higher mutation rates in female patients. The L861Q mutation in exon 21 was the most common *EGFR* mutation, while exon 19 deletions were also prevalent (Mezni Faouzi et al., 2018; Mraihi Zohra et al., 2018; Abdelmaksoud-Dammak Rania et al., 2022). *KRAS* mutations were detected in 5.73% to 9.58% of cases, and *ALK* rearrangements occurred in a smaller subset (up to 3.8%) (Mezni Faouzi et al., 2018; Dhieb, Dhoha et al., 2019; Houcine Yoldez et al., 2023). Additionally, Dhieb Dhoha et al., (2019) reported abnormal p53 expression in over 56% of patients. A unique report by Houcine, Yoldez et al., (2023) investigated the expression of PD-L1 in lung tumors, revealing notable findings. In this study, PD-L1 expression was observed to be ≥50% in only 3.8% of cases and between 1% and 49% in 4.4%, indicating a lower

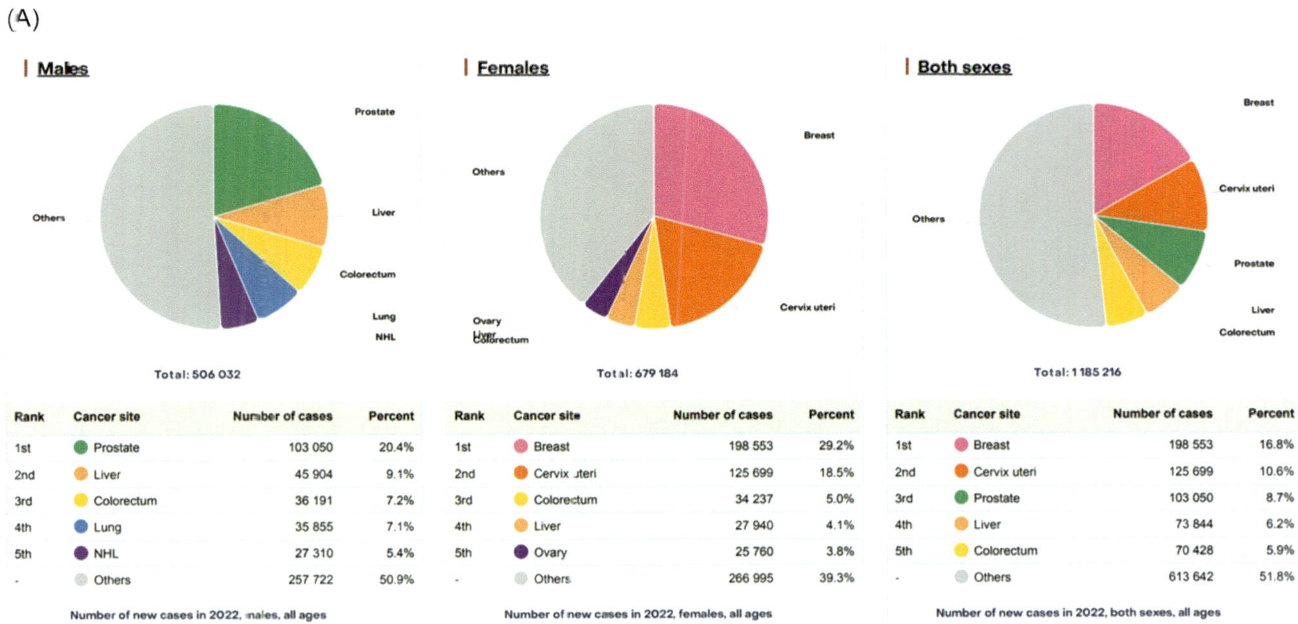

Rank	Cancer site	Number of cases	Percent
1st	Prostate	103 050	20.4%
2nd	Liver	45 904	9.1%
3rd	Colorectum	36 191	7.2%
4th	Lung	35 855	7.1%
5th	NHL	27 310	5.4%
-	Others	257 722	50.9%

Number of new cases in 2022, males, all ages

Rank	Cancer site	Number of cases	Percent
1st	Breast	198 553	29.2%
2nd	Cervix uteri	125 699	18.5%
3rd	Colorectum	34 237	5.0%
4th	Liver	27 940	4.1%
5th	Ovary	25 760	3.8%
-	Others	266 995	39.3%

Number of new cases in 2022, females, all ages

Rank	Cancer site	Number of cases	Percent
1st	Breast	198 553	16.8%
2nd	Cervix uteri	125 699	10.6%
3rd	Prostate	103 050	8.7%
4th	Liver	73 844	6.2%
5th	Colorectum	70 428	5.9%
-	Others	613 642	51.8%

Number of new cases in 2022, both sexes, all ages

Figure 10.2 (A) Top five most frequent cancers in North Africa (Globocan 2022). (B) Chord diagram of major cancer types and associated genes in North Africa.

(B)

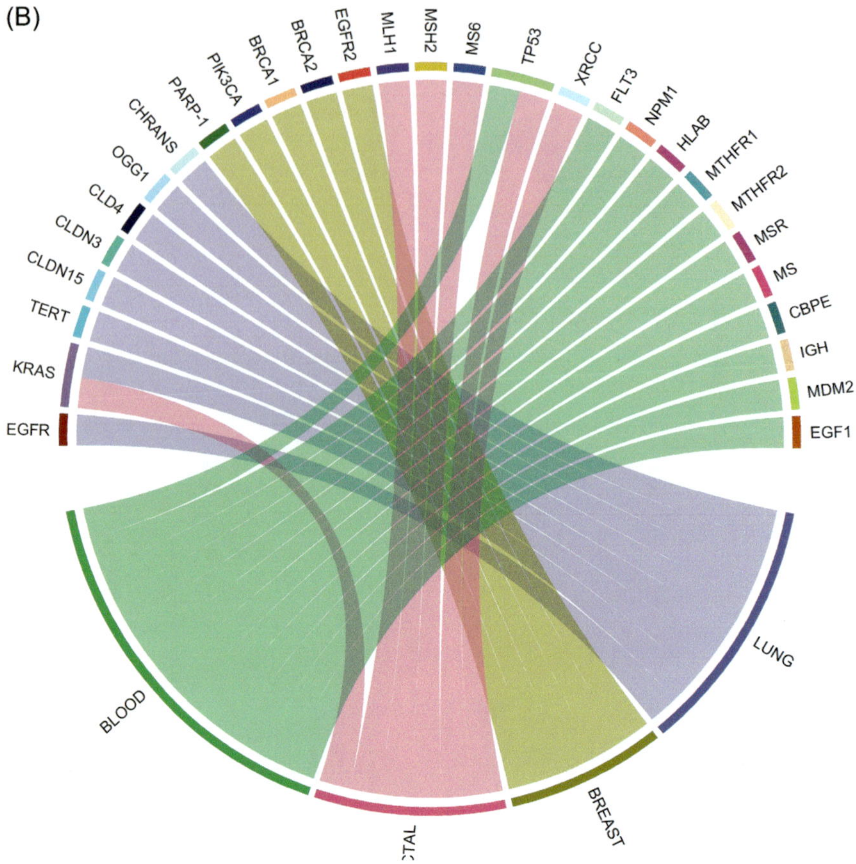

Figure 10.2 (Continued)

prevalence compared to other studies where PD-L1 positivity rates can range significantly higher, often between 20% and 54% in various cohorts (Sindhu Kilaru et al., 2024; Stephanie Tuminello et al., 2020; Konrad Pawelczyk 2019). Recent studies in Algeria have reported a high incidence and poor prognostics in lung cancer patients, particularly among those with NSCLC, who are often diagnosed at advanced stages. Adenocarcinoma has been identified as the most common histological subtype in the country (Bounedjar et al., 2020; Lahmadi et al., 2024). One of the most comprehensive studies on lung cancer in Algeria explores the association of genetic polymorphisms, such as those in the *TERT*, *OGG1*, and *CHRNA5* genes, with increased risk for lung cancer. The authors reported the *CHRNA5* rs16969968 polymorphism's association with increased risk for lung adenocarcinoma, particularly among smokers in Eastern Algeria (Mimouni et al., 2020). In Morocco, lung cancer represents the most prevalent cause of death by cancer (21.6%). In recent years, many studies have investigated NSCLC's molecular epidemiology in Morocco populations. Sara El Zaitouni et al. retrospectively analyzed next-generation sequencing (NGS) results from diagnosed NSCLC patients to identify and compare genomic alterations of NSCLC between Moroccan patients and subjects of the Cancer Genome Atlas. The results identified 234 genetic alterations in 18 genes. Targetable genetic alterations in *EGFR*, *KRAS*, *MET*, *BRAF*, *ALK*, *RET*, and *ROS1* were identified in 84.3% of tumors, and *EGFR* and *KRAS* mutations were frequently reported, occurring in 24.3% and 22.9% of cases, respectively. The untargetable genetic alterations were found in 74.3% of the specimens, in *FGFR3*, *TP53*, *ERBB2*, *PIK3CA*, *CDKN2A*, *PDL1*, *FGFR1*, *PTEN*, *CHEK2*, and *ERBB3*. There were uncommon or rare mutations in *EGFR*, *BRAF*, *RET*, and *ROS1* (Sara El Zaitouni et al., 2024). Lung cancer in the Mauritanian population has a relatively low incidence compared to other cancers. In 2022, approximately 55 new cases were reported, making it tenth in the list of countries with prevalence of cancer (Bray Freddie et al., 2024). Currently, no data have been published on the genetic markers of lung cancer within the Mauritanian population. Similarly, in Libya, although lung cancer represents a significant health concern, ranking as the most common cancer among males and the third most prevalent overall (Bray Freddie et al., 2024), molecular studies remain lacking, with no reports on key genetic markers for the disease. In a cohort study by Wail A. Eldukali et al., (2020) involving 537 NSCLC cases, only 18 (5.7%) underwent genetic testing. The results primarily identified *EGFR*-sensitizing mutations (8 cases), followed by *EGFR* wild type (7 cases), with 1 case each showing an EGFR-resistant mutation, an *ALK* rearrangement, and a combined *EGFR* and *ALK* mutation. The low rate of molecular testing in these two countries was attributed to several challenges, including the lack of local testing facilities, the need for a second biopsy, and the prohibitive cost of testing abroad. In Egypt, investigations of NSCLC cases have identified *EGFR* mutations in 22.3% to 44.1% of screened tumors (Eid SS et al., 2020; Elmetnawy Wafaa H et al., 2022; Anan Omar et al., 2022). In contrast, studies on K-RAS mutations have been less frequent. In a retrospective analysis by Anan Omar et al., K-RAS mutations were detected in 17.3% of cases, with codon 12 being the most mutated site (15.8%).

10.3.1.2 Breast Cancer

Over the past ten years, genomics research on breast cancer has made considerable progress in North Africa. The integration of advanced techniques such as exome sequencing, targeted gene panels, and genome-wide association studies (GWAS) has played a pivotal role in identifying key genetic variations associated with breast cancer susceptibility and prognosis (Hamdi et al., 2018; Riahi Aouatef et al., 2018; Mighri Najah et al., 2020; Ben Kridis-Rejeb Wala et al., 2020; Ben Ayed-Guerfali Dorra et al., 2021; Hamdi et al., 2021b; BenAyed-Guerfali Dorra et al., 2022). In Tunisia, several studies have highlighted the prevalence of recurrent mutations in key genes including c.211dupA and c.5266dupC in *BRCA1* and c.1310_1313del in *BRCA2*, which were found in multiple carriers, alongside the discovery of novel mutations specific to the Tunisian population (Mighri Najah et al., 2020; Hamdi et al., 2021b; Ben Ayed-Guerfali Dorra et al., 2021). Moreover, these studies have led to the discovery of rare and unusual genotypes, such as the cis double heterozygosity, where two cis-pathogenic mutations (c.632–1G>A and c.1310_1313DelAAGA) were detected in multiple high-risk families sharing the same geographic origins, suggesting a potential founder effect within the Tunisian population (Hamdi et al., 2021a). The use of NGS has expanded the genetic landscape beyond *BRCA* genes, offering a more comprehensive understanding of cancer susceptibility, including, notably, the identification of novel mutations in cancer susceptibility genes such as *ATM* and *RAD50* (Mighri Najah et al., 2020; Boujemaa Maroua et al., 2024), in addition to the discovery of candidate genes through exome sequencing in familial breast cancer cases including *RCC1*, *KATB6* and *MMS19* genes (Riahi Aouatef et al., 2018; Hamdi Yosr et al., 2018; BenAyed-Guerfali Dorra et al., 2022). The investigation of copy number variations (CNVs) in *BRCA*-negative families have revealed rare CNVs affecting cancer genes such as *APC2* and *PMS2*, in addition to common CNVs known to be associated with breast cancer risk affecting *APOBECA/B, UGT2B17* and *GSTT1* genes (Boujemaa Maroua et al., 2021). In recent years, multiple studies have been conducted on Algerian breast cancer (BC) patients, which profoundly influenced our understanding of the disease and investigated correlations between genetic factors, clinical features, and pathological characteristics (Uhrhammer N et al., 2008; Cherbal et al., 2015; Chaher et al, 2012; Henouda et al., 2016; Gaceb et al., 2018; Mehemmai et al., 2019). One of the most significant findings in the Algerian cancer genomic landscape is the prevalence of germline mutations in the *BRCA1* (17.43%) and *BRCA2* (7.53%), two critical DNA repair genes, with a larger mutational spectrum (Cherbal et al., 2015; Henouda et al., 2016; Gaceb et al., 2018; Mehemmai et al., 2019). In order to implement and to develop genetic testing and genomics medicine, recent studies identified distinct pathogenic variants on *BRCA1* and *BRCA2* among Algerian HBOC families with common recurrent mutations such as c.83_84del (p.Leu28fs), c.181T>G, p.(Cys61Gly), and c.798_799delTT (p.Ser267Lysfs) on *BRCA1* and c.1813dup (p.Ile605fs) and c.8485C>T (p.Gln2829Ter) on *BRCA2*, with other variants have being reported for the first time in the Algerian population, such as the exon 15 deletion on *BRCA1* gene and the c.8940del (p.Glu2981fs), as well as c.1813dupA (p. Ile605AspfsX11) on the *BRCA2* gene (Gaceb et al., 2018; Mehemmai et al., 2019).

In Morocco, a recent study by Souad Kartti et al., (2023) assessed the contribution of novel mutations to BC risk by sequencing and analyzing the main 22 genes related to BC (*ATM*, *APC*, *CHEK2*, *BARD1*, *BRCA1*, *BRCA2*, *MSH2*, *BRIP1*, *CDH1*, *PALB2*, *EPCAM*, *MLH1*, *MSH6*, *MUTYH*, *NBN*, *PMS2*, *PTEN*, *RAD50*, *RAD51C*, *RAD51D*, *STK11*, and

TP53). Mutational analysis revealed 6 likely pathogenic mutations including c.7874G>C, a heterozygous SNP in HD-OB domain of *BRCA2* gene, which led to the arginine to threonine change at codon 2625 of the protein.

The contribution of *BRCA1/2* mutations in Moroccan early onset breast cancer patients remains unknown. For this purpose, Joaira Bakkach and colleagues in 2020 assessed these genetic alterations for the first time in a cohort from north of Morocco. Three mutations were found in *BRCA1* (9%) and two within the *BRCA2* gene (6%). Three frameshift mutations have been identified (c.798_799del, c.2125_2126insA, c.5116_5119delAATA), one missense (c.116G>A), and one nonsense mutation (c.289G>T). The mutation c.5116_5119delAATA has a founder effect in North Africa. Most *BRCA* mutation carriers (80%) had no family history of breast cancer. So, these results don't support the hypothesis that *BRCA* mutations alone cause the higher frequency of breast cancer in Moroccan young women.

Studies conducted in Algeria, Morocco, and Tunisia have revealed a mix of shared mutations and several country-specific ones, underscoring the genetic heterogeneity within North African populations (Figure 10.3). These efforts led to the establishment of an ecosystem to promote genetic testing and cancer screening in North Africa, with the aim of helping both local patients and those in the diaspora to be diagnosed at an early stage and to identify high-risk family members, enabling more effective genetic counseling.

In Mauritania, breast cancer is the most prevalent cancer type among women, accounting to around 30.7% of all diagnosed cancer cases. Notably, 40.4% of these cases were classified as triple-negative breast cancer, suggesting a potential link to *BRCA* mutations, particularly *BRCA1* (Selma Mohamed Brahim et al., 2021). A unique study by Selma Mohamed Brahim and colleagues (2022), investigating 137 patients with hereditary breast cancer, identified 16 pathogenic or likely pathogenic variants among 38 patients. The most common variants reported were *BRCA1* c.815_824dup, found in 13 patients, and BRCA1 c.4986 + 6 T>C, detected in 7 patients.

Recent genetic studies in Libya have focused mainly on the *BRCA1* gene to uncover hereditary breast cancer risks among Libyan women. Key findings include the identification of multiple genetic variants, particularly in exon 11, with notable pathogenic mutations such as c.2643del and novel variants including c.1366A>G and c.2363T>G (Elmaihub Eanas Saleh et al., 2024; Elshwekh H et al., 2023).

In Egypt, research studies have documented a variety of pathogenic mutations in *BRCA1* and *BRCA2* in addition to other DNA repair genes. A recent study conducted by Auhood Nassar et al. (2023), which analyzed 101 familial cases, found that 57.4% of patients harbored 27 deleterious germline mutations across 11 cancer susceptibility genes, with 31.6% of these patients carrying more than one pathogenic mutation. The most frequently mutated genes included *BRCA2*, *BRCA1*, *ATM*, *VHL*, and *MSH6*, with significant co-occurrences noted between the *VHL* gene and others, such as *BRCA2* and *MSH2*.Additionally, a systematic review encompassing 12

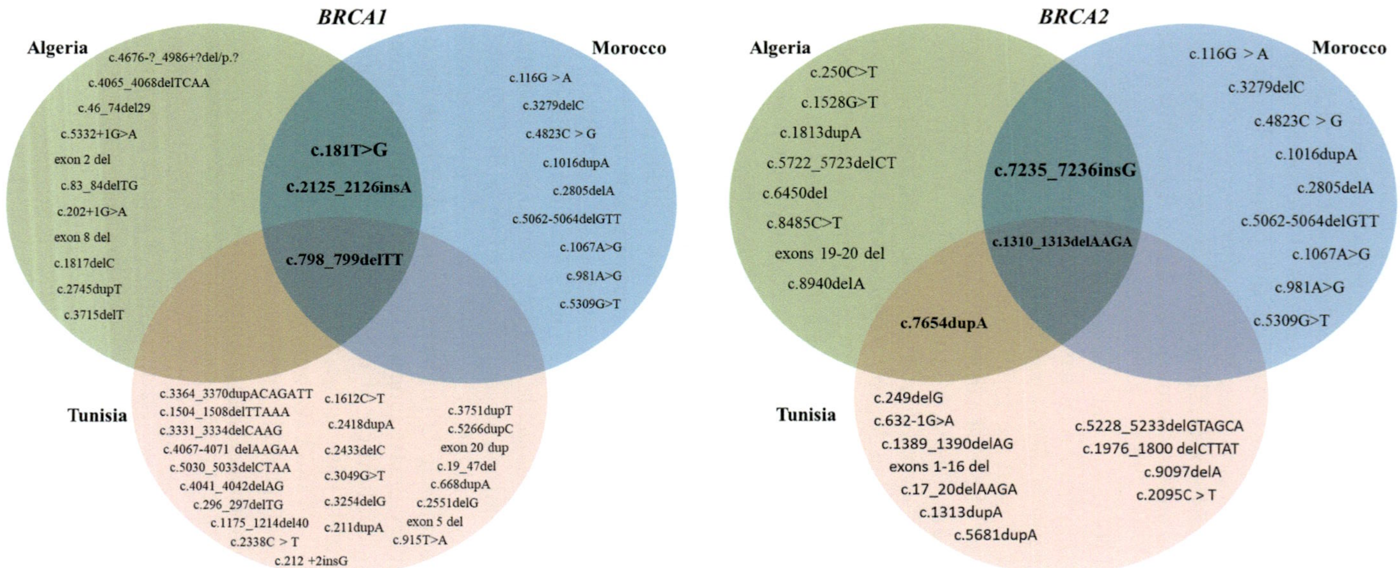

Figure 10.3 The spectrum of *BRCA1* (left panel) and *BRCA2* (right panel) mutations in Tunisia, Algeria and Morocco.

studies with a total of 610 Egyptian patients revealed a pooled prevalence of 40% for *BRCA1/2* mutations. The variability in reported mutation prevalence across these studies highlights the necessity for well-designed national studies to obtain more accurate estimates of *BRCA* mutation prevalence among Egyptian breast cancer patients (Azim Hamdy A et al., 2023).

10.3.1.3 Colorectal Cancer

Significant efforts have been made to elucidate the genetic component and molecular markers of colorectal cancer (CRC) in North Africa. The most common hereditary forms studied in the Tunisian population include Lynch syndrome (hereditary non-polyposis colorectal cancer), which is caused by mutations in mismatch repair genes and increases the risk for CRC and other malignancies like endometrial and ovarian cancers, familial adenomatous polyposis (FAP) caused by mutations in *APC* gene, and MUTYH-associated polyposis caused by mutations in *MUTYH* gene and characterized by the development of multiple polyps (Ben Sghaier Rihab et al., 2019; Jaballah-Gabteni Amira et al., 2019; Kdissa Ameni et al., 2020; Akrout Firas et al., 2023; Barbirou, Mouadh et al., 2023; Abdelmaksoud-Dammak Rania et al., 2024).

Kdissa Ameni et al., (2020) explored FAP and identified biallelic *MUTYH* variants in all the patients studied, indicating a specific polyposis phenotype without extracolonic manifestations. This study suggests that patients with fewer than 100 polyps should be considered for *MUTYH* testing to identify potential hereditary risks advocating for genetic testing even in patients with fewer than 100 polyps. All carriers harbored the same *MUTYH* mutation c.1227_1228dupGG, which is the most frequent mutation identified in previous screened Tunisian families affected by MUTYH-associated-polyposis (Rania Abdelmaksoud-Dammak et al., 2012). Recently, a WGS study conducted by Mouadh Barbirou et al., (2023) investigated a large Tunisian family with multiple members affected by CRC and revealed several relevant germline variants. These variants are located within genes known to be associated with CRC (*MLH1* and *PRH1-TAS2R14*) as well as other genes linked to various malignancies, including *PPP1R13B*, *LAMA5*, *FTO*, *FOXP1* and *NLRP14*.

In Algeria, CRC has shown a rising trend in incidence and mortality rates. Globocan data indicate that Algeria has the highest age-standardized incidence rates of CRC in Africa, with rates reported at 17.1 per 100,000 (Sung et al., 2021). Studies in the west of Algeria show that 39% of the newly diagnosed patients had cancer family history (Herrag et al., 2024). Research has uncovered various mutations, including those in the *KRAS* and *BRAF* genes, with mutation frequencies varying across the country (Jafari et al., 2022; Myer et al., 2022). Up to 50% of patients harbored a *KRAS* mutation, with G12D being the most frequently identified *KRAS* exon 2 mutation (Jafari et al., 2022; Mazouzi., 2017). The tumor suppressor gene *TP53* and its regulator *MDM2* mutations are known to be associated with CRC progression and aggressiveness. Unexpectedly, Rym Abderrahmane and colleagues found no association between *TP53 Arg 72Pro* polymorphism and CRC, and even found that *MDM2 SNP309* polymorphisms are associated with lower risks of CRC in patients from Oran, a city in the west of Algeria (Abderrahmane et al., 2015). Similarly, another study conducted in the same city found that the DNA repair genes, the *cross-complementing group 3* (*XRCC3 Thr241Met*) polymorphisms, may not be associated with CRC risk in this population (Moghtit et al., 2014). Among all CRC cases, 5% to 6% are associated with germline pathogenic variants, which is an inherited condition that increases the risk of CRC. The study of Moufid FZ et al. aimed to investigate the prevalence of *MLH1/MSH2* mutation carriers among Moroccan patients with CRC in a hospital-based cohort with Lynch syndrome (LS). LS is an autosomal dominant disorder characterized by early age of onset and an increased risk of developing extracolonic tumors. In their investigation, the authors enrolled 214 CRC patients from the COLORECFez cohort. Patients whose tumors showed mismatch repair (MMR) deficiency (MMR-D) and wild-type *BRAF* were selected to undergo mutational analysis of the *MLH1* and *MSH2* genes using Sanger sequencing. The results revealed that a total of 24 MMR-D tumors were identified (11.2%) among 214 CRC tested for MMR protein expression. The *BRAF* p.Val600Glu mutation was absent in all tumors deficient for *MLH1* protein. In addition, molecular screening showed germline *MMR* mutations (*MLH1/MSH2*) in four cases, two of which fulfilled Amsterdam criteria II and two met at least one of the revised Bethesda guidelines. The estimated frequency of *MLH1/MSH2* mutations in Moroccan CRC patients was 1.87%. The study reported a relatively high incidence of *MLH1/MSH2* (1.87%), and results confirm the contribution of *MMR* genes to CRC susceptibility in the population studied and provided evidence regarding the requirement of implementing a national screening program for LS in Morocco (Moufid FZ et al., 2018).

In Mauritania, CRC is less common compared to its prevalence in the other North African countries, with 183 new cases reported in 2022, making it the fifth most prevalent cancer in the country (Bray, Freddie et al., 2024). There is a notable lack of comprehensive genetic research on CRC in Mauritania, despite the critical role of genetic investigation in guiding therapeutic interventions. In Libya, CRC is the second most common cancer, with an increasing incidence noted over recent years (Bray Freddie et al., 2024). Genetic research has focused on *KRAS* mutations due to their implications for treatment and prognosis. A recent study by Asma Abudabous et al. (2021), analyzing frozen CRC tissues from Libyan patients, revealed *KRAS* mutations in 38.2% of the cases. Among these, codon 12 mutations were predominant (92.3%), with the most common being Gly12Asp (46.1%), followed by Gly12Val (30.8%) and Gly12Cys (15.4%).

In Egypt, CRC is particularly frequent among younger populations, representing around 42.4% of the cases studied (Noha Rashad et al., 2024). Despite this high prevalence among young patients, studies are lacking in the hereditary component of the disease. Some studies have reported microsatellite instability due to the loss of expression of one or two *MMR* genes but without confirmation at the germline level (Nahed A. Soliman et al., 2019; Neemat M. Kassem et al., 2019). *KRAS* mutations are prevalent in approximately 31.6% of Egyptian CRC patients, with significant occurrences in older individuals. The most common mutations are found at codons 12 (G12D, G12C) and 13 (G13D) (Neemat M. Kassem et al., 2019).

10.3.1.4 Hematological Malignancies

Hereditary factors can play a significant role in the development of hematological malignancies. In Tunisia, several studies have been conducted to elucidate the genetic factors that might contribute to hematological malignancies development.

Key findings include the identification of mutations in *IDH2* and *ASXL1* genes linked to non-Hodgkin lymphoma. Additionally, the analysis of *JAK2*, *CBL*, *RUNX1*, and *NPM1* genes revealed potential associations with familial cases, although some variants were classified as rare polymorphisms rather than pathogenic mutations (Hamadou Walid-Sabri et al., 2016). The investigation of *GATA2* mutations revealed its possible role as a secondary factor in malignancy development (Hamadou Walid Sabri et al., 2017a). Furthermore, significant associations were found between specific SNPs in *MTHFR*, *IKZF1*, *ARID5B*, and *CEBPE* and the risk of pediatric acute lymphoblastic leukemia (ALL) and *XPC* genes with chronic myeloid leukemia risk (Frikha Rim et al., 2021; Gharbi Hanene et al., 2016). Lastly, mutations in apoptotic genes were explored, revealing several variants with potential implications for blood cancer risk including *TP53* and *CASP8* genes (Hamadou Walid Sabri et al., 2017b; Hamadou Walid Sabri et al., 2021). Overall, these studies underscore the complex genetic landscape influencing familial hematological cancers.

Recent research conducted in Algeria have reported on the epidemiology and genetic factors linked to leukemia. Acute myeloid leukemia (AML) has been the subject of several epidemiological studies. With a median age of 46 years, the annual incidence of AML has been increasing in the region, rising from 0.91/100,000 in 2006–2010 to 1.39/100,000 in 2017. M2 is the most prevalent AML subtype, followed by M4 and M5 (Bekadja et al., 2011; Amine et al., 2018). Genetic studies have identified potential risk factors for leukemia development in the Algerian population. The *TP53* P72R allele was highly correlated with an elevated risk of chronic myeloid leukemia (CML) and may be a potential biomarker of CML susceptibility, while the *MDM2* SNP309 allele showed a low penetrance risk factor of CML susceptibility (Benseddik et al., 2022). Furthermore, it was discovered that acute leukemia patients, especially those with AML, have higher frequencies of the HLA-B*27 and HLA-B*58 alleles (Habour et al., 2024). These findings contribute to our understanding of Algerian leukemia genomics and could guide future diagnostic and therapeutic strategies.

In Morocco, hemato-oncological studies were mainly based on cytogenetic investigations, especially those related to ALL, which have been at the forefront of research in the pathogenesis of cancer related to recurring chromosomal abnormalities (either numeral or structural rearrangements), which revealed the genetic events leading to leukemia. The study of Chebihi et al. identified chromosomal abnormalities in 66% of the total cases studied, of which 70% revealed important recurrent abnormalities with high prognostic value, such as hyperdiploidy, hypodiploidy, t(9;22), t(8;14), t(1;19), and MLL rearrangements (Z Chebihi et al., 2018). Later, in a second study, Chebihi et al. reported an ALL case in which they found a t(14,21)(q11.2;q22) as a non-random chromosomal abnormality among 70 analyzed pediatric ALL cases (Chebihi et al., 2018). In Mauritania, specific data on these hematological malignancies are lacking. This might be explained by the low incidence of these cancers in the country. In 2022, non-Hodgkin lymphoma ranked as the seventh most common cancer, while leukemia was the 11th most prevalent among all cancers (Bray, Freddie et al., 2024). Research on hematological malignancies in Libya has primarily focused on leukemia, which ranks as the seventh most commonly diagnosed cancer in the country (Bray, Freddie et al., 2024). Most published studies have focused on incidence rates and immunophenotyping, with limited information available on cytogenetic or genetic testing (Elbnnani et al., 2024; Hussini et al., 2021). In one report, Hussini et al., 2021) identified three cases with the *BCR-ABL* fusion oncogene, which is associated with poor prognosis.

10.3.2 Metabolic Syndrome

Over the past three decades, the North African region has experienced significant urbanization and modernization, resulting in major changes that have had a notable impact on food consumption and lifestyle. Nearly 17.9 million deaths annually are attributed to cardiovascular diseases (CVDs), followed by cancer (9.3 million), respiratory diseases (4.1 million), and diabetes (2 million) (WHO 2023). The main risk factors contributing to CVD in North Africa are complex and include both metabolic and behavioral elements. Between 1990 and 2019, CVD in North Africa rose from the sixth to the second leading cause of death. The major causes have been notable rises in risk factors, like diabetes and hypertension, which could counteract the significant progress made in preventing communicable diseases (Minja N.W. et al., 2022; Aminorroaya et al., 2024). Rates of diabetes exhibit regional variations across the continent, with the highest prevalence recorded in North Africa. The Transition and Health Impact in North Africa project reported that a cross-sectional study involving 4,654 individuals estimated the prevalence of metabolic syndrome (MetS) to be approximately 30.0% among the population studied, with the highest prevalence in North Africa (43.9%) (Tahina 2005; Belfki et al., 2012).

10.3.2.1 Diabetes and Hypertension

An epidemiological study encompassing Algeria, Tunisia, and Morocco found an overall crude hypertension prevalence of 45.4%, indicating that Tunisia's rates are consistent with regional trends, but still highlight a critical public health issue (Nejjari C et al., 2013). A study conducted in Zaghouan revealed very high rates of type 2 diabetes (32%) and hypertension (39%) in the Tunisian population (Kheriji et al., 2023a). The Algerian population had lower rates, but still a high prevalence of diabetes (12.2%) and hypertension (24.9%) (Tahina 2005).

Diabetic diseases are significantly influenced by genetic heterogeneity. This variability of genetic defects involved more than 56 genes. A meta-analysis examining the genetic risk factors for type 2 diabetes in African populations indicated that the majority (70%) of the studies employed a candidate gene approach, with 42 out of 60 studies conducted in Tunisia and Egypt (Yako et al., 2016) from 2015 to 2024 utilizing the candidate gene-set approach in association studies between involve various genes and their variants. The most important genes studied include *ADIPOQ* (rs17300539), *MMP-2* (–790T/G; rs243864, rs243865, rs243866), *APOE* (Apoε2, Apoε3), *ELMO1* (rs7782979, rs10255208), *VEGF* (rs699947, rs1570360, rs3025020), *CDKAL1* (rs7756992), *IGF2* (rs4402960), and the *HVS1* region (Sarray S et al., 2024; Sarray S et al., 2023; Atageldiyeva KK et al., 2019; Musambil M et al., 2019; Sellami N et al., 2018; Lasram K et al., 2015; Hsouna et al., 2016). This focus underscores the complex interplay between genetic factors and the pathophysiology of diabetes, emphasizing the need for continued exploration of these genetic determinants to enhance understanding and management of the disease.

The extensive clinical spectrum of diabetes diseases can be largely attributed to genetic heterogeneity, characterized by various types of genetic defects across more than 56 genes. This diversity is dependent on the consensus of established causal genes, including those associated with rarer forms of early infancy-onset diabetes as well as the most prevalent form, type 2 diabetes. A meta-analysis exploring the genetic risk of type 2 diabetes in African populations showed that the majority (70%) of these studies employed the candidate gene approach, with 42 out of 60 studies originating from Tunisia and Egypt (Yako et al., 2016). The candidate gene-set approach from 2015 to 2024 focuses on the association of various genes and their variants, as noted earlier. Furthermore, an exome-wide association study identified *MYORG* as a gene associated with diabetes development in Tunisia (Dallali et al., 2023). In Algeria, the Apolipoprotein E (*APOE)* is the gene most studied in patients with metabolic disorders. The polymorphism of *APOE* (rs439401, rs4420638) was assessed in subjects in the western Algerian city Oran, and was associated with MetS, hypertension, and diabetes (Boulenouar et al., 2013; Sapkota et al., 2015). Another study conducted in Oran, Algeria, assessed four polymorphisms located in *APOE* gene (rs7412, rs429358, rs4420638, and rs439401) and found the rs4420638 variant to be associated to obesity while the rs439401 was associated to hypertension, but not to MetS (Boulenouar et al., 2019). Other studies conducted in the same city showed no association between hypertension and nitric oxide synthase 3 (*NOS3* –786T/C), angiotensinogen (*AGT* M235T), angiotensin II type 1 receptor (*AT1R* +1166A/C), or the *ACE* I/D polymorphisms (Amrani et al., 2023; Meroufel et al., 2014). However, arginase 1 (*ARG1*, rs2781667C/T) polymorphism was associated with a significantly decreased systolic blood pressure (Meroufel et al., 2014).

Recent studies have identified that the antioxidant system alteration is associated with diabetes in the Algerian population (Eddaikra et al., 2018; Rabhi et al., 2022). The polymorphism of two antioxidant enzymes, catalase (*CAT* –262C/T, rs1001179) and superoxide dismutase (*MnSOD* 47C/T, rs4880), showed a strong association with type 1 diabetes (T1D) in the Arab–Berber population (Eddaikra et al., 2018). T1D was also shown to be associated with the human leukocyte antigens (HLA) DR4 and DR3DR4 variants in western Algerian populations (Aribi et al., 2004). This genetic susceptibility, combined with unhealthy dietary habits, favors the development of diabetes in the Algerian population. Several studies have shown that genetic polymorphisms of the transcription factor 7-like 2 (*TCF7L2*) are highly associated with the development of type 2 diabetes mellitus and its associated complications in several populations. Sarah Elhourch et al. investigated the association of the rs7903146 (C/T) and rs12255372 (G/T) polymorphism in the *TCF7L2* gene with the risk of developing type 2 diabetes mellitus in the Moroccan population. The study confirmed a significant association of the *TCF7L2* gene (rs7903146 (C/T) and rs12255372 (G/T) polymorphisms with a higher risk for type 2 diabetes mellitus in Morocco. No significant difference with respect to anthropometric and metabolic parameters between different genotypes was reported. Furthermore, type 2 diabetes mellitus development can be influenced by mutations in the aldosterone synthase (*CYP11B2*) gene. Youssef Nouhi et al. (2024) explored the link between two SNPs in the *CYP11B2* gene—-344T/C and *K173R* and type 2 diabetes mellitus development. The results suggest a significant association between the *CYP11B2-K173R* polymorphism and type 2 diabetes mellitus development in the Moroccan population.

The maturity-onset diabetes of the young (MODY) is a monogenic form of diabetes characterized by autosomal dominant inheritance. To offer adequate patient management and therapeutic treatment, it is relevant to set an accurate molecular diagnosis. In 2021, Said Trhanint et al. completed a molecular and clinical assessment of MODY, with the aim of determining the frequency of *HNF1A* and *GCK* genes among Moroccan-suspected MODY patients. They enrolled 20 suspected MODY patients, which were screened for *HNF1A* and *GCK* mutations using Sanger sequencing and MLPA methods. Segregation analysis of the mutations identified was performed among family members, and the pathogenic nature of missense variants was predicted. The results showed a total of two mutations were revealed among all patients, raising the diagnostic rate to 10%. In addition, the authors identified a large novel *GCK* deletion (c.209-?_1398+?del) by MLPA in one patient and a previously reported missense substitution (c.92G > A) in the *HNF1A* gene (Said Trhanint et al., 2021). Whole mitochondrial genome screening of a family with maternally inherited diabetes and deafness associated with retinopathy identified two de novo variations: A novel mutation, m.8241T>G (p.F219C), in the *MT-CO2* gene, and a known mutation, m.13276G>A (p.M314V), in the *MT-ND5* gene (Tabebi et al., 2020; Finsterer et al., 2017). According to Tabebi et al., (2020), a mutational screening of mitochondrial DNA in a Tunisian family also identified known substitutions, such as the m.3243A>G mutation in the *MT-TL1* gene. Furthermore, a study examining polymorphisms in genes linked to apoptosis and mitochondrial diabetes found important variations in the following genes: *p53* (rs1042522 and rs17878362), *BCL2* (rs2279115), *BAX* (rs1805419), *BAK1* (rs210132 and rs2227925), *CASP3* (rs105937), *CASP7* (rs2227310), *CASP8* (rs1045485), and *CASP10* (rs13006529) (Tabebi et al., 2018).

10.3.2.2 Cardiovascular Diseases

Shaboodien et al. (2020) investigated the molecular genetics and incidence of cardiomyopathies across the African continent. The study focused on inherited forms such as dilated cardiomyopathy (DCM), hypertrophic cardiomyopathy (HCM), arrhythmogenic cardiomyopathy, restrictive cardiomyopathy, and left ventricular noncompaction (LVNC) within familial/genetic and non-familial/non-genetic causes. These categories are recognized as significant contributors to sudden cardiac death and heart failure during childhood and early adulthood. In Tunisia, high-throughput sequencing of nine sarcomeric genes has revealed a mutation detection rate of 27% among patients diagnosed with DCM, with mutations predominantly affecting the *MYBPC3* and *MYH7* genes. Similarly, a mutation detection rate of 27% has been reported in a cohort of 45 Tunisian patients with HCM, where mutations were primarily identified in the *MYBPC3* and *MYH7* genes (Jaafar N et al., 2016).

In 2018, a study utilizing NGS for DCM on a Moroccan family targeted the resequencing of 50 cardiomyopathy genes and identified a pathogenic *LMNA* p.R54C mutation as the disease cause (Adadi et al., 2018). The Egyptian Collaborative Cardiac Genomics Project sequenced a targeted panel of 174 genes associated with inherited cardiac conditions, identifying 1,262 variants across 27 cardiomyopathy genes. Notably, 15.1% of these variants were not found in current global genetic reference databases. The distribution of rare *TTN* loss-of-function and *MYH7* variants was also characterized within various cohorts. The findings underscore the imperative for comprehensive genetic testing and research to deepen the understanding of the genetic landscape of cardiomyopathies in Africa. Such efforts are essential for enhancing clinical management strategies for populations affected by these conditions.

10.4 THE GENETIC LANDSCAPE OF RARE DISEASES IN NORTH AFRICA

Despite their rarity, rare disease phenotypes are diverse in Morocco, and their frequencies and incidences vary according to the disease.

10.4.1 Xeroderma Pigmentosum

Xeroderma pigmentosum (XP) is a group of eight rare autosomal recessive disorders that predispose to early sunlight-induced cancers of the skin and eyes. XP is caused by mutations in DNA repair genes that protect cells from UV-induced DNA damage (Tamura et al., 2010). *XPC* and *XPA* complementation groups are the most prevalent in North Africa, with 85% of families expressing a mutated *XPC*, and 12% expressing a mutated *XPA* (Soufir et al., 2010). The c.1643_1644delTG (p.Val548Alafs*25) was found in 74% of XP patients of Maghreb region and has been suggested to be a founder mutation for the Mediterranean region (Bensenouci et al., 2016; Schäfer et al., 2013; Soufir et al., 2010). In Tunisia, XP is relatively common, with an estimated prevalence of 1 in 10,000, likely due to high rates of consanguineous marriage and endogamy (Chikhaoui A et al., 2022). Besides *XPA* and *XPC* mutations, Tunisian studies identified five other major mutations: A homozygous *XPF* mutation (c.1762G>T, p.V588F), mildly affecting protein stability; two homozygous *XPE* mutations (c.613T>C, p.C205R and c.618C>A, p.S206R), with the second variant mildly affecting protein structure; the *XPE* p.Lys381Argfs*2 mutation, present in a homozygous state and associated with recent relatedness; a homozygous *XPG* mutation (c.2333T>C, p.Leu778Pro); and, finally, the *XPD* p.Arg683Gln mutation in an homozygous state, linked to ancient relatedness (medium ROH region). Khalat N et al. (2023) found that besides *XPC* c.1643_1644delTG (p.Val548Alafs*25), the Libyan population expresses two other mutations: The *XPC* c.658C>T (p.Arg220*) and the *XPA* c.682C>T (p.Arg228*). More than 50% of patients from northwest, middle-west, and southwest Algeria with two-pair base deletion c.1643_1644delTG had skin cancer and/or ocular cancer. Another study also conducted in the west of Algeria revealed no association between *XPD* Lys751Gln polymorphisms and CRC (Moghtit et al., 2014). The *XPC* c.1643_1644delTG mutation was found to be the major cause (76.19%) of XP in Moroccan families (Senhaji et al., 2013). The frequency of c.1643_1644delTG heterozygous mutation was found in 1/250 newborn (Doubaj et al., 2012).

10.4.2 Ataxia

Autosomal recessive cerebellar ataxias (ARCA), like most neurodegenerative disorders, represent a complex group with variable clinical expressions, high phenotypic properties, and genotypic heterogeneity involving more than 30 genes. Their prevalence varies across ethnicities, ranging from 0.0 to 7.2 per 100,000 individuals (Ruano et al., 2014). Several studies have been conducted on Algerian patients suffering from ARCA, one of which—a cohort of 110 patients—helped define its genetic context (Hamza et al., 2015). The study revealed 23 different molecular alterations, including 6 novel, in 9 different genes responsible for ARCA. Another study analyzed 19 ataxia with oculo-motor apraxia type 2 (AOA2) patients with *SETX* mutations, identifying it as the third most frequent form of ARCA in Algeria, after Friedreich ataxia and ataxia with vitamin E deficiency (Tazir et al., 2009). In the Moroccan context, a study was conducted on the mutation *ATM* gene in patients with ataxia and ocular telangiectasia. The authors screened 27 patients, out of 22 unrelated families, and found 14 different mutations in 19 unrelated families, of which 7 were not been reported before. The study showed that the c.5644C>T (p.Arg1882*) mutation was the most common cause of ataxia in Morocco (Jeddane et al., 2013).

10.4.3 Usher Syndrome

Usher syndrome (USH) is an autosomal recessive disorder characterized by a combination of hearing loss and subsequent development of retinitis pigmentosa and constitutes approximately 50% of all hereditary deaf-blindness cases. Vestibular dysfunction, retinitis pigmentosa, and profound congenital sensorineural deafness are the characteristics of Usher syndrome type I (USH1), the most severe version of this illness. Ben-Rebeh I et al. (2016) found a significant genetic heterogeneity in Tunisian families afflicted by USH1, including three novel mutations. These mutations included two new nonsense mutations (p.Arg3X and p.Arg134X) in the *USH1C* and *PCDH15* genes, respectively, one frameshift mutation (p.Lys615Asnfs*6) in

Myosin VIIA (*MYO7A/USH1B*), and one previously documented splicing error (c.470+1G>A) in *MYO7A* gene. The *MYO7A*, *Cadherin Related 23* (*CDH23*) and *Protocadherin 15* (*PCDH15*) mutations were the most relevant genes in USH type 1 pathogenesis and diagnosis in the Algerian population (Abdi et al., 2016; Ammar-Khodja et al., 2009; Roux et al., 2006). Furthermore, the pathogenic *DFNB1* genotype is identified in 40% of USH1 cases in Algeria, with the 35delG mutation being the most prevalent alteration (Ammar-Khodja et al., 2009). More recently, researchers identified in the Algerian population two homozygous missense variants: p.(Arg84Trp) in the *TMIE* gene, linked to deafness, and p.(His337Arg) in the *PDE6B* gene, associated with retinitis pigmentosa. When coexisting, these two variants mimic the USH (Abdi et al., 2024). Concurrently, *Adhesion G Protein-Coupled Receptor V1* (*GPR98*) frame-shift mutation was found in Algerian families suffering from USH type 2 (Ammar-Khodja et al., 2015).

Numerous investigations related to hearing loss disease were made in Morocco; more recently in 2024, Assia Idyahia et al., investigated the genetic etiology of sensorineural hearing loss in three Moroccan patients using whole exome sequencing. The results revealed in two families Perrault syndrome caused by LARS2, p. Asn153His; p. Thr629Met compound heterozygous variants in two siblings in one family and p. Thr522Asn, a homozygous variant, in two sisters in another. In the third family, the patient was diagnosed with D-bifunctional protein deficiency, linked to compound heterozygous mutations p. Asn457Tyr and p. Val643Argfs*5 in HSD17B4. Molecular dynamic simulation results showed that Val643Argfs*5 does not prevent HSD17B4 protein from binding to the PEX5 receptor, but further studies are recommended to verify its effect on HSD17B4 protein functionality (Assia Idyahia et al., 2024).

10.5 METAGENOMICS

The development of NGS technology has led to rapid advances in microbiome studies to explore various aspects of human diseases in North Africa. In the Algerian population, several metagenomic studies were conducted. An international work gathering multiple populations analyzed the gut microbiota of traditional rural populations versus urban/semi-urban populations across different regions. Forty-seven Tuaregs individuals living in traditional rural societies in southern Algeria were analyzed. The study reports that the Tuareg population tends to have more diverse gut microbiomes enriched in anaerobic bacteria like *Treponema Succinifaciens*, compared to urban populations (Ouarabi et al., 2021). Another interesting research work has been conducted on the vaginal microbiome of Algerian women. The study indicates that the vaginal microbiome in Algerian women is characterized by a diverse array of bacterial species, but predominantly composed of *Lactobacillus*, which are known to play a crucial role in maintaining vaginal health by producing lactic acid and lowering the pH, which is a particular profile compared to other ethnic groups. The researchers also correlated hormone levels to changes in microbial composition for menopausal women (Ouarabi et al., 2021). The key findings from the majority of studies characterizing metagenomics in Tunisia include associations with diabetes and cancer. Significant alterations in gut microbiota composition were observed among Tunisian diabetic patients compared to healthy subjects. Specifically, individuals with type 1 diabetes (T1D) exhibited higher levels of *Fusobacteria* and a decline in the TM7 phylum, while those with type 2 diabetes (T2D) showed a notable decrease in *Verrucomicrobia*. At the genus level, T2D patients experienced more pronounced changes, including reduced relative abundances of *Faecalibacterium*, *Akkermansia*, *Clostridium*, *Blautia*, and *Oscillibacter*. In contrast, the T1D group primarily demonstrated a decrease in *Blautia*. (Hanachi et al., 2022). Zidi O et al. (2021) strongly suggested that chemotherapy treatment plays a key role in modulating the fecal metabolomic profile of breast cancer patients, enhancing the feasibility of identifying specific fecal metabolic profiles reflecting biochemical changes that occur during chemotherapy treatment. Mathlouthi et al,. (2023) reported a significant association between *Halobacteria* and tumor tissues from formalin-fixed paraffin-embedded tissues in Tunisian patients with CRC. Similarly, in Morocco, the study of Allali I et al. (2018) compared the stool microbiome of 11 CRC patients with 12 healthy individuals from Morocco, and results showed that cancer samples had higher proportions of *Firmicutes*, specifically of *Clostridia*, and *Fusobacteria*, especially of Fusobacteriia, while *Bacteroidetes* were enriched in healthy samples particularly the class *Bacteroidia*. *Porphyromonas*, *Clostridium*, *Ruminococcus*, *Selenomonas*, and *Fusobacterium* were significantly overrepresented in diseased patients. These studies employ metagenomic approaches to understand microbial diversity and its implications for health and environmental management. Despite the significance of gut microbiota in health, most studies have historically concentrated on populations from Asia, North America, and Europe, leaving African populations—including those from North Africa—largely underrepresented in microbiome research.

10.6 PHARMACOGENOMICS

Pharmacogenomics is the study of the influence of interindividual genetic variations on the response to drugs, with the aim of improving patient care using a personalized medicine approach. The pharmacogenomics field is pivotal to understanding the differences in drug metabolism among individuals, especially concerning the cytochrome P450 (*CYP450*) enzymes, which are essential for drug metabolism. Genetic variability in the genes encoding these enzymes can significantly influence drug efficacy and adverse reactions in Africa (Rajman et al., 2017; Zhou and Lauschke 2022).

The involvement of variants annotated as *very important pharmacogene* variants in the outcomes of anticoagulants, oral anti-diabetic medications, and statins has been reported by Jmel et al. (2018). In the Tunisian population, a total of 1,056 variants across 24 pharmacogenes were identified. A significant proportion of ethnic variability in response to anticoagulants has been attributed to differential frequencies of variants in the *CYP2C9*, *CYP2C19*, and *VKORC1* genes (Abdelhedi et al., 2015; Jmel et al., 2018). The genetic landscape of these pharmacogenes has shown that a cluster of the Tunisian population is distinguished from Asian and sub-Saharan African populations, aligning more closely with North African populations (Algeria, Morocco, Egypt) and Tuscan and European population groups, including Utah residents with Northern and Western European ancestry from the CEPH collection (CEU). The *CYP2D6* enzyme is essential for the metabolism of approximately 25% of all clinically used medications (Alsanosi et al., 2014; Nahid and Johnson, 2022). Studies denote that a large part of the Algerians are ultra-metabolizers for *CYP2D6*, indicating that drug posology should be adjusted in consequence (Sistonen et al., 2007; Alali et al., 2022). The frequency of *CYP3A5 rs776746* and *ABCB1 rs1045642* polymorphism in the Algerian population was notably different from that in Caucasian and African populations (Spierings et al., 2013; Ammour et al., 2024). Researchers have also found that, unlike *CYP2E1*, *CYP2A6*, *GSTM1*, and *GSTT1*, the *CYP1A1* polymorphism plays a role in the susceptibility to nasopharyngeal cancer in the Algerian population (Ghania et al., 2020; Alami et al., 2024).

In Morocco, pharmacogenetics research is starting to attract the interest of researchers with a few studies. The first study occurred in 1986, on the acetylation of isoniazid in the Moroccan population, which was followed by two others in 2011, focusing on the metabolism of tacrolimus and anti-vitamin K. So, the hope now is to identify the majority of the genetic polymorphisms affecting Moroccan patients, in order to provide them with appropriate care (Janati Idrissi et al., 2013).

10.7 CHALLENGES AND OPPORTUNITIES

10.7.1 Challenges

10.7.1.1 Language Barriers and Proximity to Europe

Geopolitical dynamics, especially those concerning Europe, play a significant role in shaping research priorities and funding for genetic studies in North Africa. On the one hand, populations located closer to Europe often receive more focus in research initiatives due to historical connections, the availability of funding, and perceived relevance to European demographics.

This preferential treatment can lead to insufficient representation of more isolated or politically marginalized populations within the region, but it has also contributed to fostering collaborative research partnerships with sub-Saharan African countries. On the other hand, European funding for research often prioritizes excellence, which can lead to a concentration of resources in select countries and/or institutions, which can lead at the mid-term to the marginalization of North African countries due to insufficient research infrastructure and well-established academic networks. Also, North African countries are not eligible for some funding opportunities designed to enhance research capacity. Additionally, the emphasis on information technology over other genomics-based areas can hinder the capacity of startups acting in this field for fundraising.

10.7.1.2 Regulatory Landscape in Genomics

The ethical challenges in North African countries are quite similar, and similar solutions/structure have been adopted. Regional/institutional ethics committees have been identified in all North African countries, in addition various laws on ethics have been enacted in each county. The ethics committees oversee biomedical research within the countries, by monitoring and approving the validity of research projects in each country, at its national level. Their role is to review study protocols, evaluate their validity, and ensure that biomedical research is compliant with the principles of fairness, transparency, and responsibility. On the legal and regulatory level, there are appropriate conditions in which biomedical research is carried out in a way that protects the privacy of the participants. There is, however, a clear need for an improved legal framework, like the European GDPR (General Data Protection Regulation).

10.7.1.3 Training

In terms of genomics/genetics and related research (bioinformatics), some North African countries have a well-established tradition to ensure skills development needed, while others need to develop appropriate curricula to enable interdisciplinary workforce. The list of required skills includes specialists in genetics, omics, statistics, genetic/genomic counseling, bioinformatics, computer systems administration, as well as engineers and technicians capable of operating and maintaining high technology equipment. There is also a deficit in soft skills, including leadership, management, and governance. The genomics innovation ecosystem should therefore engage a close collaboration between the different stakeholders, from patients and clinicians to policymakers and civil society, which needs developing specific skills and training activities. The shortage of trained geneticists, bioinformaticians, and other skilled researchers significantly hinders North Africa's ability to conduct thorough genetic studies. Additionally, the lack of advanced training and mentorship opportunities exacerbates these challenges by creating gaps in the talent pipeline. This deficiency in expertise not only restricts research output but also impedes the overall advancement of genetic science in the region. Addressing these issues is essential for enhancing the capacity for impactful genetic research and fostering a new generation of skilled professionals in Africa.

10.7.1.4 Community Involvement and Ethical Considerations

Successful genomics research in North Africa depends on community involvement and ethical oversight. Engaging with local health services and understanding cultural contexts is essential for building trust and facilitating participation in research projects (Bekada et al., 2015). It is critical to recognize the hierarchies within communities and to establish *rules of engagement* through dialogue, ensuring that expectations are clearly communicated, and that informed consent is tailored to various educational levels and cultural backgrounds. Moreover, researchers must ensure that consent processes are not only informed but also respectful of community norms and practices (Joseph et al., 2021).

10.7.1.5 Capacity Building for Sustainable Research

Advancing genomics research in North Africa relies greatly on capacity buildings, including individual skills enhancement, institutional resources improvement, and good research infrastructures establishment. Knowledge transfer through collaborative efforts, such as staff exchanges between institutions, is also a crucial role. Additionally, funding applications must strategically outline plans for capacity building, ensuring access to advanced technologies and updated research methodologies to maximize the impact of research initiatives (Zeghbib et al., 2021; Khoury and Holt, 2021).

10.7.1.6 Data Management and Data Sharing

An essential element of successful research collaborations is the development of open and transparent data management and sharing plans. Such plans should adhere to the Findable, Accessible, Interoperable, Reusable (or FAIR) principles, thus ensuring access to genomics data, maximizing the chance of a return of benefit to the population and implementing the fundamental ethical principles of fairness, transparency, and accountability.

10.7.1.7 Public Health Genomics (Importance of COVID-19)

The major role of genomics showed up during the COVID-19 pandemic, underscoring the need for a genomic surveillance strategy of infectious diseases (Knyazev et al., 2022; Zeghbib et al., 2021). Moreover, the integration of genomics technologies, such as tumor genome analysis, non-invasive prenatal screening, and genomic tests in medical practices, is expected to provide better diagnostic and treatment options resulting in an improved healthcare system (Khoury and Holt, 2021).

10.7.1.8 Brain Drain

Talent migration of highly skilled scientists and health professionals from North African countries to the high-income European Union member states is increasingly contributing to a brain drain in the region. This represents a big challenge for the North African region, which will very likely hinder scientific and technological progress. According to the World Bank's World Development Report 2021, the main cause of this phenomenon is economic, linked to the huge wage gaps that exist between high- and low-income countries. To face this crisis, there is a need for mechanisms and regulations that stimulate the digital economy and start-up creation in high technology sectors (including genomics) and encourage private sector investment in research, engineering, health, and data science.

10.7.2 Opportunities

Over the past decade, research studies in North Africa have built a strong foundation for precision medicine implementation. By uncovering the genetic factors involved in both communicable and non-communicable diseases, researchers are helping to develop personalized treatments based on individual genetic profiles. This leads to better outcomes and a more focused approach to healthcare and prevention in the region. Despite this progress, there remain significant challenges, such as limited access to advanced genomic technologies, the need for larger, more diverse patient and healthy cohorts, the integration of multi-omics approaches, such as transcriptomics and epigenomics, and data science tools, which are critical for capturing the full genetic landscape of several diseases in North Africa.

10.7.2.1 Human Genome Projects

Following are some major human genome projects in the region.

Genome Tunisia: The main goal of the Genome Tunisia Project is to develop a healthcare system capable of incorporating omics data for use in routine medical practice, enabling medical doctors to better treat, prevent, and diagnose patients (Hamdi Y. et al., 2024) The project also aims to sequence 10,000 healthy genomes and thousands of clinical exomes (4,000 per year). So far, national resources have been redeployed, and an urgent update was proposed for the current state of the genomic investigation platforms that would be implemented. A multidisciplinary partnership involving Tunisian experts has come to discerning all requirements that would be of high priority to fulfill the project goals. One of the most urgent priorities is to determine the reference sequence of the Tunisian genome to reinforce R&I in the field of genomics and to strengthen discoverability in the health sector. Extensive situation analysis and revision of the education programs, community awareness, training activities, appropriate infrastructure including NGS platforms and biobanking, and ethical and regulatory framework have been undertaken toward building sufficient capacity to effectively integrate genomics medicine and molecular diagnostics in the Tunisian healthcare system. An ecosystem with all engaged stakeholders has been implemented. This includes healthcare providers, clinicians, researchers, IT and bioinformaticians, biotechnology and pharma companies, policymakers, advocacy groups, and patients.

Genome Egypt: The Genome Egypt project has taken advantage of technology advancements to sequence and de novo assemble the genome of an Egyptian individual. They integrated the sequences of an additional 109 Egyptian individuals to generate an Egyptian Reference Human Genome. Comparing Egyptian genetic variant data and African and European variants showed that Egyptians constitute an individual genetic group that does carry specific "Egyptian" gene variants. The generation of the first Egyptian reference genome and the identification of population-specific variation will support genetic research and *precision* medicine in Egypt and North Africa. The next stage of the project is to extend sequencing efforts to better characterize *Egyptian* genetic variations and evaluate their relevance for common diseases such as cancer (Elmonem et al., 2024).

Genome Morrocco: The Moroccan Genome Project was launched with the goal of characterizing the genetic diversity of the Moroccan population and establishing a Moroccan reference genome. In its first phase, the project sequenced 109 high-coverage Moroccan genomes, leading to the identification of more than 27 million genetic variants, including around 1.4 million novel variants not previously reported in global databases. Among these, 15,378 variants were found to be common within the Moroccan population.

To better reflect the local genomic landscape, researchers constructed the Moroccan Major Allele Reference Genome—a population-specific reference that incorporates the most frequent alleles observed in Moroccan individuals.

The study highlights Morocco's unique genetic mosaic shaped by centuries of migration and admixture and underscores the importance of regional genomics data in advancing precision medicine, disease risk prediction, and population-specific therapeutic strategies (El Fahime et al.,2025).

Advancements in genomics research have greatly contributed to our understanding of hereditary diseases in North Africa and enhanced opportunities from genetic resources. Autosomal analyses have revealed that genetic variation exists within populations, further informed by comparative studies that include various regional and ethnic groups. The significant overlap observed between the genetic profiles of Arabs and Berbers underscores the complex interactions that have shaped the demographic landscape of North Africa over the centuries. Ongoing research continues to explore these genetic connections and the implications for understanding human history in North Africa. In addition, genomic technologies are increasingly utilized in clinical practice in North Africa, providing significant advancements in diagnostic and treatment options. These technologies include tumor genome analysis for targeted therapies, non-invasive prenatal screening, and genomic tests for rare genetic disorders. Notably, exome sequencing has proven invaluable for diagnosing previously unexplained rare disorders, such as Miller syndrome and Charcot Marie-Tooth disease, revealing causal mutations that guide clinical decisions. Moreover, newborn screening for treatable inherited conditions exemplifies a successful application of genomics in public health, with ongoing efforts to expand population screening for various genetic conditions. The integration of genomics data into healthcare practices faces numerous challenges, including ethical, legal, and technical considerations. Addressing these challenges is crucial for effectively utilizing genomic information in the diagnosis and clinical management of diseases. Research continues to explore how genomic information can complement existing models of *one health* approaches, which incorporate a range of genetic, environmental, and social determinants (Khoury and Holt, 2021; Huan Zhang et al., 2019). North Africa's economy may benefit greatly from its genetic resources, through economic impact, employment generation, and technology transfer. The long-term solution to capitalize these opportunities relies on a comprehensive national legal, regulatory, and institutional framework for access and benefit sharing. This framework will not only enhance fair and equitable access to genetic resources but also ensure that local populations benefit appropriately from these resources (Dine et al., 2024).

10.7.2.2 Genomics Entrepreneurship/Startups

The ecosystem of genomics startups in North Africa is experiencing rapid transformation, driven by initiatives such as the Genome Tunisia Project and collaborative efforts like the Egyptian Genome Project, both aimed at enhancing research capabilities, R&I, and R&D. By addressing challenges such as training deficits and funding shortages, these initiatives have the potential to positively impact healthcare systems, scientific research, and the economy of the region. As they expand, they may also contribute valuable insights to global genomics databases, thereby helping to close the gap in genetic data for underrepresented populations. Recent initiatives have focused on providing seed funding to early-stage startups. For instance, the Sawari Ventures North Africa Fund I aims to support 300 companies across Egypt, Tunisia, and Morocco, with a goal of creating or sustaining approximately 5,000 jobs. This fund has received substantial backing from the European Investment Bank and other investors, indicating a growing interest in the region's entrepreneurial potential. Indeed, North Africa is emerging as a hub for deep-tech innovations. A notable example is InstaDeep, a Tunisian AI startup recently acquired by BioNTech for $683 million. InstaDeep developed a system to predict high-risk coronavirus variants and has raised over $100 million prior to its acquisition. Moreover, Tunisia launched a Start-Up Act to build an entrepreneurial ecosystem and to help startups, investors, and startup support organizations develop a vibrant tech-economy. Morocco is also among the most recognized in Africa in terms of digital pathology. The site of the Moroccan Startup of digital pathology "Data Pathology" (https://www.datapathology.ma/) shows that their main mission is to fight cancer in Africa, through the development of digital solutions for rapid and effective diagnosis allowing better patient care. In Algeria, startups are now a key focus of government policy. A specialized ministry for startups was formally created by executive order named the Ministry of Knowledge Economy, Startups, and Micro-enterprises. This shows the government's commitment to making a supportive environment for innovation and entrepreneurship in the country, promoting long-lasting development for the future. In order to help businesses and make project implementation easier, it launched the Moukawil.dz platform. Overall, the combination of targeted funding initiatives and innovative startups positions North Africa as a promising landscape for advancements in genomics and related fields.

Efforts are underway to advance genomics research and genomics medicine in North African, but significant gaps remain in understanding its unique genetic diversity and its implications in health (Omotoso et al., 2022).

Recently, the ministries of health and ministries of scientific research from North African countries have shown a significant political commitment to advance genomics medicine in the region. Indeed, a *National Genome Center*, cofunded by the Tunisian Ministry of Health and the World Bank, was launched in Tunisia. This reference center will host the Genome Tunisia Project, other genomics research activities, and bioinformatics and biobanking services. It will also centralize all genetic testing activities in the country.

In Egypt, the government along with the scientific community has invested in genomics research mainly in the Egypt Center for Research and

Regenerative Medicine, which hosts the Egypt Genome Project and several other genomics medicine activities, including bioinformatics analysis, biobanking, and data science. Ain Shams and Cairo Universities are also considered as the main reference centers for human genomics and precision medicine research in Egypt.

National strategies for precision medicine implementation are also underway in Tunisia, Algeria, and Morocco as an outcome of the Personalized Medicine in North Africa Project. This regional initiative aims to lay the groundwork for integrating personalized medicine approaches into national health systems by fostering collaboration among stakeholders, assessing existing capacities, and identifying strategic priorities.

High-level discussions, stakeholder engagement, and capacity-building activities in these countries have led to the development of country-specific road maps. These road maps address critical components such as the following:

- *Governance and policy frameworks* to guide implementation.
- *Infrastructure and data systems* to support genomics research and clinical application.
- *Training and workforce development* in genomics, bioinformatics, and clinical translation.
- *Public engagement and ethical considerations* tailored to local contexts.

The national precision medicine strategies also emphasized the importance of regional cooperation and knowledge exchange, helping to align national efforts with global standards while taking into account local health priorities and population-specific genetic diversity.

As a result, Tunisia, Algeria, Morocco, and Egypt are now positioning themselves as emerging leaders in precision public health within the African continent. Collaboration and knowledge exchange with neighboring countries, including Libya and Mauritania, are key components of the national precision medicine strategies of the four countries. These regional partnerships aim to foster shared learning, harmonize approaches, and strengthen collective capacities for implementing precision medicine research across North Africa.

10.8 CONCLUSION

The recent findings related to the genetic structure of North African populations and the genomics underlying several non-communicable diseases have highlighted some specific genomic features of those populations. The hundreds of whole genome sequences generated recently by ongoing genome projects in Tunisia, Libya, and Egypt will bring new and valuable insights into population diversity and especially into medically relevant variants. These data will constitute an unprecedented resource to improve the healthcare system by implementing relevant genomics-based prevention, diagnosis, prognosis, and therapeutic interventions and to pave the way for personalized medicine.

REFERENCES

1000 Genomes Project Consortium et al. A global reference for human genetic variation. Nature. 2015;526(7571):68–74. https://doi.org/10.1038/nature15393

Abdelhedi, R., Bouayed, N. A., Alfadhli, S., Abid, L., Rebai, A., & Kharrat, N. Characterization of drug-metabolizing enzymes CYP2C9, CYP2C19 polymorphisms in Tunisian, Kuwaiti and Bahraini populations. J Genet. 2015;94(4):765–770.

Abdelmaksoud-Dammak, R. et al. High prevalence of the c.1227_1228dup (p.Glu410GlyfsX43) mutation in Tunisian families affected with MUTYH-associated-polyposis. Famil Cancer. 2012;11(3):503–508. https://doi.org/10.1007/s10689-012-9543-5

Abdelmaksoud-Dammak, R. et al. Predominance of the rare EGFR mutation p.L861Q in Tunisian patients with non-small cell lung carcinoma. Genes. 2022 Aug 22;13(8):1499. https://doi.org/10.3390/genes13081499

Abdelmaksoud-Dammak, R. et al. Strategies for diagnosis and management of CMMRD in low-resource countries: Report of a Tunisian family. Famil Cancer. 2024 Apr 30. https://doi.org/10.1007/s10689-024-00386-z

Abderrahmane, R., Louhibi, L., Moghtit, F. Z., Boubekeur, A., Benseddik, K., Boudjema, A., Benrrahal, F., Aberkane, M., Fodil, M., & Saidi-Mehtar, N. TP53 Arg 72Pro and MDM2 SNP309 polymorphisms and colorectal cancer risk: A west Algerian population study. Pathol Oncol Res. 2015;21(3):629–635. https://doi.org/10.1007/s12253-014-9867-6

Abdi, S., Bahloul, A., Behlouli, A., Hardelin, J.-P., Makrelouf, M., Boudjelida, K., Louha, M., Cheknene, A., Belouni, R., & Rous, Y. Diversity of the genes implicated in Algerian patients affected by Usher syndrome. PLoS One. 2016;11(9):e0161893.

Abdi, S., Makrelouf, M., Rous, I. N., Ounnoughi, K., Zenati, A., Petit, C., & Bonnet, C. (2024). Biallelic variants in TMIE and PDE6B genes mimic Usher syndrome. Gene Reports, 101954.

Abudabous, A., Drah, M., Aldehmani, M., Parker, I., & Alqawi, O. KRAS mutations in patients with colorectal cancer in Libya. Mol Clin Oncol. 2021;15(4):197. https://doi.org/10.3892/mco.2021.2359

Adadi, N., Radi, F. Z., Lahrouchi, N., Hara, L., Ratbi, I., Elalaoui, S. C., Alders, M., Zarzur, J., Bezzina, C., & Sefiani, A. Anatol Inherited dilated cardiomyopathy in a large Moroccan family caused by LMNA mutation. J Cardiol. 2018;20(1):65–68.

Akrout, Firas et al. Constitutional mismatch repair deficiency syndrome with atypical features caused by a homozygous MLH1 missense variant (c.1918C>A, p.(Pro640Thr)): A case report. Front Oncol. 2023 Aug 17;13:1195814. https://doi.org/10.3389/fonc.2023.1195814

Alali, M., Ismail Al-khalil, W., Rijjal, S., Al-Salhi, L., Saifo, M., & Youssef, L. A. Frequencies of CYP2D6 genetic polymorphisms in Arab populations. Hum Genomics. 2022;16(1):6.

Alami, I. E., Khaali, W., Jalbout, M., Gihbid, A., Ayoub, W. B., Benider, A., Brahim, S. M., Cherif, M. H., Benchakroun, N., & Mzibri, M. E. (2024). Genetic variations in CYP2A6, CYP2E1, GSTM1, GSTT1 genes and the risk of Nasopharyngeal carcinoma in North African population. Ann Hum Genet. 2024;89(4):158–168.

Ali, J., Cohn, B., Mwaka, E. et al. A scoping review of genetics and genomics research ethics policies and guidelines for Africa. BMC Med Ethics. 2021;22:39. https://doi.org/10.1186/s12910-021-00611-9

Allali, I., Boukhatem, N., Bouguenouch, L., Hardi, H., Boudouaya, H. A., Cadenas, M. B., Ouldim, K., Amzazi, S., Azcarate-Peril, M. A., & Ghaza, H. Gut microbiome of Moroccan colorectal cancer patients. Med Microbiol Immunol. 2018 Aug;207 (3–4):211–225. https://doi.org/10.1007/s00430-018-0542-5. Epub 2018 Apr 23.

Alsanosi, Safaa Mohammed M., Craig Skiffington, & Sandosh Padmanabhan. Pharmacokinetic Pharmacogenomics, Academic Press, Handbook of Pharmacogenomics and Stratified Medicine, Chapter 17, 341–364, 2014.

Amine, B. M., Krim, A., Bouhass, R. A., Rekab, M., bouchakor Moussa, Y., Taoussi, S., Abad, M. T,. Bradai, M., Graine, M., & Ait Ali, H. J. B. Epidemiological, clinical, and biological characterization of newly diagnosed acute Myeloid Leukemia in Algeria. Report on Behalf of the Algerian Acute Leukemia Study Group. 2018;132:5172.

Aminorroaya, A., Saeedi Moghaddam, S., Tavolinejad, H., Aryan, Z., Heidari, B., Ebrahimi, H., Naderian, M., Shobeiri, P., Ghanbari, A., Rezaei, N., Malekpour, M. R., Haghshenas, R., Rezaei, N., Larijani, B., & Farzadfar, F. Burden of ischemic heart disease and its attributable risk factors in North Africa and the Middle East, 1990 to 2019: Results from the GBD Study 2019. J Am Heart Assoc. 2024 Jan 16;13(2):e030165. https://doi.org/10.1161/JAHA.123.030165.

Ammar-Khodja, F., Bonnet, C., Dahmani, M., Ouhab, S., Lefèvre, G. M., Ibrahim, H., Hardelin, J. P., Weil, D., Louha, M., & Petit, C. Diversity of the causal genes in hearing impaired Algerian individuals identified by whole exome sequencing. Mol Genet Genomic Med. 2015;3(3):189–196.

Ammar-Khodja, F., Faugère, V., Baux, D., Giannesini, C., Léonard, S., Makrelouf, M., Malek, R., Djennaoui, D., Zenati, A., & Claustres, M. Molecular screening of deafness in Algeria: High genetic heterogeneity involving DFNB1 and the Usher loci, DFNB2/USH1B, DFNB12/USH1D and DFNB23/USH1F. Eur J Med Genet. 2009;52(4):174–179.

Ammour, A., Aberkane, M., Boudjema, A., Boughrara, W., & Benchekor, S. M. The frequencies distribution of CYP3A5 rs776746 and ABCB1 rs1045642 polymorphisms in the west Algerian population and relationships with pharmacogenetics. African Health Sci. 2024;24(1):307–312.

Amrani, A., Kiando, S. R., Treard, C., Jeunemaitre, X., & Bouatia-Naji, N. Genetic association study between T-786C NOS3 polymorphism and essential hypertension in an Algerian population of the Oran city. J Hypertens. 2023;41(Suppl 3):e235.

Arauna, L. R., Mendoza-Revilla, J., Mas-Sandoval, A., et al. Recent historical migrations have shaped the gene pool of Arabs and Berbers in North Africa. Mol Biol Evol. 2017;34(2):318–329.

Aribi, M., Moulessehoul, S., Benabadji, A.-B., & Kendoucitani, M. HLA DR phenotypic frequencies and genetic risk of Type 1 diabetes in west region of Algeria, Tlemcen. BMC Genet. 2004;5:1–7.

Atageldiyeva, K. K., Nemr, R., Echtay, A., Racoubian, E., Sarray, S., Almawi, W. Y. Apolipoprotein E genetic polymorphism influence the susceptibility to nephropathy in type 2 diabetes patients. Gene. 2019 5;715:144011.

Azim, Hamdy, A. et al. The landscape of BRCA mutations among Egyptian women with breast cancer. Oncol Ther. 2023;11(4):445–459. https://doi.org/10.1007/s40487-023-00240-9

Badache, H., Boussetta, S., Elgaaeid, A. B., Cherni, L., & El-Khil, H. K. Investigation of the genetic structure of Kabyle and Chaouia Algerian populations through the polymorphism of Alu insertion markers. Ann Hum Biol. 2019;46(2):150–159. https://doi.org/10.1080/03014460.2019.1588994

Badro, D. A. et al. Y-chromosome and mtDNA genetics reveal significant contrasts in affinities of modern Middle Eastern populations with European and African populations. PloS One. 2013;8(1):e54616. https://doi.org/10.1371/journal.pone.0054616

Bahri, R., Esteban, E., Moral, P., & Chaabani, H. New insights into the genetic history of Tunisians: Data from Alu insertion and apolipoprotein E gene polymorphisms. Ann Hum Biol. 2008;35(1):22–33. https://doi.org/10.1080/03014460701753729

Bahri, R. et al. Genetic differentiation and origin of the Jordanian population: An analysis of Alu insertion polymorphisms. Genet Test Mol Biomarkers. 2012;16(5):324–329. https://doi.org/10.1089/gtmb.2011.0197

Bakkach, J., Mansouri, M., Derkaoui, T. et al. Contribution of BRCA1 and BRCA2 germline mutations to early onset breast cancer: A series from north of Morocco. BMC Cancer. 2020;20:859. https://doi.org/10.1186/s12885-020-07352-9

Barbirou, M. et al. Variant characterization of a representative large pedigree suggests "variant risk clusters" convey varying predisposition of risk to lynch syndrome. Cancers. 2023 Aug 12;15(16):4074. https://doi.org/10.3390/cancers15164074

Barrou, T. (1982). Ancient history of Arabs. Edition Dar El Feker, Damas, Syria.

Bekada, A., Arauna, L. R., Deba, T., Calafell, F., Benhamamouch, S., & Comas, D. Genetic heterogeneity in Algerian human populations. PLoS One. 2015;10(9):e0138453. https://doi.org/10.1371/journal.pone.0138453

Bekada, A., Fregel, R., Cabrera, V. M., Larruga, J. M., Pestano, J., Benhamamouch, S., & González, A. M. Introducing the Algerian mitochondrial DNA and Y-chromosome profiles into the North African landscape. PLoS One. 2013;8(2):e56775. https://doi.org/10.1371/journal.pone.0056775

Bekadja, M. A., Hamladji, R. M., Belhani, M., Ardjoun, F. Z., Abad, M. T., Touhami, H., Ait-Ali, H., Zouaoui, Z., Sidimansour, N., Hamdi, S., Grifi, F., Mesli, N., Saidi, M., Mehalhal, N., Bachiri, A., Bouhass, R., Said, Y. M., & Midoun, N. A population-based study of the epidemiology and clinical features of adults with acute myeloid leukemia in Algeria: Report on behalf of the Algerian Acute Leukemia Study Group. Hematol Oncol Stem Cell Ther. 2011;4(4):161–166. https://doi.org/10.5144/1658-3876.2011.161

Belfki, H., Ben Ali, S., Aounallah-Skhiri, H., Traissac, P., Bougatef, S., Maire, B., Delpeuch, F., Achour, N., & Ben Romdhane, H. Prevalence and determinants of the metabolic syndrome among Tunisian adults: Results of the Transition and Health Impact in North Africa (TAHINA) project. Public Health Nutr. 2013

Apr;16(4):582–590. https://doi.org/10.1017/S1368980012003291. Epub 2012 Aug 13. PMID: 22883486; PMCID: PMC10271567.

Ben Ayed-Guerfali, D. et al. The identification by exome sequencing of candidate genes in BRCA-negative Tunisian patients at a high risk of hereditary breast/ovarian cancer. Genes. 2022 Jul 22;13(8):1296. https://doi.org/10.3390/genes13081296

Ben Ayed-Guerfali, D. et al. Novel and recurrent BRCA1/BRCA2 germline mutations in patients with breast/ovarian cancer: A series from the south of Tunisia. J Transl Med. 2021 Mar 16;19(1):108. https://doi.org/10.1186/s12967-021-02772-y

Ben Halim, N., Hsouna, S., Lasram, K., Chargui, M., Khemira, L., Saidane, R.,. . .& Kefi, R. Mitochondrial DNA structure of an isolated Tunisian Berber population and its relationship with Mediterranean populations. Ann Hum Biol. 2018;45(1):86–97.

Ben Halim, N. et al. Consanguinity, endogamy, and genetic disorders in Tunisia. J Community Genet. 2013;4(2):273–284. https://doi.org/10.1007/s12687-012-0128-7

Ben Halima, A., Bahri, R., Esteban, E., Moral, P., & Chaabani, H. Variation of Rhesus haplotype frequencies in North Africans and in worldwide population analyses. Int J Hum Genet. 2015;15(1):21–31.

Ben Kridis-Rejeb, W. et al. Identification of novel candidate genes by exome sequencing in Tunisian familial male breast cancer patients. Mol Biol Rep. 2020;47(9):6507–6516. https://doi.org/10.1007/s11033-020-05703-0

Benseddik, K., Abderrahmane, R., Boudjemaa, A., Meriem, A., Samia, D., Zohra, T. C., & Naima, M. D. Association of TP53 P72R and MDM2 SNP309 variants with susceptibility to chronic myeloid leukaemia development in Algerian population: TP53 P72R C allele is a strong risk marker and MDM2 SNP309 G allele has a protective effect. South Asian Journal of Experimental Biology. 2022;12(1):28–34.

Ben Sghaier, R. et al. Targeted next generation sequencing screening of Lynch syndrome in Tunisian population. Fam Cancer. 2019;18(3):343–348. https://doi.org/10.1007/s10689-019-00130-y

Bensenouci, S., Louhibi, L., De Verneuil, H., Mahmoudi, K., & Saidi-Mehtar, N. Diagnosis of xeroderma pigmentosum Groups A and C by detection of two prevalent mutations in West Algerian population: A rapid genotyping tool for the frequent XPC mutation c. 1643_1644delTG. BioMed Res Int. 2016;2016(1):2180946.

Bentayebi, K., Abada, F., Ihzmad, H., & Amzazi, S. Genetic ancestry of a Moroccan population as inferred from autosomal STRs. Meta Gene. 2014 Jun 21;2:427–438. https://doi.org/10.1016/j.mgene.2014.04.011. eCollection 2014 Dec. PMID: 25606427 PMCID: PMC4287812

Boujemaa, M. et al. Germline copy number variations in BRCA1/2 negative families: Role in the molecular etiology of hereditary breast cancer in Tunisia. PloS One. 2021 Jan 27;16(1):e0245362. https://doi.org/10.1371/journal.pone.0245362

Boujemaa, M. et al. Uncovering the clinical relevance of unclassified variants in DNA repair genes: A focus on BRCA negative Tunisian cancer families. Front Genet. 2024 Jan 19;15:1327894. https://doi.org/10.3389/fgene.2024.1327894

Boulenouar, H., Mediene Benchekor, S., Meroufel, D. N., Lardjam Hetraf, S. A., Ouhaibi Djellouli, H., Hermant, X., Grenier-Boley, B., Hamani Medjaoui, I., Saidi Mehtar, N., Amouyel, P., Houti, L., Meirhaeghe, A., & Goumidi, L. Impact of APOE gene polymorphisms on the lipid profile in an Algerian population. Lipids Health Dis. 2013;12:155. https://doi.org/10.1186/1476-511x-12-155

Boulenouar, H., Mediene Benchekor, S., Ouhaibi Djellouli, H., Larjam Hetraf, S. A., Houti, L., & Hammani-Medjaoui, I. Association study of APOE gene polymorphisms with diabetes and the main cardiometabolic risk factors, in the Algerian population. Egyptian J Med Hum Genet. 2019;20(1):5.

Bounedjar, A., Benkali, R., Badoui, L., Abada, M., Moumeni, A., Marouani, A., Sedkaoui, C., Ammour, H., Dib, A., Braikia, S., Reggad, R., Zidane, H., Taieb, L., Mahfouf, H., Belacel, A., Mechiat, Z., Maachou, Y., Souilah, S., Gamaz, M.,. . . Kedar, M. Incidence of lung cancer in males and females in Algeria: The lung cancer registry in Algeria (LuCaReAl). Cancer Epidemiol. 2020;69:101799. https://doi.org/10.1016/j.canep.2020.101799

Brahim, S. M., Hamed, C. T., EkhtElbenina, Z. E. I. N., Salame, M., Veten, F., Zein, M. V.,. . . & Tolba, A. (2021). Epidemiological and clinicopathological features of breast cancer in Mauritania.

Brahim, S. M. et al. Screening of BRCA1/2 variants in Mauritanian breast cancer patients. BMC Cancer. 2022 Jul 20;22(1):802. https://doi.org/10.1186/s12885-022-09903-8

Bray, F. et al. Global cancer statistics 2022: GLOBOCAN estimates of incidence and mortality worldwide for 36 cancers in 185 countries. CA Cancer J Clin. 2024;74(3):229–263. https://doi.org/10.3322/caac.21834

Chaabani, H. GM polymorphism and the evolutionary history of modern humans. Ann Genet. 2002;45(4):197–206. https://doi.org/10.1016/s0003-3995(02)01140-1

Chaabani, H. The Tunisian Revolution "The Free, Youth Revolution" from an anthropological perspective. Int J Mod Anthropol. 2017;1(10):50–72. http://dx.doi.org/10.4314/ijma.v1i10.1

Chaâbani, H., & Cox, D. W. Genetic characterization and origin of Tunisian Berbers. Hum Hered. 1988;38(5):308–316. https://doi.org/10.1159/000153804

Chaabani, H. et al. Genetic differentiation of Yemeni people according to rhesus and Gm polymorphisms. Ann Genet. 2000;43(3–4):155–162. https://doi.org/10.1016/s0003-3995(00)01023-6

Chaabani, H. et al. Restriction fragment length polymorphisms associated with immunoglobulin heavy chain gamma genes in Tunisians. Hum Genet. 1986;73(2):110–113. https://doi.org/10.1007/BF00291597

Chaher, N. et al. Molecular and epidemiological characteristics of inflammatory breast cancer in Algerian patients. Breast Cancer Res Treat. 2012;131(2):437–444. https://doi.org/10.1007/s10549-011-1422-5

Chemlal, N. (2024, September). Pharmacogenomics and oncology in Morocco. Conference: Pharmacoviligance Annual Conference: Mohammed V University of Rabat. https://doi.org/10.13140/RG.2.2.23136.08968

Cherbal, F. et al. BRCA1 and BRCA2 germline mutations screening in Algerian breast/ovarian cancer families. Disease Markers. 2010;28(6):377–384. https://doi.org/10.3233/DMA-2010-0718

Cherbal, F. et al. Distribution of molecular breast cancer subtypes among Algerian women and correlation with clinical and tumor characteristics: A population-based study. Breast Disease. 2015;35(2):95–102. https://doi.org/10.3233/BD-150398

Chikhaoui, A., Jones, M., Režen, T., Ben Ahmed, M., Naouali, C., Komel, R., Zghal, M., Boubaker, S., Abdelhak, S., & Yacoub-Youssef, H. Inflammatory landscape in Xeroderma pigmentosum patients with cutaneous melanoma. Sci Rep. 2022;16;12(1):13854.

Choudhury, A., Hazelhurst, S., Meintjes, A., Achinike-Oduaran, O., Aron, S., Gamieldien, J., Jalali Sefid Dashti, M., Mulder, N., Tiffin, N., & Ramsay, M. Population-specific common SNPs reflect demographic histories and highlight regions of genomic plasticity with functional relevance. BMC Genomics. 2014;15(1):437. https://doi.org/10.1186/1471-2164-15-437

Coudray, C. et al. Study of GM immunoglobulin allotypic system in Berbers and Arabs from Morocco. Am J Hum Biol. 2006;18(1):23–34. https://doi.org/10.1002/ajhb.20465

Dallali, H., Boukhalfa, W., Kheriji, N., Fassatoui, M., Jmel, H., Hechmi, M., Gouiza, I., Gharbi, M., Kammoun, W., Mrad, M., Taoueb, M., Krir, A., Trabelsi, H., Bahlous, A., Jamoussi, H., Messaoud, O., Abid, A., & Kefi, R. The first exome wide association study in Tunisia: Identification of candidate loci and pathways with biological relevance for type 2 diabetes. Front Endocrinol (Lausanne). 2023;19(14):1293124.

Dhieb, D. et al. Analysis of genetic alterations in Tunisian patients with lung adenocarcinoma." Cells. 2019 May 28;8(6):514. https://doi.org/10.3390/cells8060514

Dine, R. D., Elkheir, L. Y. M., Raimi, M. O., Alemayehu, M., Mohamed, S. Y., Turzin, J. K., et al. Ten simple rules for successful and sustainable African research collaborations. PLoS Comput Biol. 2024;20(6):e1012197. https://doi.org/10.1371/journal.pcbi.1012197, https://doi.org/10.1016/j.lrr.2018.11.001. eCollection 2019

Doubaj, Y., Laarabi, F. Z., Chafai Elalaoui, S., Barkat, A., & Sefiani, A. Carrier frequency of the recurrent mutation c. 1643_1644delTG in the XPC gene and birth prevalence of the xeroderma pigmentosum in Morocco. The J Dermatol. 2012;39(4):382–384.

Eddaikra, A., Amroun, H., Raache, R., Galleze, A., Abdallah-Elhadj, N., Azzouz, M., Mecabih, F., Mechti, B., Abbadi, M., & Touil-Boukoffa, C. Clinical variables and ethnicity may influenced by polymorphism of CAT– 262C/T and MnSOD 47C/T antioxidant enzymes in Algerian type1 diabetes without complications. Gene, 670, 182–192.

Eid, S. S. M., Ahmed, A. R. H., Abdelgawad, M. I. K., Mohamed, A. E. S., Mohamed, A. O., Mohamed, D. O.,. . .& Saad-Eldeen, H. M. Mutation of EGFR in non-small cell lung cancer, a regional study in Upper Egypt. Cancer Res J. 2020;8:1.

El Fahime, E., Kartti, S., Chemao-Elfihri, M. W. et al. Moroccan genome project: Genomic insight into a North African population. Commun Biol. 2025;8:584. https://doi.org/10.1038/s42003-025-08020-z

El Moncer, W., Esteban, E., Bahri, R., Gayà-Vidal, M., Carreras-Torres, R., Athanasiadis, G.,. . .& Chaabani, H. Mixed origin of the current Tunisian population from the analysis of Alu and Alu/STR compound systems. J Hum Genet. 2010;55(12):827–833.

El Moncer, W. et al. Research of the origin of a particular Tunisian group using a physical marker and Alu insertion polymorphisms. Genet Mol Biol 2011;34(3):371–376. https://doi.org/10.1590/S1415-47572011005000021

El Zaitouni, S. et al. "Genetic profiling of non-small cell lung cancer in Moroccan patients by targeted next-generation sequencing. Technol Cancer Res Treat. 2024;23:15330338241288907.https://doi.org/10.1177/15330338241288907

Elbnnani, A. S. et al. Flow cytometric detection of leukemic blasts in Libyan pediatric patients with acute lymphoblastic leukemia. Libyan J Med. 2024;19(1):2319895. https://doi.org/10.1080/19932820.2024.2319895

Eldukali, W., Omran, K., & Azzuz, R. Trends, and tumor characteristics of lung cancer and malignant pleural mesothelioma in the East of Libya. Ibnosina J Med Biomed Sci. 2020;12(04):272–279.

Elkamel, S., Cherni, L., Alvarez, L., Marques, S. L., Prata, M. J., Boussetta, S., et al. The orientalisation of North Africa: New hints from the study of autosomal STRs in an Arab population. Ann Hum Biol. 2017;44:180–190.

Elloumi-Zghal, H., & Chaabouni Bouhamed, H. Genetics and genomic medicine in Tunisia. Mol Genet Genomic Med. 2018;6(2):134–159.

Elmaihub, E. S. et al. Analysis of BRCA1 germline variants (exons 5, 11 and 20) in breast cancer families from Libya. The Libyan J Med. 2024;19(1):2356906. https://doi.org/10.1080/19932820.2024.2356906

Elmetnawy, W. H., Qenawi, M., Sabet, S., & Bassiony, H. Mutation patterns of epidermal growth factor receptor gene in non-small cell lung cancer among Egyptian patients. Egypt J Basic Appl Sci. 2022;9(1):77–90.

Elmonem, M. A., Soliman, N. A., Moustafa, A., Gad, Y. Z., Hassan, W. A., Taha, T., El-Feky, G., Sakr, M., & Amer, K. The Egypt Genome Project. Nat Genet. 2024;56(6):1035–1037. https://doi.org/10.1038/s41588-024-01739-1

Elshwekh, H., Alhudiri, I., Abdul Jalil, H. M., Mohamed, A. A., Elkikli, A., Jornaz, N., Enattah, N. Investigation of BRCA1 Exon 11 Genetic Variations in Breast Cancer among Libyan Women. medRxiv. 2023.

Fadhlaoui-Zid, K., Martinez-Cruz, B., Khodjet-el-khil, H., Mendizabal, I., Benammar-Elgaaied, A., & Comas, D. Genetic structure of Tunisian ethnic groups revealed by paternal lineages. Am J Phys Anthropol. 2011;146(2):271–280.

Fadhlaoui-Zid, K., Plaza, S., Calafell, F., Ben Amor, M., Comas, D., El Gaaied, A. B. Mitochondrial DNA heterogeneity in Tunisian Berbers. Ann Hum Genet. 2004;68:222–233.

Finsterer, J., Frank, M. Maternally inherited diabetes and deafness is a mitochondrial multiorgan disorder syndrome (MIMODS). Acta Diabetol. 2017 Oct;54(10):979–980. https://doi.org/10.1007/s00592-017-1029-3.

Fregel, R., Méndez, F. L., Bokbot, Y., Martín-Socas, D., Camalich-Massieu, M. D., Santana, J., et al. Ancient genomes from North Africa evidence prehistoric migrations to the Maghreb from both the Levant and Europe. Proc Natl Acad Sci USA. 2018;115:6774–6779.

Frigi, S., Mota-Vieira, L., Cherni, L., van Oven, M., Pires, R., Boussetta, S., & El-Gaaied, A. B. A. Mitochondrial DNA analysis of Tunisians reveals a mosaic genetic structure with recent population expansion. Homo. 2017;68(4):298–315.

Frikha, R. et al. Involvement of MTHFR rs1801133 in the Susceptibility of Acute Lymphoblastic Leukemia: A Preliminary Study. J Pediatr Hematol Oncol. 2021;43(6):e816–e818. https://doi.org/10.1097/MPH.0000000000001970

Gaceb, H. et al. Clinicopathological and molecular study of triple-negative breast cancer in Algerian patients. Pathol Oncol Res. 2018;24:297–308.

Garrigan, D., & Hammer, M. F. Reconstructing human origins in the genomic era. Nat. Rev. Genet. 2006;7:669–680. https://doi.org/10.1016/j.mgene.2014.04.011

Ghania, D., Katia, B., & Dalila, S. Cyp1a1 gene polymorphism and nasopharyngeal carcinoma in Algerian population. Curr Trends Nat Sci. 2020;9(18):218–225.

Gharbi, H. et al. Association of genetic variation in IKZF1, ARID5B, CDKN2A, and CEBPE with the risk of acute lymphoblastic leukemia in Tunisian children and their contribution to racial differences in leukemia incidence. Pediat Hematol Oncol. 2016;33(3):157–167. https://doi.org/10.3109/08880018.2016.1161685

Goodspeed, G. S. (2007) A history of the Babylonians and Assyrians. Gorgias Press, Piscataway, NJ, p 54.

Guidoumi, M., Kefi, R., Abdelhak, S., & Bouslama, Z. Consanguinity and endogamy of a Northeastern Algerian population (population of El-Kala). Adv Environ Biol. 2015;9(3):457–465.

Gurdasani, D., et al. The African genome variation project shapes medical genetics in Africa. Nature. 2015;517(7534):327–332.

Habour, N., Yafour, N., Youcef, B. Y., Bouhass, R., Chekkal, M., Brahimi, M., Bekadja, M. A., & Sahraoui, T. HLA-B*58 and HLA-B*27 play a role in the development of acute leukemia: A case control study. Asian Pac J Cancer Prev APJCP. 2024;25(1):169–173. https://doi.org/10.31557/APJCP.2024.25.1.169

Haghighatdoost, F., Mehrabani-Zeinabad, K., Hajihashemi, P. et al. Burden of colorectal cancer and its risk factors in the North Africa and Middle East (NAME) region, 1990–2019: A systematic analysis of the global burden of disease study. BMC Public Health. 2024;24:557. https://doi.org/10.1186/s12889-024-18027-6

Hamadou, W. S. et al. Familial hematological malignancies: ASXL1 gene investigation. Clin Transl Oncol.: Official Publication of the Federation of Spanish Oncology Societies and of the National Cancer Institute of Mexico. 2016;18(4):385–390. https://doi.org/10.1007/s12094-015-1379-7

Hamadou, W. S. et al. Familial hematological malignancies: New IDH2 mutation. Ann Hematol. 2016;95(12):1943–1947. https://doi.org/10.1007/s00277-016-2813-9

Hamadou, W. S. et al. GATA2 gene analysis in several forms of hematological malignancies including familial aggregations. Ann Hematol. 2017a;96(10):1635–1639. https://doi.org/10.1007/s00277-017-3076-9

Hamadou, W. S. et al. Mutational analysis of apoptotic genes in familial aggregation of hematological malignancies. Bull Cancer. 2021;108(9):798–805. https://doi.org/10.1016/j.bulcan.2021.04.009

Hamadou, W. S. et al. Mutational analysis of JAK2, CBL, RUNX1, and NPM1 genes in familial aggregation of hematological malignancies. Ann Hematol. 2016;95(7):1043–1050. https://doi.org/10.1007/s00277-016-2678-y

Hamadou, W. S. et al. Mutational analysis of TP53 gene in Tunisian familial hematological malignancies and sporadic acute leukemia cases. Fam Cancer. 2017b;16(1):153–157. https://doi.org/10.1007/s10689-016-9931-3

Hamdi, Y., Trabelsi, M., Ghedira, K., Boujemaa, M., Ben Ayed, I., Charfeddine, C., Souissi, A., Rejeb, I., Kammoun Rebai, W., Hkimi, C., Neifar, F., Jandoubi, N., Mkaouar, R., Chaouch, M., Bennour, A., Kamoun, S., Chaker Masmoudi, H., Abid, N., Mezghani Khemakhem, M., GTCA Consortium,. . .Rebai, A. Genome Tunisia project. Paving the way for precision medicine in North Africa. Genome Med. 2024;16(1):104. https://doi.org/10.1186/s13073-024-01365-w

Hamdi, Y. et al. Family specific genetic predisposition to breast cancer: Results from Tunisian whole exome sequenced breast cancer cases. J Transl Med. 2018 Jun 7;16(1):158. https://doi.org/10.1186/s12967-018-1504-9

Hamdi, Y. et al. Identification of BRCA2 Cis double heterozygous breast cancer cases using whole exome sequencing: Phenotypic expression and impact on personalized oncology. Front Genet. 2021a Aug 12;12:674990. https://doi.org/10.3389/fgene.2021.674990

Hamdi, Y. et al. Identification of eleven novel BRCA mutations in Tunisia: Impact on the clinical management of BRCA related cancers. Front Oncol. 2021b Aug 20;11:674965. https://doi.org/10.3389/fonc.2021.674965

Hamza, W., Ali Pacha, L., Hamadouche, T., Muller, J., Drouot, N., Ferrat, F., Makri, S., Chaouch, M., Tazir, M., Koenig, M., & Benhassine, T. Molecular and clinical study of a cohort of 110 Algerian patients with autosomal recessive ataxia. BMC Med Genet. 2015;16.

Hanachi, M., Maghrebi, O., Bichiou, H., Trabelsi, F., Bouyahia, N. M., Zhioua, F., Belghith, M., Harigua-Souiai, E., Baouendi, M., Guizani-Tabbane, L., Benkahla, A., & Souiai, O. Longitudinal and comparative analysis of gut microbiota of Tunisian newborns according to delivery mode. Front Microbiol. 2022 Apr 25;13:780568.

Henn, B. M. et al. Genomic ancestry of North Africans supports back-to-Africa migrations. PLoS Genet. 2012;8:e1002397.

Henouda, S., Bensalem, A., Reggad, R., Serrar, N., Rouabah, L., & Pujol, P. Contribution of BRCA1 and BRCA2 germline mutations to early algerian breast cancer. Dis Markers. 2016;7869095. https://doi.org/10.1155/2016/7869095

Herrag, S. E. E., Moulessehoul, S., Mehadji, D. E., Yekrou, D., & Khaled, M. B. Epidemiological insights into colorectal cancer in northwestern Algeria. Ecancermedicalscience. 2024;18:1684. https://doi.org/10.3332/ecancer.2024.1684

Houcine, Y. et al. PDL1 and molecular biomarkers expression in non-small cell lung cancer in Tunisian patients. Monaldi Arch Chest Dis = Arch Monaldi Mal Torace. 2023 Nov 3. https://doi.org/10.4081/monaldi.2023.2778

Hsouna, S., Ben Halim, N., Lasram, K., Meiloud, G., Arfa, I., Kerkeni, E., Romdhane, L., Jamoussi, H., Bahri, S., Ben Ammar, S., Abid, A., Barakat, A., Houmeida, A., Abdelhak, S., & Kefi, R. Study of the T16189C variant and mitochondrial lineages in Tunisian and overall Mediterranean region. Mitochondrial DNA A DNA Mapp Seq Anal. 2016;27(2):1558–1563.

Hussini, H., Elmahdi, K., & Nasef, A. Diagnostic pattern of adult acute leukemia in Benghazi Medical Center/Libya. J Clin Exp Immunol. 2021;6(2):300–304.

Idyahia, A., Redouan, S., Amalou, G., Charoute, H., Harmak, H., Bonnet, C., Petit, C., Benrahma, C., & Barakat, A. Exome sequencing reveals pathogenic mutations in the LARS2 and HSD17B4 genes associated with Perrault syndrome and D-bifunctional protein deficiency in Moroccan families. Mol Biol Rep. 2024 Jul 25;51(1):850. https://doi.org/10.1007/s11033-024-09740-x.

Ivanova, R., Astrinidis, A., Djoulah, S., Lepage, V., Wijnen, E., Hors, J., & Charron, D. Mitochondrial DNA polymorphisms of a west Algerian population (Oran region). Biomed Pharmacother. 1999;53(8):386–392. https://doi.org/10.1016/s0753-3322(99)80110-3

Jaafar, N., Gómez, J., Kammoun, I., Zairi, I., Amara, W. B., Kachboura, S., Kraiem, S., Hammami, M., Iglesias, S., Alonso, B., & Coto, E. Spectrum of mutations in hypertrophic cardiomyopathy genes among Tunisian patients. Genet Test Mol Biomarkers. 2016;20(11):674–679.

Jaballah-Gabteni, A. et al. Identification of novel pathogenic MSH2 mutation and new DNA repair genes variants: Investigation of a Tunisian Lynch syndrome family with discordant twins. J Transl Med. 2019 Jun 27;17(1):2. https://doi.org/10.1186/s12967-019-1961-9

Jafari, M., Laraqui, A., Baba, W., Benmokhtar, S., Zaitouni, S. E., Ali, A. A., Bounaim, A., Moujahid, M., Tanz, R., Mahfoud, T., Sbitti, Y., Annaz, H. E., Abi, R., Tagajdid, M. R., Kochri, S. E., Lahlou, I. A., Hsaini, H. E., Belayachi, L., Benjouad, A.,. . .Ennibi, K. Prevalence and patterns of mutations in RAS/RAF/MEK/ERK/MAPK signaling pathway in colorectal cancer in North Africa. BMC Cancer. 2022;22(1):1142. https://doi.org/10.1186/s12885-022-10235-w

Janati Idrissi, M., Ouldim, K., Amarti, A., Mohammed, E. H., & Khabbal, Y. Pharmacogenetics: what about Morocco?. Pan Afr Med J. 2013;14:143.

Jeddane, L., Ailal, F., Dubois-d'Enghien, C., Abidi, O., Benhsaien, I., Kili, A., Chaouki, S., Kriouile, Y., El Hafidi, N., Fadil, H., Abilkassem, R., Rada, N., Bousfiha, A. A., Barakat, A., Stoppa-Lyonnet, D., & Bellaoui, H. Molecular defects in Moroccan patients with ataxia-telangiectasia. Neuromolecular Med. 2013 Jun;15(2):288–294. https://doi.org/10.1007/s12017-013-8218-1. Epub 2013 Jan 16.

Jmel, H., Romdhane, L., Ben Halima, Y., Hechmi, M., Naouali, C., Dallali, H., Hamdi, Y., Shan, J., Abid, A., Jamoussi, H., Trabelsi, S., Chouchane, L., Luiselli, D., Abdelhak, S., Kefi, R. Pharmacogenetic landscape of Metabolic Syndrome components drug response in Tunisia and comparison with worldwide populations. PLoS One. 2018;13(4):e0194842.

Kartti, S., Bouricha, E. M., Zarrik, O., Aghlallou, Y., Mounjid, C., ELJaoudi, R., Belyamani, L., Ibrahimi, A., & El Khannoussi, B. Targeted gene panel sequencing unveiled new pathogenic mutations in patients with breast cancer. Bioinformatics Biol Insights. 2023;17:11779322231182054. https://doi.org/10.1177/11779322231182054

Kassem, N. M., Emera, G., Kassem, H. A. et al. Clinicopathological features of Egyptian colorectal cancer patients regarding somatic genetic mutations especially in KRAS gene and microsatellite instability status: A pilot study. Egypt J Med Hum Genet. 2019;20(20). https://doi.org/10.1186/s43042-019-0028-z

Kdissa, A. et al. c.1227_1228dupGG (p.Glu410Glyfs), a frequent variant in Tunisian patients with MUTYH associated polyposis. Cancer Genet. 2020;240:45–53. https://doi.org/10.1016/j.cancergen.2019.10.007

Khalat, N., Messaoud, O., Ben Rekaya, M., Chargui, M., Zghal, M., Zendah, B., Saqer, N., Mokni, M., Abdelhak, S., & Mohamed, O. A. First genetic characterization of Xeroderma pigmentosum in Libya. High frequency of XP-C founder mutation. Mol Genet Genomic Med. 2023 Jun;11(6):e2158.

Kheriji, N., Dakhlaoui, T., Kamoun Rebai, W., Maatoug, S., Thabet, M. T., Mellah, T., Mrad, M., Trabelsi, H., Soltani, M., Kabbage, M., Hassine, H. B., Hadj Salah Bahlous, A., Mahjoub, F., Jamoussi, H., Abid, A., Abdelhak, S., & Kefi, R. Prevalence and risk factors of diabetes mellitus and hypertension in North East Tunisia calling for efficient and effective actions. Sci Rep. 2023a Aug 5;13(1):12706.

Kheriji, N., Dallali, H., Gouiza, I., Hechmi, M., Mahjoub, F., Mrad, M., Krir, A., Soltani, M., Trabelsi, H., Hamdi, W., Bahlous, A., Ben Ahmed, M., Jamoussi, H., & Kefi, R. Whole-exome sequencing reveals novel variants of monogenic diabetes in Tunisia: Impact on diagnosis and healthcare management. Front Genet. 2023b Dec 14;14:1224284.

Khoury, M. J., & Holt, K. E. The impact of genomics on precision public health: Beyond the pandemic. Genome Med. 2021;13(1):67. https://doi.org/10.1186/s13073-021-00886-y

Knyazev, S., Chhugani, K., Sarwal, V. et al. Unlocking capacities of genomics for the COVID-19 response and future pandemics. Nat Method. 2022;19:374–380. https://doi.org/10.1038/s41592-022-01444-z

Kilaru, S., Panda, S. S., Moharana, L. et al. PD-L1 expression and its significance in advanced NSCLC: Real-world experience from a tertiary care center. J Egypt Natl Canc Inst. 2024;36(3). https://doi.org/10.1186/s43046-024-00207-5

Lahmadi, M., Beddar, L., Ketit, S., Makhbouche, T., Laouar, N., & Filali, T. Clinicopathological characteristics and prognosis of non-small cell lung cancer in Algeria: A single-center retrospective study. BMC Cancer. 2024;24(1):946. https://doi.org/10.1186/s12885-024-12709-5

Lasram, K., Ben Halim, N., Benrahma, H., Mediene-Benchekor, S., Arfa, I., Hsouna, S., Kefi, R., Jamoussi, H., Ben Ammar, S., Bahri, S., Abid, A., Benhamamouch, S., Barakat, A., & Abdelhak, S. Contribution of CDKAL1 rs7756992 and IGF2BP2 rs4402960 polymorphisms in type 2 diabetes, diabetic complications, obesity risk and hypertension in the Tunisian population. J Diabetes. 2015;7(1):102–113.

Lucas-Sánchez, M., Fadhlaoui-Zid, K. & Comas, D. The genomic analysis of current-day North African populations reveals the existence of trans-Saharan migrations with different origins and dates. Hum Genet. 2023;142(2):305–320. https://doi.org/10.1007/s00439-022-02503-3

Lucas-Sánchez, M., Serradell, J. M., & Comas, D. Population history of North Africa based on modern and ancient genomes. Hum Mol Genet. 2021;30(R1):R17–R23. https://doi.org/10.1093/hmg/ddaa261

Mathlouthi, N. E. H, Oumarou Hama, H., Belguith, I., Charfi, S., Boudawara, T., Lagier, J. C., Ammar Keskes, L., Grine, G., & Gdoura, R. Colorectal Cancer Archaeome: A Metagenomic Exploration, Tunisia. Curr Issues Mol Biol. 2023 19;45(9):7572–7581.

Mazouzi, K. Abstract A20: Genomic study of KRAS/NRAS mutations of metastatic colorectal cancer in eastern Algeria. Cancer Res. 2017;77(22_Supplement):A20–A20.

Mehemmai, C. et al. BRCA1 and BRCA2 Germline mutation analysis in hereditary breast/ovarian cancer families from the Aures region (Eastern Algeria): First report. Pathol Oncol Res. 2019;1–12.

Meroufel, D. N., Médiène-Benchekor, S., Dumont, J., Benhamamouch, S., Amouyel, P., & Brousseau, T. A study on the polymorphisms of the renin–angiotensin system pathway genes for their effect on blood pressure levels in males from Algeria. J Renin-Angiotensin-Aldosterone Syst. 2014;15(1):1–6.

Mezni, F. et al. About molecular profile of lung cancer in Tunisian patients. J Immunoassay Immunochem. 2018;39(1):99–107. https://doi.org/10.1080/15321819.2017.1407339

Mezzi, N., Messaoud, O., Mkaouar, R., Zitouna, N., Romdhane, S., Abdessalem, G., Charfeddine, C., Maazoul, F., Ouerteni, I., Hamdi, Y., Zaouak, A., Mrad, R.,

Abdelhak, S., & Romdhane, L. Spectrum of genetic diseases in Tunisia: Current situation and main milestones achieved. Genes (Basel). 2021;12(11):1820.

Mighri, N. et al. Identification of novel BRCA1 and RAD50 mutations associated with breast cancer predisposition in Tunisian patients. Front Genet. 2020 Nov 6;11:552971. https://doi.org/10.3389/fgene.2020.552971

Mimouni, A., Rouleau, E., Saulnier, P., Marouani, A., Abdelali, M. L., Filali, T., Beddar, L., Lakehal, A., Hireche, A., Boudersa, A., Aissaoui, M., Ramtani, H., Bouhedjar, K., Abdellouche, D., Oudjehih, M., Boudokhane, I., Abadi, N., & Satta, D. Association of TERT, OGG1, and CHRNA5 Polymorphisms and the Predisposition to Lung Cancer in Eastern Algeria. Pulm Med. 2020;7649038. https://doi.org/10.1155/2020/7649038

Minja, N. W., Nakagaayi, D., Aliku, T., Zhang, W., Ssinabulya, I., Nabaale, J., Amutuhaire, W., de Loizaga, S. R., Ndagire, E., Rwebembera, J., Okello, E., & Kayima, J. Cardiovascular diseases in Africa in the twenty-first century: Gaps and priorities going forward. Front Cardiovasc Med. 2022 Nov 10;9:1008335. https://doi.org/10.3389/fcvm.2022.1008335.

Moghtit, F. Z., Aberkane, M. S., Le Morvan, V., Louhibi, L., Bellot, R., Bousahba, A., Megaiz, A., Fodil, M., Mediene-Benchekor, S., Zemani-Fodil, F., Boudjema, A., Robert, J., & Saidi-Mehtar, N. No association between XRCC3 Thr241Met and XPD Lys751Gln polymorphisms and the risk of colorectal cancer in West Algerian population: A case-control study. Med Oncol. 2014;31(5):942. https://doi.org/10.1007/s12032-014-0942-3

Moufid, F. Z., Bouguenouch, L., El Bouchikhi, I., Chbani, L., Iraqui Houssaini, M., Sekal, M., Belhassan, K., Bennani, B., & Ouldim, K. The first molecular screening of MLH1 and MSH2 genes in Moroccan colorectal cancer patients shows a relatively high mutational prevalence. Genet Test Mol Biomarkers. 2018 Aug;22(8):492–497. https://doi.org/10.1089/gtmb.2018.0067. Epub 2018 Jul 25.PMID: 30044143

Mraihi, Z. et al. EGFR mutation status in Tunisian non-small-cell lung cancer patients evaluated by mutation-specific immunohistochemistry. BMC Pulm Med. 2018 Aug. 9;18(1):132. https://doi.org/10.1186/s12890-018-0706-5

Musambil, M., & Siddiqui, K. Genetics and genomics studies in type 2 diabetes: A brief review of the current scenario in the Arab region. Diabetes Metab Syndr. 2019;13(2):1629–1632.

Nahid, N. A., & Johnson, J. A. CYP2D6 pharmacogenetics and phenoconversion in personalized medicine. Expert Opin Drug Metab Toxicol, 18(11), 769–785.

Nassar, A., Zekri, A.-R. N., Kamel, M. M., Elberry, M. H., Lotfy, M. M., Seadawy, M. G., Hassan, Z. K., Soliman, H. K., Lymona, A. M., Youssef, A. S. E.-D. Frequency of pathogenic germline mutations in early and late onset familial breast cancer patients using multi-gene panel sequencing: An Egyptian study. Genes. 2023;14:106. https://doi.org/10.3390/genes14010106

National Institute of Public Health. (2005). National health survey (Epidemiological transition and health impact in North Africa) TAHINA Project.

Nejjari, C., Arharbi, M., Chentir, M. T., Boujnah, R., Kemmou, O., Megdiche, H., Boulahrouf, F., Messoussi, K., Nazek, L., & Bulatov, V. Epidemiological Trial of Hypertension in North Africa (ETHNA): An international multicentre study in Algeria, Morocco and Tunisia. J Hypertens. 2013;31(1):49–62.

Omar, A., Bahnassy, A., Gaafer, R., El Desouky, E., & Medhat, A. M. Assessing mutations in treatment response-related genes in Egyptian patients with non-small cell lung cancer. Asian Pac J Cancer Biol. 2021;7(2):125–132.

Omotoso, O. E., Teibo, J. O., Atiba, F. A. et al. Bridging the genomic data gap in Africa: Implications for global disease burdens. Global Health. 2022;18:103. https://doi.org/10.1186/s12992-022-00898-2

Ouarabi, L., Drider, D., Taminiau, B., Daube, G., Bendali, F., & Lucau-Danila, A. Vaginal microbiota: Age dynamic and ethnic particularities of Algerian women. Microb Ecol. 2021;82:1020–1029.

Pawelczyk, K. et al. Role of PD-L1 expression in non-small cell lung cancer and their prognostic significance according to clinicopathological factors and diagnostic markers. Int J Mol Sci. 2019 Feb 14;20(4):824. https://doi.org/10.3390/ijms20040824

Rabhi, H., Guermouche, B., Merzouk, H., & Merzouk, S. A. The Mediterranean diet biodiversity impact on metabolic and oxidative stress parameters in type 2 diabetes. Genet Biodiver J. 2022;6(2):87–102.

Rajman, I., Knapp, L., Morgan, T., & Masimirembwa, C. African genetic diversity: Implications for cytochrome P450-mediated drug metabolism and drug development. EBioMedicine, 17, 67–74.

Rashad, N., Eid Salem, S., Meheissen, M. A. M., Refaat, G., Mahmoud Sami, H., Temerik, A., Kordy, N., Daniel, M. A., El-Kaffas, M., Esam, M., El Mansy, H., Elkerm, Y., Abdelaziz Elsaid, A., Attia Ismail, A., Saber Abdelhalim, M., Moustafa Ahmad, L., Akram Mahmoud, M., & El Desouky, E. D. Early-onset colorectal cancer in Egypt: Pathological characters, patterns of care, and survival compared to average-age onset colorectal cancer: A retrospective multicenter study. JCO Global Oncol. 2024;10:e2300372. https://doi.org/10.1200/GO.23.00372

Riahi, A. et al. Exome sequencing and case-control analyses identify RCC1 as a candidate breast cancer susceptibility gene. Int J Cancer. 2018;142(12):2512–2517. https://doi.org/10.1002/ijc.31273

Riou, S., El Younsi, C., & Chaabouni, H. Consanguinity in the population of northern Tunisia (French). Tunis Med. 1989;67:167–172.

Romdhane, L., Mezzi, N., Hamdi, Y., El-Kamah, G., Barakat, A., & Abdelhak, S. Consanguinity and Inbreeding in health and disease in North African populations. Annu Rev Genomics Hum Genet. 2019 31;20:155–179.

Roux, A. F., Faugere, V., Le Guedard, S., Pallares-Ruiz, N., Vielle, A., Chambert, S., . . . & Claustres, M. Survey of the frequency of USH1 gene mutations in a cohort of Usher patients shows the importance of cadherin 23 and protocadherin 15 genes and establishes a detection rate of above 90%. J Med Genet. 2006;43(9):763–768.

Ruano, L., Melo, C., Silva, M. C., & Coutinho, P. The global epidemiology of hereditary ataxia and spastic paraplegia: A systematic review of prevalence studies. Neuroepidemiology. 2014;42(3):174–183. https://doi.org/10.1159/000358801

Sapkota, B., Subramanian, A., Priamvada, G., Finely, H., Blackett, P. R., Aston, C. E., & Sanghera, D. K. Association of APOE polymorphisms with diabetes and cardiometabolic risk factors and the role of APOE genotypes in response to antidiabetic therapy: Results from the AIDHS/SDS on a South Asian population.

J Diabetes Complic. 2015;29(8):1191–1197. https://doi.org/10.1016/j.jdiacomp.2015.07.025

Sarray, S., Ezzidi, I., Moussa, S., Abdennebi, H. B., & Mtiraoui, N. Association study between adiponectin gene variants, serum levels and the risk of type 2 diabetes in Tunisian women: Insights from BMI stratification. Cytokine. 2024;181:156695.

Sarray, S., Lamine, L. B., Dallel, M., Ezzidi, I., Sellami, N., Turki, A., Moustafa, A. E. E, Mtiraoui, N. Association of matrix metalloproteinase-2 gene variants with diabetic nephropathy risk. J Gene Med. 2023;25(11):e3553.

Schäfer, A., Hofmann, L., Gratchev, A., Laspe, P., Schubert, S., Schürer, A., Ohlenbusch, A., Tzvetkov, M., Hallermann, C., & Reichrath, J. Molecular genetic analysis of 16 XP-C patients from Germany: Environmental factors predominately contribute to phenotype variations. Exp Dermatol. 2013;22(1):24–29.

Sellami, N., Lamine, L. B., Turki, A., Sarray, S., Jailani, M., Al-Ansari, A. K., Ghorbel, M., Mahjoub, T., & Almawi, W. Y. Association of VEGFA variants with altered VEGF secretion and type 2 diabetes: A case-control study. Cytokine. 2018;106:29–34.

Senhaji, M. A., Abidi, O., Nadifi, S., Benchikhi, H., Khadir, K., Ben Rekaya, M., Eloualid, A., Messaoud, O., Abdelhak, S., & Barakat, A. c.1643_1644delTG XPC mutation is more frequent in Moroccan patients with xeroderma pigmentosum. Arch Dermatol Res. 2013 Jan;305(1):53–57. https://doi.org/10.1007/s00403-012-1299-0. Epub 2012 Nov 11. PMID: 23143338. https://doi.org/10.1007/s00403-012-1299-0

Serradell, J. M., Lorenzo-Salazar, J. M., Flores, C. et al. Modelling the demographic history of human North African genomes points to a recent soft split divergence between populations. Genome Biol. 2024;25:201. https://doi.org/10.1186/s13059-024-03341-4

Shawky, R. M., El-Awady, M. Y., Elsayed, S. M., & Hamadan, G. E. Consanguineous matings among Egyptian population. Egyptian J Med Hum Genet. 2011;12(2).

Sistonen, J., Sajantila, A., Lao, O., Corander, J., Barbujani, G., & Fuselli, S. CYP2D6 worldwide genetic variation shows high frequency of altered activity variants and no continental structure. Pharmacogenet Genomics. 2007;17(2):93–101.

Skhoun, H., Khattab, M., Takki Chebihi, Z., Belkhayat, A., Dakka, N., & El Baghdadi, J. B/T mixed phenotype acute leukemia with high hyperdiploidy and lineage switch to B-cell acute leukemia. Leuk Res Rep. 2022 Jan 15;17:100289. https://doi.org/10.1016/j.lrr.2022.100289. eCollection 2022.

Soliman, N. A., Morsia, D. F., & Helmy, N. A. H. Immunohistochemical expression of MMR proteins with clinicopathological correlation in colorectal cancer in Egypt. Open Access Maced J Med Sci. 2019;7(10):1608–1617. https://doi.org/10.3889/oamjms.2019.357

Soufir, N., Ged, C., Bourillon, A., Austerlitz, F., Chemin, C., Stary, A., Armier, J., Pham, D., Khadir, K., & Roume, J. A prevalent mutation with founder effect in xeroderma pigmentosum group C from north Africa. J Investig Dermatol. 2010;130(6):1537–1542.

Spierings, N., Holt, D., & MacPhee, I. A. CYP3A5 genotype had no impact on intrapatient variability of tacrolimus clearance in renal transplant recipients. Therapeutic Drug Monit 2013;35(3):328–331.

Sung, H., Ferlay, J., Siegel, R. L., Laversanne, M., Soerjomataram, I., Jemal, A., & Bray, F. Global cancer statistics 2020: GLOBOCAN estimates of incidence and mortality worldwide for 36 cancers in 185 countries. CA Cancer J Clin. 2021;71(3):209–249. https://doi.org/10.3322/caac.21660

Tabebi, M., Khabou, B., Boukadi, H., Ben Hamad, M., Ben Rhouma, B., Tounsi, S., Maalej, A., Kamoun, H., Keskes-Ammar, L., Abid, M., Mnif, M., & Fakhfakh, F. Association study of apoptosis gene polymorphisms in mitochondrial diabetes: A potential role in the pathogenicity of MD. Gene. 2018 10;639:18–26.

Tabebi, M., Safi, W., Felhi, R., Alila Fersi, O., Keskes, L., Abid, M., Mnif, M., & Fakhfakh, F. The first concurrent detection of mitochondrial DNA m.3243A>G mutation, deletion, and depletion in a family with mitochondrial diabetes. Mol Genet Genomic Med. 2020;8(7):e1292.

Tadmouri, G. O., Nair, P., Obeid, T., Al Ali, M. T., Al Khaja, N., Hamamy, H. A. Consanguinity and reproductive health among Arabs. Reprod Health. 2009;6(1):17. https://doi.org/10.1186/1742-4755-6-17

Takki Chebihi, Z., Belkhayat, A., Chadli, E., Hessissen, L., El Khorassani, M., El Kababri, M., Kili, A., Khattab, M., Bakri, Y., & Dakka, N. The rare translocation t(14;21)(q11;q22) detected in a Moroccan patient with T-cell acute lymphoblastic leukemia. Leuk Res Rep. 2018 Nov 22;11:1–4.

Takki Chebihi, Z., Belkhayat, A., Chadli, E., Hilal, L., Skhoun, H., Hessissen, L., El Khorassani, L., El Kababri, M., Kili, A., Khattab, M., Bakri, Y., & Dakka, N. Cytogenetic profile of Moroccan pediatric acute lymphoblastic leukemia: Analysis of 155 cases with a review of the literature. Clin Lymphoma Myeloma Leuk. 2018 Jun;18(6):e241–e248. https://doi.org/10.1016/j.clml.2018.04.004. Epub 2018 Apr 25.

Talbi, J. et al. Etude de la consanguinité dans la population marocaine. Impact sur le profil de la santé. Anthropologie. 2007;1–11.

Tamura, D., DiGiovanna, J. J., & Kraemer, K. H. Founder mutations in xeroderma pigmentosum. J Investig Dermatol. 2010;130(6):1491–1493.

Tazir, M., Ali-Pacha, L., M'zahem, A., Delaunoy, J., Fritsch, M., Nouioua, S., Benhassine, T., Assami, S., Vallat, J.-M., & Hamri, A. Ataxia with oculomotor apraxia type 2: A clinical and genetic study of 19 patients. J Neurol Sci. 2009;278(1–2):77–81.

Trhanint, S., Bouguenouch, L., Abourazzak, S., El Ouahabi, H., Latrech, H., Benyakhlef, S., Bennani, B., El Bouchikhi, I., Moufid, F. Z., Ouldim, K., El Ghadraoui, L., & Maazouzi, N. Molecular and clinical assessment of maturity-onset diabetes of the young revealed low mutational rate in Moroccan families. Int J Pediatr Adolesc Med. 2022 Jun;9(2):98–103. https://doi.org/10.1016/j.ijpam.2021.03.006. Epub 2021 Mar 22.

Triki-Fendri, S. et al. Genetic structure of the Kuwaiti population revealed by paternal lineages. Am J Hum Biol. 2016;28(2):203–212. https://doi.org/10.1002/ajhb.22773

Tuminello, S., Sikavi, D., Veluswamy, R., Gamarra, C., Lieberman-Cribbin, W., Flores, R., & Taioli, E. PD-L1 as a prognostic biomarker in surgically resectable non-small cell lung cancer: A meta-analysis. Transl Lung Cancer Res. 2020;9(4):1343–1360. https://doi.org/10.21037/tlcr-19-638

Uhrhammer, N., Abdelouahab, A., Lafarge, L., Feillel, V., Ben Dib, A., Bignon, & Y. J. BRCA1 mutations in Algerian breast cancer patients: High frequency in young, sporadic cases. Int J Med Sci. 2008 Jul 8;5(4):197–202. https://doi.org/10.7150/ijms.5.197.

Vilà-Valls, L., Abdeli, A., Lucas-Sánchez, M. et al. Understanding the genomic heterogeneity of North African Imazighen: From broad to microgeographical perspectives. Sci Rep. 2024;14:9979. https://doi.org/10.1038/s41598-024-60568-8

World Health Organization. (2023). Noncommunicable diseases. World Health Organization. Retrieved 1 November 2024, from https://www.who.int/news-room/fact-sheets/detail/noncommunicable-diseases

Yako, Y. Y., Guewo-Fokeng, M., Balti, E. V., Bouatia-Naji, N., Matsha, T. E., Sobngwi, E., Erasmus, R. T., Echouffo-Tcheugui, J. B., Kengne, A. P. Genetic risk of type 2 diabetes in populations of the African continent: A systematic review and meta-analyses. Diabetes Res Clin Pract. 2016;114:136–150.

Zalloua, P. A., Platt, D. E., El Sibai, M., Khalife, J., Makhoul, N., Haber, M., Xue, Y., Izaabel, H., Bosch, E., Adams, S. M., Arroyo, E., López-Parra, A. M., Aler, M., Picornell, A., Ramon, M., Jobling, M. A., Comas, D., Bertranpetit, J., Wells, R. S., Tyler-Smith, C., & Genographic Consortium. Identifying genetic traces of historical expansions: Phoenician footprints in the Mediterranean. Am J Hum Genet. 2008 Nov;83(5):633–642. https://doi.org/10.1016/j.ajhg.2008.10.012. Epub 2008 Oct 30. PMID: 18976729; PMCID: PMC2668035

Zeghbib, S., Somogyi, B. A., Zana, B., Kemenesi, G., Herczeg, R., Derrar, F., & Jakab, F. The Algerian chapter of SARS-CoV-2 pandemic: An evolutionary, genetic, and epidemiological prospect. Viruses. 2021;13(8):1525. https://doi.org/10.3390/v13081525

Zhang, H., Klareskog, L., Matussek, A., Pfister, S. M., & Benson, M. Translating genomic medicine to the clinic: Challenges and opportunities. Genome Med. 2019;11(1):9. https://doi.org/10.1186/s13073-019-0622-1

Zhou, Y., & Lauschke, V. M. The genetic landscape of major drug metabolizing cytochrome P450 genes—an updated analysis of population-scale sequencing data. Pharmacogenomics J. 2022;22(5):284–293.

Chapter 11

Genomics and Infectious Diseases in Africa

Gerald Mboowa[1*]

[1]Africa Centres for Disease Control and Prevention, Addis Ababa, Ethiopia

11.1 INTRODUCTION

In 2015, the United Nations General Assembly adopted the Sustainable Development Goals (SDGs), committing to end the epidemics of HIV/AIDS, tuberculosis (TB), malaria, and other communicable diseases by 2030 [1]. Achieving this ambitious target requires innovative and data-driven approaches—one of which is public health genomics. This field enables researchers and practitioners to investigate how genetic factors influence disease progression across populations, generating insights that can inform health promotion, disease prevention, and the precise design and delivery of population-based interventions [2].

A central tool in public health genomics is the genome-wide association study (GWAS), which scans the genome to identify genetic variants associated with specific disease outcomes. By uncovering these associations, GWAS contributes to a deeper understanding of disease progression, treatment response, and prevention strategies. To date, GWAS has identified thousands of variants linked to complex traits, revealing the highly polygenic architecture of many common diseases—including HIV/AIDS, TB, and malaria.

This expanding body of genetic knowledge holds significant promise for improving clinical outcome prediction based on individual genetic profiles. In recent years, both GWAS and candidate gene studies have increasingly been used to examine how host genetic variation influences susceptibility or resistance to infectious diseases. These efforts have led to the discovery of important associations with malaria [3–5], HIV infection [6], and TB [7], particularly involving immune response genes. Together, these findings highlight the complex, polygenic nature of disease susceptibility and progression across diverse populations [6].

Although individual genetic variants typically contribute only modestly to variation in disease outcomes, their combined effects can offer a more

***Corresponding author:** gmboowa@gmail.com (Gerald Mboowa)

 DOI: 10.1201/9781003436584-11

comprehensive view of an individual's genetic risk. By summing the number of risk alleles—each weighted by its estimated effect size—researchers can calculate a cumulative genetic score that reflects a person's likelihood of experiencing specific outcomes, such as disease susceptibility, protection, or progression. When applied to binary disease outcomes, these cumulative scores are known as polygenic risk scores (PRSs); for broader quantitative traits, they are referred to more generally as polygenic scores [8]. PRSs capture an individual's genetic predisposition to complex diseases and are emerging as powerful tools in precision medicine [9]. However, their generalizability remains limited: Most PRSs have been developed using data from populations of predominantly European ancestry, raising concerns about their accuracy and equity in underrepresented populations—particularly those of African ancestry—thus exacerbating existing disparities in health research and healthcare [8]. To address this limitation, researchers are developing multi-ancestry PRS methods—such as CT-SLEB—that integrate genomics data across diverse populations. These approaches aim to improve both the accuracy and the fairness of genetic risk prediction, ensuring that the benefits of precision medicine are equitably distributed across global populations [10].

Pathogens have long exerted significant selective pressure on the human genome, particularly influencing genes involved in immune responses [11, 12]. This evolutionary pressure is especially intense for pathogens that impair reproductive fitness through morbidity or mortality. Over time, such pressures can alter the frequency of genetic variants that confer resistance or susceptibility to infection. This process is most evident for long-standing human pathogens—such as *Plasmodium falciparum* (malaria), *Mycobacterium tuberculosis* (TB), *Mycobacterium leprae* (leprosy) [13], and HIV—where prolonged exposure has left detectable signatures of natural selection in host populations.

These evolutionary insights are essential for understanding population-level differences in disease susceptibility and for guiding precision public health interventions. Ancient diseases like malaria, which dates back at least to 2,700 BC [14], and TB, which has afflicted humans for 15,000 to 20,000 years [15], have imposed sustained selective pressure on the human genome over millennia. In contrast, HIV/AIDS—a comparatively recent disease identified only around 40 years ago [16]—has had limited evolutionary time to influence host genetic variation at the population level. This difference in evolutionary timelines helps explain why genomic adaptations to older pathogens are more pronounced and widespread.

The three major infectious diseases—HIV, tuberculosis (TB), and malaria—exert a unique and interrelated influence on the genetic architecture of human populations, particularly in Africa, where the burden of all three is highest. Genetic adaptations that evolved in response to one pathogen can influence susceptibility or resistance to others, with important implications for disease dynamics and control strategies. For instance, genetic variants associated with resistance to TB and malaria may also impact vulnerability to emerging or re-emerging infections.

A well-studied example is the Duffy antigen receptor for chemokines (DARC), a red blood cell surface protein that serves as the entry receptor for *Plasmodium vivax*. The DARC –46C/C genotype, common in African populations, produces a DARC-negative phenotype that confers innate resistance to *P. vivax* malaria. Interestingly, this same variant influences circulating levels of chemokines such as CCL5/RANTES, which are involved in immune responses and have been shown to suppress HIV-1 replication. This illustrates how long-standing evolutionary pressures from endemic pathogens can shape host immunity and may result in cross-protective benefits—or potential trade-offs—against other infections [17].

Although HIV-1 is a relatively recent pathogen compared to malaria and TB, it too interacts with host genetic factors such as DARC. HIV-1 can bind to red blood cells via DARC, facilitating trans-infection of target immune cells. Among African Americans, the DARC –46C/C genotype has been associated with a 40% increased risk of HIV-1 acquisition. Extrapolating these findings suggests that up to 11% of the HIV-1 burden in Africa could be attributable to this genotype. Intriguingly, once infected, individuals with the DARC-negative phenotype often experience slower disease progression. Moreover, the effect of the disease-accelerating CCL5 –471A/A polymorphism appears to be modulated by DARC expression—it is observed only in DARC-positive individuals, not in those who are DARC negative.

Together, these findings underscore the complex interplay between host genetic factors and HIV-1 pathogenesis and highlight how genetic adaptations driven by ancient pathogens may influence susceptibility, progression, and immune responses to contemporary infectious diseases in African populations [17].

People living with HIV are 20 to 30 times more likely to develop active TB following infection with *M. tuberculosis*, highlighting the profound impact of HIV-induced immunosuppression. This increased susceptibility is primarily driven by the depletion of $CD4^+$ T cells, which are essential for mounting an effective immune response against TB. Since the emergence of the HIV epidemic, sub-Saharan Africa (SSA) has experienced a sharp rise in TB incidence, underscoring the synergistic interaction between these two infections. HIV compromises host immunity, enabling the reactivation of latent TB infections and accelerating progression to active disease, thereby increasing the likelihood of transmission. Conversely, TB can exacerbate HIV disease progression by promoting chronic immune activation.

The evolutionary success—or fitness—of pathogens such as HIV and *M. tuberculosis* is closely tied to their ability to transmit between hosts. This co-epidemic dynamic in high-burden regions like SSA calls for integrated approaches to surveillance, diagnosis, and management. Genomic surveillance can provide critical insights into transmission dynamics and inform public health strategies tailored to co-infected populations [18, 19].

While HIV infection is a well-established risk factor for active TB, recent findings suggest a more complex interaction. A study conducted in SSA revealed a counterintuitive observation: HIV/TB co-infection is associated with reduced transmissibility of *M. tuberculosis*. This may be due to lower bacterial loads in individuals with advanced HIV infection and atypical clinical presentations that generate fewer infectious aerosols. These findings suggest that although HIV facilitates disease progression in individuals, it may also reduce the efficiency of TB transmission at the population level [20].

Supporting this, a recent study using a phylodynamic model applied to *M. tuberculosis* genomic sequences from HIV-positive and HIV-negative individuals across multiple African countries found that HIV/TB co-infection was associated with a reduced effective reproductive number (R_0) for TB. Stratification by $CD4^+$ T-cell count confirmed these trends, with HIV-positive TB patients generating significantly fewer secondary infections. These patients functioned more as epidemiological sinks, in contrast to HIV-negative individuals, who acted as primary drivers of TB transmission within communities.

Together, these insights underscore the complex and sometimes paradoxical effects of HIV on TB transmission dynamics. They highlight the need for precision public health strategies that account for host–pathogen interactions, co-infection biology, and the use of microbial genomics to guide interventions in high-burden settings [20].

Elucidating host genetic determinants of susceptibility and protection in infectious diseases—including emerging and re-emerging pathogens—is essential for advancing precision public health, particularly in Africa, where the infectious disease burden is highest. Understanding how genomic variants influence clinical outcomes can inform the design and evaluation of targeted interventions, such as vaccines and therapeutics, by helping to interpret diverse clinical endpoints across genetically heterogeneous populations. The disproportionate impact of HIV/AIDS, tuberculosis, and malaria in Africa underscores the urgency of applying advances in infectious disease genomics to the continent's highly diverse genetic landscape.

Importantly, discoveries from African populations may also yield insights with global relevance, particularly for less genetically diverse populations. For instance, the CCR5Δ32 variant—known to confer resistance to HIV infection in individuals homozygous for the mutation (Δ32/Δ32)—was first identified in European populations [21] and is rare among native African populations [22]. This example highlights the importance of inclusive genomics research in uncovering population specific mechanisms of disease resistance, which can inform both regional and global prevention and treatment strategies. The identification of CCR5Δ32 has catalyzed several translational advances in HIV therapeutics, including the development of CCR5 antagonists such as Maraviroc and the dual CCR5/CCR2 inhibitor

Cenicriviroc [23], both of which show promise in blocking HIV entry into host cells.

Furthermore, allogeneic hematopoietic stem cell transplantation from donors homozygous for CCR5Δ32 has provided proof-of-concept for a functional HIV-1 cure in select clinical cases [24]. In contrast, for malaria and TB, no similarly transformative host genetic markers have yet been identified and translated into broadly applicable clinical interventions. This gap underscores the need for continued research focused on population-specific genetic determinants of disease resistance or susceptibility and their clinical utility.

The anticipated reduction in the cost of next-generation sequencing (NGS) to approximately $100 per genome [25, 26], combined with growing investments in sustainable public health genomics, international collaborations, and capacity-building initiatives in Africa, presents a transformative opportunity. These advancements will enable large-scale identification of genetic markers that confer protection, elevate disease risk, or influence progression in African populations. Conducting genetic studies in those most affected by HIV, TB, and malaria is critical for identifying key genes and biological pathways that mediate host–pathogen interactions. Such knowledge holds immense potential to accelerate the development of affordable, targeted interventions that can reduce disease burden and mortality across the region.

11.2 HOST GENOMIC FACTORS ASSOCIATED WITH TUBERCULOSIS

In 2021, an estimated 10.6 million people worldwide developed TB, resulting in 1.6 million deaths—including 187,000 among people living with HIV [27]. In the post–COVID-19 era, TB has re-emerged as the leading cause of death from a single infectious agent globally, surpassing HIV/AIDS. The African region bears a disproportionate burden, accounting for 25% of all TB-related deaths [28]. Ending the TB epidemic by 2030 remains a key target under the United Nations SDGs [27].

Host genetic factors have been increasingly recognized as important determinants of susceptibility and resistance to TB. Numerous studies have documented associations between genetic polymorphisms and TB risk, including variants in genes such as *SLC11A1* (*NRAMP-1*), vitamin D receptor, MHC genes, *TLRs*, *NOD2*, *IL-1*, *IL-10*, *CCL2/MCP-1*, *SP110*, *P2X7*, and *IFN-γ* [29]. These genes are involved in pathways critical to innate and adaptive immunity, antigen processing, and macrophage function.

A meta-analysis of 36 peer-reviewed studies reported significant associations between SLC11A1 polymorphisms and increased TB risk, with pooled odds ratios of 1.35 (95% CI: 1.17–1.54), 1.25 (95% CI: 1.04–1.50),

1.23 (95% CI: 1.04–1.44), and 1.31 (95% CI: 1.08–1.59) for the 3′ UTR, D543N, INT4, and 5′ (GT)n variants, respectively [30]. These associations remained consistent across population subgroups, reinforcing the hypothesis that SLC11A1 plays a key role in host defense against TB.

In a landmark GWAS, researchers analyzed TB cohorts from Ghana and The Gambia and, after replication in 11,425 individuals, identified rs4331426—a single nucleotide polymorphism (SNP) located in a gene-poor region on chromosome 18q11.2—as a susceptibility locus for TB in African populations (combined $P = 6.8 \times 10^{-9}$; OR = 1.19; 95% CI: 1.13–1.27) [31]. This same group also identified a –362C variant in the MCP-1 promoter, which was associated with protection from pulmonary TB in a Ghanaian cohort [32].

Further evidence for the role of host genetics in TB susceptibility comes from analysis of ASAP1 (ArfGAP with SH3 Domain, Ankyrin Repeat, and PH Domain 1), a gene highly expressed in dendritic cells (DCs). In cohorts from Ghana (971 cases, 988 controls) and The Gambia (1,306 cases, 1,372 controls), seven ASAP1 SNPs were analyzed, revealing associations with TB risk [33]. Notably, the rs10956514 SNP was linked to reductions in ASAP1 expression following *M. tuberculosis* infection. ASAP1 plays a critical role in actin cytoskeleton remodeling and the formation of podosomes essential for DC migration. DCs depleted of ASAP1 exhibited impaired matrix degradation and reduced mobility, suggesting a novel mechanism by which impaired DC function contributes to TB susceptibility.

These findings highlight the importance of continued investigation into host genomic factors that influence TB resistance, susceptibility, and disease progression—particularly in African populations that bear a high burden of disease. Identifying population-specific genetic variants not only provides insight into the biological mechanisms underlying host–pathogen interactions but also supports the development of targeted interventions. Table 11.1 summarizes key findings from selected studies that have explored host genomic determinants of TB in African cohorts.

11.3 HOST GENOMIC FACTORS ASSOCIATED WITH HIV IN AFRICA

HIV susceptibility and disease progression result from a complex interplay of viral characteristics, environmental exposures, host genetic factors [71], and epigenetic modifications. An early theoretical model proposed in 2001 explored how HIV-resistant population haplotypes could become enriched in a susceptible population—such as in SSA—through delayed progression to AIDS. The model suggested that individuals with genetic variants associated with slower disease progression would experience extended survival and increased reproductive success, thereby promoting natural selection for

Table 11.1 Selected Host Genomic Factors Associated with Tuberculosis in Africa

Main Description/Outcome of the Study	*Study Design*	*Number of Individuals Analyzed*	*Infectious Condition and Associated Phenotype*	*Gene(s) Variants Implicated*	*Population*	*Reference*
Variations in the *NRAMP1* gene and susceptibility to tuberculosis in West Africans.	Case–control	827	Susceptibility to tuberculosis	*NRAMP1*	West Africans	[34]
Polymorphisms in *TICAM2* and *IL1B* are associated with tuberculosis.	Family studies	835	Susceptibility to pediatric tuberculosis	*TICAM2* and *IL1B*	Ugandans	[35]
Investigating the role of gene–gene interactions in tuberculosis susceptibility.	Case–control	1,476	Susceptibility to tuberculosis	*NRG1-NRG3, GRIK1–GRIK3* and *IL23R–ATG4C*	South Africans	[36]
Genetic analysis of tuberculosis susceptibility variants in Ghana reveals Candidate protective loci in *SORBS2* and *SCL11A1* genes.	Case–control	426	Protective against tuberculosis	*SORBS2* and *SCL11A1*	Ghanaians	[37]
Polymorphisms in the *RANTES* gene increase susceptibility to active tuberculosis in Tunisia.	Case–control	373	Susceptibility to tuberculosis	*RANTES*	Tunisians	[38]
Common polymorphisms in *TLR4* gene associated with susceptibility to pulmonary tuberculosis in the Sudanese.	Case–control	602	Susceptibility to tuberculosis	*TLR4*	Sudanese	[39]
Allelic variation in *BTNL2* and susceptibility to tuberculosis in a South African population.	Case–control	914	Susceptibility to tuberculosis	*BTNL2*	South Africans	[40]

(*Continued*)

Table 11.1 (Continued)

Main Description/Outcome of the Study	*Study Design*	*Number of Individuals Analyzed*	*Infectious Condition and Associated Phenotype*	*Gene(s) Variants Implicated*	*Population*	*Reference*
Putative pathogen-selected polymorphisms in the *PKLR* gene are associated with mycobacteria susceptibility in Brazilian and African populations.	Case–control	471	Susceptibility to tuberculosis	*PKLR*	Mozambicans	[41]
The *CYP7A1* gene rs3808607 variant is associated with susceptibility to tuberculosis in a Moroccan population studied.	Case–control	324	Susceptibility to tuberculosis	*CYP7A1*	Moroccans	[42]
A regulatory variant in the *C1Q* gene cluster is associated with tuberculosis susceptibility and C1qA plasma levels in a South African population.	Case–control	904	Susceptibility to tuberculosis	*C1Q*	South Africans	[43]
Investigation of chromosome 17 candidate genes in susceptibility to tuberculosis in a South African population.	Case–control	800	Susceptibility to tuberculosis	*NOS2A promoter*	South Africans	[44]
Vitamin D receptor polymorphisms and susceptibility to tuberculosis in West Africa: A case–control and family study	Case–control and family study	1,521	Susceptibility to tuberculosis	*VDR haplotypes*	West Africans	[45]
Polymorphisms in *CARD8* and *NLRP3* are associated with extrapulmonary tuberculosis and poor clinical outcome in active tuberculosis in Ethiopia.	Case–control	3,180	Susceptibility to extrapulmonary tuberculosis	*CARD8 and NLRP3*	Ethiopians	[46]

(*Continued*)

Table 11.1 (Continued)

Main Description/Outcome of the Study	*Study Design*	*Number of Individuals Analyzed*	*Infectious Condition and Associated Phenotype*	*Gene(s) Variants Implicated*	*Population*	*Reference*
Age- and gender-specific effects on *NRAMP1* gene polymorphisms and risk of the development of active tuberculosis in Tunisian populations.	Case–control	373	Susceptibility to tuberculosis	*NRAMP1 3'-UTR, D543N, P2X7 1513A/C, IFN-γ +874T/A*	Tunisians	[47]
Variants in toll-like receptors 2 and 9 influence susceptibility to pulmonary tuberculosis in Caucasians, African Americans, and West Africans.	Case–control	667	Susceptibility to tuberculosis	*TLR2* and *TLR9*	West Africans	[48]
Deciphering genetic susceptibility to tuberculous meningitis.	Case–control	903	Tuberculous meningitis susceptibility	*CCP110*	South Africans	[49]
Association of *HLA-DR, -DQ*, and *vitamin D receptor* alleles and haplotypes with tuberculosis in the Venda of South Africa.	Case–control	212	Susceptibility to tuberculosis	*HLA-DRB1* and *−DQB1*	South Africans	[50]
Interleukin-18 and interferon-γ single nucleotide polymorphisms in Egyptian patients with tuberculosis.	Case–control	211	Susceptibility to tuberculosis	*INF-γ+874*	Egyptians	[51]
Genetic diversity of *TLR2, TLR4*, and *VDR* loci and pulmonary tuberculosis in Moroccan patients.	Case–control	546	Protection against pulmonary tuberculosis	*TLR2* and *TLR4*	Moroccans	[52]
New variant identified in major susceptibility locus to tuberculosis on chromosomal region 8q12-q13 in a Moroccan population: A case–control study.	Case–control	269	Susceptibility to tuberculosis	*NSMAF*	Moroccans	[53]

(Continued)

Table 11.1 (Continued)

Main Description/Outcome of the Study	*Study Design*	*Number of Individuals Analyzed*	*Infectious Condition and Associated Phenotype*	*Gene(s) Variants Implicated*	*Population*	*Reference*
CLEC4E (Mincle) genetic variation associates with pulmonary tuberculosis in Guinea-Bissau (West Africa).	Case–control	611	Susceptibility to tuberculosis	*CLEC4E*	Bissau-Guineans	[54]
TLR1 variant H305L associated with protection from pulmonary tuberculosis.	Case–control	4,588	Protection against pulmonary tuberculosis	*TLR1 Variant H305L*	Ghanaians	[55]
Genetic susceptibility to tuberculosis associated with cathepsin Z haplotype in a Ugandan household contact study.	Case–control	565	Susceptibility to tuberculosis	*Cathepsin X*	Ugandans	[56]
Low vitamin-D levels combined with *PKP3-SIGIRR-TMEM16J* Host Variants is associated with tuberculosis and death in HIV-infected and -exposed infants.	Case–control	466	Susceptibility to tuberculosis	*PKP3-SIGIRR-TMEM16J*	South Africans	[57]
Influence of *interferon-gamma receptor 1* gene polymorphisms on the susceptibility to pulmonary tuberculosis among a Sudanese population studied.	Case–control	150	Susceptibility to tuberculosis	*IFN-γR1*	Sudanese	[58]
Association of *PTPN22* gene functional variants with development of pulmonary tuberculosis in a Moroccan population studied.	Case–control	278	Susceptibility to tuberculosis	*PTPN22*	Moroccans	[59]

(*Continued*)

Table 11.1 (Continued)

Main Description/Outcome of the Study	*Study Design*	*Number of Individuals Analyzed*	*Infectious Condition and Associated Phenotype*	*Gene(s) Variants Implicated*	*Population*	*Reference*
Age-dependent association between pulmonary tuberculosis and common *TOX* variants in the 8q12–13 linkage region.	Case–control and family study	1,210	Susceptibility to tuberculosis	*TOX variants*	Moroccans and Madagascan	[60]
Association of toll-like receptors with susceptibility to tuberculosis suggests sex-specific effects of *TLR8* polymorphisms.	Case–control	1,216	Susceptibility to tuberculosis	*TLR8*	South Africans	[61]
A locus at 5q33.3 confers resistance to tuberculosis in highly susceptible individuals.	Case–control	581	Resistance to tuberculosis	*IL12B*	Ugandans and Tanzanians	[62]
MCP-1 promoter variant −362C associated with protection from pulmonary tuberculosis in Ghana, West Africa.	Case–control and family study	4,632	Resistance to tuberculosis	*MCP-1 promoter variant −362C*	Ghanaians	[32]
Interferon-γ (+874 T/A) and *interleukin-10* (−1082 G/A) gene polymorphisms are associated with active tuberculosis in the Algerian population of Oran's city.	Case–control	245	Susceptibility to tuberculosis	*IFNG+874 TA and IL10–1082G**	Algerian	[63]
Association of *IL-10* and *CCL5* single nucleotide polymorphisms with tuberculosis in the Sudanese population.	Case–control	438	Susceptibility to tuberculosis	*IL-10 and CCL5*	Sudanese	[64]

(Continued)

Table 11.1 (Continued)

Main Description/Outcome of the Study	*Study Design*	*Number of Individuals Analyzed*	*Infectious Condition and Associated Phenotype*	*Gene(s) Variants Implicated*	*Population*	*Reference*
ALOX5 variants associated with susceptibility to human pulmonary tuberculosis.	Case–control	4,185	Susceptibility to tuberculosis	*ALOX5 variants*	Ghanaians	[65]
Association between human leukocyte antigen class II and pulmonary tuberculosis due to *Mycobacterium tuberculosis* in Uganda.	Case–control	85	Resistance to tuberculosis	*HLA-DQB1*03:03*	Ugandans	[66]
A functional haplotype in the 3′ untranslated region of *TNFRSF1B* is associated with tuberculosis in two African populations.	Case–control	2,709	Susceptibility to tuberculosis	*3'untranslated region ofTNFRSF1B*	Ghanaians and South Africans	[67]
The *MCP-1* (*CCL2*) −2518 GG genotype is associated with protection against pulmonary tuberculosis in Moroccan patients.	Case–control	541	Protection against tuberculosis	*MCP-1 (CCL2) −2518 GG*	Moroccans	[68]
Association study of genes controlling IL-12-dependent IFN-γ immunity: *STAT4* alleles increase risk of pulmonary tuberculosis in Morocco.	Case–control and family study	1,563	Susceptibility to tuberculosis	*STAT4 alleles*	Moroccans	[69]
IL-10R1 S138G loss-of-function polymorphism is associated with extrapulmonary tuberculosis risk development in Tunisia.	Case–control	373	Susceptibility to tuberculosis	*IL-10R1 S138G*	Tunisians	[70]

these protective haplotypes [72]. The authors argued that the high prevalence of HIV in some African countries created conditions conducive to strong selective pressure on host genetic variants that modulate the interval between infection and AIDS onset.

Two major public health milestones have since altered this evolutionary landscape. First, the widespread rollout of antiretroviral therapy (ART) in 2003 led to a tenfold increase in treatment access across low- and middle-income countries by 2008 [73, 74]. Second, the World Health Organization's 2015 recommendation of HIV universal test and treat (UTT) marked a pivotal shift in care, advocating for immediate ART initiation for all individuals diagnosed with HIV, regardless of CD4 count or clinical staging [75]. This approach has improved health outcomes by enabling early treatment and has contributed to reduced HIV incidence through population-level viral suppression.

However, the success of UTT has also introduced new challenges for research. Early ART initiation, while beneficial for individuals, limits opportunities to observe the natural course of HIV infection, making it increasingly difficult to identify long-term non-progressors and study genetic markers of delayed AIDS progression. This presents a need for innovative study designs that can account for early treatment effects while still capturing host genomic contributions to disease outcomes.

Addressing this need is especially critical in SSA, which accounts for over 70% (23.5 million) of the global population of people living with HIV [76]. Within SSA, eastern and southern Africa continue to experience the highest global rates of HIV-1 prevalence and incidence. Notably, SSA populations harbor the greatest levels of genetic and phenotypic diversity worldwide, offering unparalleled opportunities to identify novel genomic variants associated with HIV susceptibility, disease progression, and treatment response [77].

Despite more than 40 years since the discovery of HIV [78] no effective protective vaccine has been successfully developed. While previous GWASs have identified key genes involved in HIV control among populations of European ancestry, individuals of African ancestry remain significantly underrepresented in human genomics research, limiting the generalizability and equity of these findings [79]. Host alleles and haplotypes such as CCR5Δ32 and CCR5m303 [80], which have been identified in European populations as conferring resistance to HIV-1 infection, are rarely detected or completely absent in African populations. Nevertheless, there are documented cases of native African individuals who appear to have remained uninfected despite repeated exposure to HIV-1, suggesting the existence of alternative, yet-to-be-identified protective genetic variants within these populations [81–83].

Interestingly, nearly 30% of children in Southern Africa are classified as HIV-exposed but uninfected (HEU)—meaning they were born to HIV-positive mothers but did not acquire the virus themselves [84]. This large and growing population underscores the critical need to identify genetic and immunological markers associated with both resistance and susceptibility to HIV, particularly among vulnerable groups such as children. South Africa alone is estimated to have approximately 3.5 million HEU children—the highest number globally—highlighting the urgency of understanding the biological mechanisms underpinning [their resistance [85]]. However, the genetic basis for this outcome remains poorly defined.

A population-based study in Uganda examining CCR5 gene variants found no association between non-Δ32 polymorphisms and HIV/AIDS disease risk [86]. Nevertheless, follow-up research within the same cohort identified a subset of individuals who naturally controlled HIV despite being ART naive. These individuals maintained high $CD4^{+}$ T cell counts, partially due to significantly reduced surface expression of CCR5 on their $CD4^{+}$ T cells [87]. In addition, certain human leukocyte antigen (HLA) alleles in African individuals have been linked to delayed progression from HIV infection to AIDS [88]. The HLA complex is essential for immune defense, facilitating the recognition and presentation of viral antigens, including those from HIV to cytotoxic T lymphocytes. Together, these findings underscore the importance of host genetic variation in shaping HIV disease outcomes in African populations.

A 2023 GWAS involving over 3,800 people living with HIV (PLWH) from genetically diverse African populations identified a novel locus on chromosome 1 significantly associated with reduced HIV viral load. Notably, this association appeared specific to individuals of African ancestry, reinforcing the value of inclusive genomic studies in uncovering ancestry-specific protective variants [89]. This discovery has implications for therapeutic development and supports global efforts—such as the UNAIDS 2030 target—to end the HIV epidemic.

The natural control of HIV is a complex phenomenon driven in part by host genetics, which influence both susceptibility to infection and disease progression. Identifying these genetic determinants, particularly in underrepresented populations, is essential for developing equitable and effective HIV interventions [90]. Numerous host genes have been studied across African populations (summarized in Table 11.2), with growing evidence that epistatic interactions—interplay between different genetic loci—may contribute to variability in resistance, susceptibility, and clinical outcomes [91].

Despite the high HIV burden in SSA, the region offers unique opportunities to uncover therapeutic targets. Of particular interest are rare individuals who, despite repeated exposure to replication-competent HIV-1, remain uninfected without ART. Understanding the genetic and immunological mechanisms that confer protection in these individuals is crucial to advancing preventive strategies.

Clinically, HIV disease progression varies widely among PLWH in the absence of ART. Based on progression patterns, individuals are typically categorized into rapid progressors (RPs), who develop AIDS within 3 years of infection; typical progressors (TPs), who progress within 3 to 10 years; and long-term non-progressors (LTNPs), who remain asymptomatic for more than 10 years without treatment. Epidemiological data suggest that approximately 70–80% of PLWH are TPs, 10–15% are RPs, and fewer than 5% qualify as LTNPs [92, 93] (Figure 11.1). Understanding the genetic underpinnings of these phenotypic categories remains a key frontier in HIV research.

Figure 11.1 illustrates the variation in disease progression among individuals infected with HIV following exposure to replication-competent virus. On the left, the schematic highlights the outcomes of HIV exposure: While many individuals become infected, a subset remains uninfected despite repeated exposure, likely due to innate or acquired resistance mechanisms. Once infected, individuals show a spectrum of clinical phenotypes based on the time to HIV/AIDS progression. The bell-shaped distribution on the right side of the figure represents the frequency of different HIV disease progression phenotypes among PLWH:

- *Rapid progressors* (green) represent a minority who progress to AIDS within 3 years of infection.
- *Typical progressors* (blue) constitute the majority, showing disease progression within 3–10 years.
- *Long-term non-progressors* (red) are rare individuals who maintain stable CD4+ T cell counts and low viral loads without ART for over 10 years.

The x-axis denotes the estimated years from infection to AIDS (a proxy for disease progression), while the y-axis shows the relative frequency of each phenotype among PLWH. This distribution underscores the heterogeneity in host response to HIV and highlights the value of studying extreme phenotypes—such as long-term non-progressors and individuals resistant to infection—for identifying protective genetic variants and immune mechanisms. These insights are critical for informing HIV cure and vaccine research. GWAS and host–pathogen interaction research often focus on these extreme phenotypes to identify protective or susceptibility-related genetic markers. However, such designs face several challenges.

First, identifying well-characterized cohorts—particularly HIV-exposed uninfected individuals or those with extreme phenotypes—is logistically and ethically complex. Establishing the precise timing of HIV exposure, infection, and progression remains difficult in routine settings. Second, the widespread implementation of UTT policies across SSA, where ART is offered regardless of CD4 count, has further complicated phenotypic stratification, as early treatment initiation alters natural disease trajectories. These constraints underscore the need for innovative study designs and longitudinal

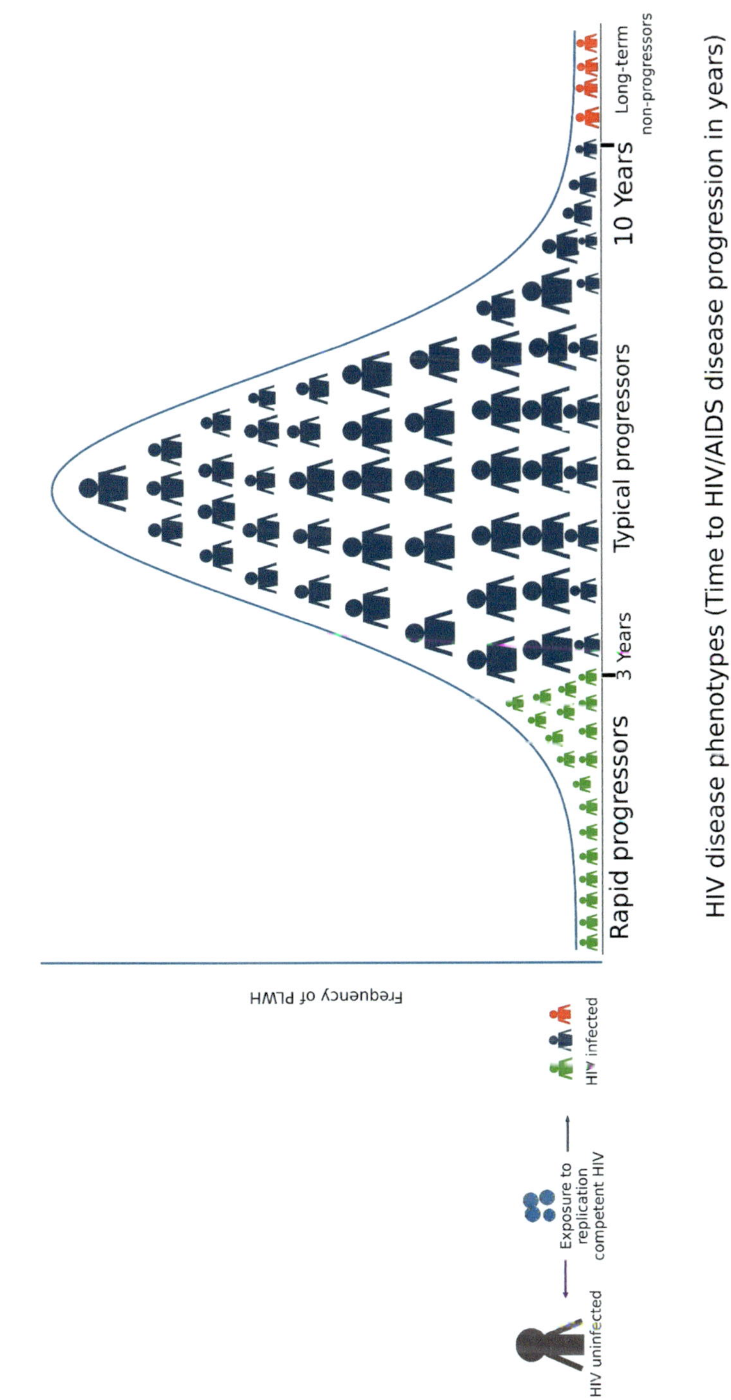

Figure 11.1 Distribution of clinical phenotypes among people living with HIV and implications for host genetic studies. Colors indicate clinical phenotypes: rapid progressors (green), typical progressors (blue), and long-term non-progressors (red).

cohorts to elucidate host genomic determinants of HIV acquisition, resistance, and disease progression. Despite these challenges, such research remains critical to informing vaccine development and curative strategies tailored to genetically diverse African populations.

11.3.1 Strategies to Design Clinical Studies to Identify Markers in HIV Disease Progression

During the initial years of ART rollout across Africa, in the early 2000s, a critical opportunity to study well-characterized, ART-naïve, HIV-exposed or -infected cohorts—particularly those with extreme disease progression phenotypes—was not fully leveraged. This gap largely stemmed from the limited availability of large-scale genomics research infrastructure, biobanking facilities, and trained expertise in many African settings at the time. Consequently, Africa missed a key window to generate insights from individuals exhibiting rare phenotypes, such as those who remained uninfected despite repeated exposure to replication-competent HIV-1—a phenomenon that defies conventional epidemiological expectations. While such individuals have provided essential clues in other global populations, similar discoveries in Africa remain scarce. To date, homozygosity for the CCR5Δ32 deletion [94] remains the only consistently replicated host genetic variant strongly associated with resistance to HIV-1 infection. However, this mutation is virtually absent in native African populations, underscoring the need for expanded efforts to identify and study alternative resistance mechanisms within Africa's genetically diverse populations. Although the CCR5Δ32 mutation is a well-known protective factor, it accounts for only a minority of individuals exhibiting the HEU clinical phenotype. Ongoing research aims to uncover additional genetic mechanisms underlying this rare and informative phenotype [95].

Study designs aimed at identifying genetic factors associated with HIV disease progression—particularly those leveraging diverse HIV-exposed phenotypes such as exposed-uninfected individuals and extreme cases like rapid and long-term non-progressors—must be critically re-evaluated in light of the widespread adoption of UTT policies, as well as evolving pre- and post-exposure prophylaxis guidelines. To overcome these new challenges, recent advancements in genomics research emphasize the importance of integrating large-scale, well-annotated phenotypic datasets that account for social, environmental, and clinical factors alongside genomic information [96] with multi-omics information, benefiting from spectacular advances in sequencing technology and data science [97]. Integrating rich genomic and phenotypic data holds significant potential for advancing HIV research in both prospective cohorts and cross-sectional studies. However, such efforts would require large sample sizes and may take a generation to yield comprehensive insights. To date, most HIV host genetics studies have concentrated on clinically defined outcomes, such as resistance to infection, susceptibility, or variation in disease progression. While valuable, these approaches must evolve to incorporate more nuanced and longitudinal data that reflect the complex interplay of genetics, environment, and clinical interventions [97]. However, intermediate phenotypes—such as typical progressors, who represent the majority of PLWH—have proven valuable in identifying host genetic association signals that may not be detectable when using more complex or extreme clinical outcomes alone [97].

Case–control studies are particularly valuable when investigating rare phenotypes with long latency periods between exposure and disease outcome. In the context of HIV, outcomes such as repeated exposure without infection (HIV-exposed uninfected, HEU) or long-term non-progression are uncommon. In such studies, individuals exhibiting the phenotype of interest (cases—for example, HEU or long-term non-progressors) are compared to individuals without the phenotype (controls— for example, rapid or typical progressors). Case–control designs offer several advantages: (1) they are more cost effective than cohort studies; (2) they can efficiently assess the relative importance of predictor variables; (3) they enable the estimation of odds ratios, which often approximate relative risks; and (4) they generate substantial data from relatively small sample sizes [98]. This type of study is therefore particularly useful for generating hypotheses that can be further evaluated through longitudinal cohort designs or experimental approaches. The discovery and widespread roll-out of rapid, affordable HIV point-of-care diagnostics—alongside the availability of effective ART—have significantly reduced HIV-related mortality and morbidity. Consequently, there is a growing population of individuals requiring lifelong, viral-suppressive treatment, underscoring the importance of continued research into host–pathogen interactions and long-term outcomes [99]. These advances have been accompanied by evolving clinical guidelines and recommendations for HIV treatment and viral load monitoring, which are routinely updated in response to emerging scientific evidence. Sustained viral suppression through antiretroviral therapy not only significantly improves individual health outcomes but also reduces the risk of HIV transmission, forming the cornerstone of treatment-as prevention strategies.

In the early years of the HIV epidemic, before the availability of ART, the virus exerted strong directional selection pressure on African populations. Individuals categorized as rapid or typical progressors often experienced reduced survival and reproductive potential due to AIDS-related mortality, while long-term non-progressors had a survival advantage. Today, the widespread availability of ART has transformed HIV into a manageable chronic condition, effectively equalizing life expectancy across different disease progression phenotypes. This shift presents a unique opportunity to establish longitudinal cohorts for HIV host genomics research in Africa.

Given the higher likelihood of encountering rapid and typical progressors—individuals more prone to symptomatic HIV and therefore more likely to present at health facilities—Africa is uniquely positioned to investigate the genetic determinants of disease progression. However, with the widespread adoption of the UTT strategy across SSA, HIV prognosis and long-term quality of life will increasingly be shaped by a combination of factors. These include the strength of healthcare systems (e.g., access to drug resistance testing, antiretroviral therapy, and routine viral load monitoring), socio-demographic variables, and the burden of comorbidities. The successful identification and longitudinal follow-up of informative cohorts will depend on the following factors: (1) the implementation of routine, large-scale HIV testing integrated with electronic health records; (2) the establishment of robust biobanking infrastructure; and (3) leveraging continental initiatives such as the African Population Cohort Consortium [100] and the forthcoming African Genomics Centres of Excellence network [101, 102].

11.4 HOST GENOMIC FACTORS ASSOCIATED WITH MALARIA IN AFRICA

Malaria continues to be one of the most significant and widespread parasitic infectious diseases in Africa. In 2021, the continent accounted for 95% of global malaria cases and 96% of malaria-related deaths, underscoring the disproportionate burden it bears [118]. For generations, malaria has exerted a distinct and intense selection pressure on children and pregnant women in Africa. High mortality rates in these vulnerable groups have profound demographic and evolutionary consequences, potentially shaping the genetic architecture of surviving populations. In 2021, children under 5 years of age accounted for approximately 80% of all malaria deaths in the African region. Alongside pregnant women, individuals living with HIV/AIDS also face significantly elevated risks, highlighting the intersection of malaria with other health vulnerabilities and the urgent need for targeted interventions [118]. Four species of *Plasmodium* (*P. falciparum*, *P. vivax*, *P. malariae*, and *P. ovale*) are responsible for almost all human malaria infections [119]. Human populations with a long history of exposure to *Plasmodium* infections have undergone evolutionary adaptations, developing genetic mechanisms that confer protection against severe malaria. These protective traits primarily counteract the consequences of infected erythrocyte lysis and the ensuing host immune and inflammatory responses that drive disease severity [119]. Malaria's high mortality and pervasive burden have made it one of the strongest evolutionary selective pressures in recent human history. Host genetic variants that confer resistance to malaria—such as those affecting hemoglobin and red blood cell function—represent some of the most well-documented examples of strong positive selection

Table 11.2 Selected Host Genomic Factors Associated with HIV/AIDS in Africa

Main Description/Outcome of the Study	*Study Design*	*Number of Individuals Analyzed*	*Infectious Condition and Associated Phenotype*	*Gene(s) Variants Implicated*	*Population*	*Reference*
Lack of *RH2* gene expression may have influenced the HIV pandemic in sub-Saharan Africa.	Case–control	1,600	Reduced susceptibility to HIV infection	Rhesus system *RH2*	Batswana	[103]
Chemokine coreceptor-2 gene polymorphisms among HIV-1 infected individuals in Kenya.	Cross-sectional	118	Delay HIV disease progression	*CCR-2*	Kenyans	[104]
genome-wide analyses reveal gene influence on HIV disease progression and HIV-1C acquisition in Southern Africa.	Cross-sectional	556	Reduces HIV-1C acquisition and delays HIV disease progression	*HCG22* (*HLA* Complex Group 22)	Batswana	[105]
Association of polymorphisms in the LEDGF/p75 gene (PSIP1) with susceptibility to HIV-1 infection and disease progression.	Longitudinal study cohort	776	Susceptibility to HIV-1 infection and disease progression	*LEDGF/p75* gene (*PSIP1*)	Black South Africans	[106]
Association of polymorphisms in the regulatory region of the cyclophilin a gene (*PPIA*) with gene expression and HIV/AIDS disease progression.	Longitudinal study cohort	539	Rapid HIV disease progression	Regulatory Region of the Cyclophilin a Gene (*PPIA*)	Black South Africans	[107]

(*Continued*)

Table 11.2 (Continued)

Main Description/Outcome of the Study	*Study Design*	*Number of Individuals Analyzed*	*Infectious Condition and Associated Phenotype*	*Gene(s) Variants Implicated*	*Population*	*Reference*
The impact of bone marrow stromal antigen-2 (*BST2*) gene variants on HIV-1 control in Black South African individuals.	Case–control	143	Delay HIV disease progression	Bone marrow stromal cell antigen 2 (*BST2* or *tetherin*)	Black South Africans	[108]
Associations of human leukocyte antigen *DRB* with resistance or susceptibility to HIV-1 infection in the Pumwani sex worker cohort.	Case–control	1,090	Resistance to HIV infection	*DRB1**010101, *DRB1**010201, *DRB1**1102 and haplotype DRB1*1102-*DRB3**020201	Kenyans	[109]
			Susceptibility to HIV infection	*DRB1**030201, *DRB1**070101, *DRB1**1503, *DRB5**010101, haplotypes *DRB1**070101-*DRB4**01010101 and *DRB1**1503-*DRB5**01010101		
African infants' *CCL3* gene copies influence perinatal HIV transmission in the absence of maternal nevirapine.	Case–control	849	Reduced HIV susceptibility	Infant *CCL3L1* (*CCR5* ligand) gene copies	South Africans	[110]
CD4 intragenic SNPs associate with HIV-2 plasma viral load and CD4 count in a community-based study from Guinea-Bissau, West Africa.	Case–control	396	Increased HIV-2 infection	CD4 intragenic SNPs	Bissau-Guineans	[111]

(Continued)

Table 11.2 (Continued)

Main Description/Outcome of the Study	*Study Design*	*Number of Individuals Analyzed*	*Infectious Condition and Associated Phenotype*	*Gene(s) Variants Implicated*	*Population*	*Reference*
Polymorphisms in *IRF-1* associated with resistance to HIV-1 infection in highly exposed uninfected Kenyan sex workers.	Cross-sectional	687	Resistance to HIV-1 infection	Polymorphisms in *IRF-1*	Kenyans	[112]
Distribution of *CCR5*-delta32, *CCR5* promoter 59029 A/G, *CCR2–64I* and *SDF1*-3′: A genetic polymorphisms in HIV-1 infected and uninfected patients in the west region of Cameroon.	Case–control	179	Susceptibility to HIV-1 infection Resistance to HIV-1 infection	*CCR5* promoter 59029 G/G *CCR2*-64I A/A	Cameroonians	[113]
The influence of human leukocyte antigen-types on disease progression among HIV-2 infected patients in Guinea-Bissau.	Retrospective cohort	437	Protects against HIV-2 disease progression	*HLA-B**58:01, *HLA*-DPB1*10:01 and *HLA*-DRB1*11:01	Bissau-Guineans	[114]
A *TRIM5alpha* exon 2 polymorphism is associated with protection from HIV-1 infection in the Pumwani sex worker cohort.	Retrospective cohort	1,032	Protects against HIV-1	*TRIM5*alpha	Kenyans	[115]
Risk for HIV-1 infection associated with a common *CXCL12* (*SDF1*) polymorphism and *CXCR4* variation in an African population.	Case–control	370	Increased risk for HIV-1 infection	*SDF1–3'A*	South Africans	[116]
Interleukin-1 gene cluster polymorphisms and susceptibility to clinical malaria in a Gambian case–control study.	Case–control	376	Susceptibility to clinical malaria	Interleukin-1 gene	Gambians	[117]

in modern human populations [3]. Studies have shown that the severity of malaria—including severe malarial anemia and cerebral malaria—varies considerably among individuals and across populations, reflecting differences in both host genetic susceptibility and environmental factors [5, 120].

The best-characterized human genetic polymorphism associated with malaria resistance is the sickle hemoglobin (HbS) variant, resulting from a single nucleotide mutation in the β-globin gene (HBB), which confers protection against severe forms of *Plasmodium falciparum* malaria in heterozygous individuals) [121]. The high prevalence of the HbS allele in SSA and other malaria-endemic regions is widely attributed to the selective advantage conferred to heterozygous individuals (HbAS), who are partially protected against severe *Plasmodium falciparum* malaria [121]. HbS is estimated to have emerged approximately 4,000 to 5,000 years ago, and its evolutionary interaction with malaria remains one of the most well-characterized examples of natural selection in the human genome, reflecting a persistent selective pressure exerted by *Plasmodium falciparum* over millennia [122]. The HbS balanced polymorphism exemplifies a classic case of heterozygote advantage, where the homozygous genotype (HbSS) leads to sickle cell disease—a severe condition responsible for the birth of at least 225,000 affected infants annually in SSA [123]. Despite this significant health burden, the persistence of the HbS allele in malaria-endemic regions is driven by the protective effect of the heterozygous genotype (HbAS). Individuals with HbAS exhibit partial resistance to severe *Plasmodium falciparum* malaria, resulting in higher survival rates and reproductive fitness. The mechanism of protection involves impaired parasite growth under low oxygen tension and altered trafficking of parasite-encoded exported proteins, which disrupts the parasite's ability to remodel host red blood cells. This evolutionary trade-off underscores the powerful role of malaria in shaping human genetic variation [124–126].

HbS is a structural variant of normal adult hemoglobin (HbAA), consisting of two α- and two β-globin chains, where a single amino acid substitution in the β-chain distinguishes it from the wild-type [121]. HbS arises from a single nucleotide mutation (GAG → GTG) in the sixth codon of the β-globin gene (HBB), resulting in an amino acid substitution of glutamic acid with valine (Glu6Val). Individuals homozygous for this mutation (HbSS) develop sickle cell disease, characterized by the polymerization of hemoglobin under low oxygen conditions. This leads to the deformation of red blood cells into a sickle shape, promoting vascular occlusion and impaired blood flow. The resulting vaso-occlusive events can affect multiple organs and tissues, contributing to significant morbidity and increased mortality [121]. Individuals with the HbAS genotype—carriers of the sickle cell trait—are typically asymptomatic and are largely protected against severe forms of malaria. While they do not experience the clinical complications associated with sickle cell disease, their heterozygous genotype confers a significant survival advantage in malaria-endemic regions. Notably, recent studies indicate that approximately one third of the variation in susceptibility to severe and complicated malaria can now be attributed to additive host genetic factors, highlighting the complex interplay of multiple loci beyond HbS in mediating resistance [127]. Beyond this protective genetic mechanism, the host immune response plays a critical role in controlling *Plasmodium* infection. The balance between pro-inflammatory (Th1-type) and anti-inflammatory (Th2-type) cytokines is particularly important—Th1 responses aid in the clearance of intracellular parasites, while Th2 responses contribute to the elimination of extracellular stages and support humoral immunity. This cytokine equilibrium influences both parasite control and the severity of clinical symptoms, and its dysregulation has been implicated in malaria pathogenesis [121]. Studies from African children have also found evidence for enhanced humoral response and increased levels of gamma globulin in subjects with HbAS compared to HbAA [121]. Therefore, host genetic variations outside the α (HBA) and β (HBB) loci influence malaria susceptibility in Africa.

In support of this, a nested case–control study conducted in Nanoro, Burkina Faso, involving 656 infants, found that genetic polymorphisms in genes involved in innate immune responses were associated with malaria susceptibility during the first year of life—likely through modulation of inflammatory mediator production [128]. A recent case–control study of severe malaria (n = 983) conducted in Tanga, Tanzania—an area of intense malaria transmission—identified genetic polymorphisms in the *USP38*, *FREM3*, *SDC1*, *DDC*, and *LOC727982* genes as putatively associated with differential host susceptibility to severe malaria [129]. Elsewhere in Africa, malaria risk has been associated with a SNP in the MyD88-adaptor-like (MAL) gene, where individuals carrying the rs8177374 mutant variant were found to have a reduced likelihood of developing severe malaria [130]. An increasing number of reports have linked various genetic disorders to either susceptibility or resistance to malaria. Heritability studies suggest that approximately 25% of the risk of progressing to severe malaria is attributable to human genetic factors [131]. Numerous genetic loci have been identified in African populations that appear to confer either protection against or susceptibility to malaria (Table 11.3) but the overall impact of host genetics on malarial risk remains unknown [123]. Recent advances in genome sequencing and data science hold great promise for identifying additional genetic factors that can inform the development of more effective vaccines and targeted treatments.

11.5 ADVANCING DISEASE CONTROL WITH PATHOGEN GENOMICS IN AFRICA

Traditional diagnostic and surveillance methods, while foundational, often fall short in the face of rapidly evolving pathogens, diverse transmission dynamics, and emerging drug resistance patterns. Pathogen genomics offers a transformative toolkit that enables more precise, timely, and localized

Table 11.3 Selected Host Genomic Factors Associated with Malaria in Africa

Main Description/Outcome of the Study	*Study Design*	*Number of Individuals Analyzed*	*Infectious Condition and Associated Phenotype*	*Gene(s) Variants Implicated*	*Population*	*Reference*
A variant in the gene *FUT9* is associated with susceptibility to placental malaria infection.	Case–control	360	Placental malaria infection	*FUT9*	Mozambican	[132]
USP38, *FREM3*, *SDC1*, *DDC*, and *LOC727982* gene polymorphisms and differential susceptibility to severe malaria in Tanzania,	Case–control	983	Differential susceptibility to severe malaria	*USP38, FREM3, SDC1, DDC,* and *LOC727982*	Tanzanians	[129]
Type I interferon receptor variants in gene regulatory regions are associated with susceptibility to cerebral malaria in Malawi.	Observational	212	Susceptibility to Cerebral Malaria	*Type I interferon (IFN) receptor 1 (IFNAR1)*	Malawians	[133]
G6PD and *HBB* polymorphisms in the Senegalese population: Prevalence, correlation with clinical malaria.	Case–control	437	Protective effect against severe malaria	*HbS*	Senegalese	[134]
Haplotype of non-synonymous mutations within *IL-23R* is associated with susceptibility to severe malaria anemia in a *P. falciparum* holoendemic transmission area of Kenya.	Cross-sectional	369	Pediatric severe malarial anemia	*IL-23R*	Kenyans	[135]
Common variation in the *ABO* glycosyltransferase is associated with susceptibility to severe *Plasmodium falciparum* malaria.	Case–control and Family-based study	9,000	Susceptibility to severe *Plasmodium falciparum* malaria	*ABO glycosyltransferase*	Gambians, Malawians, Kenyans	[136]

Table 11.3 (Continued)

Main Description/Outcome of the Study	*Study Design*	*Number of Individuals Analyzed*	*Infectious Condition and Associated Phenotype*	*Gene(s) Variants Implicated*	*Population*	*Reference*
Association of a single nucleotide polymorphism in the C-reactive protein gene (−286) with susceptibility to *Plasmodium falciparum* malaria.	Longitudinal study	192	Susceptibility to *Plasmodium falciparum* malaria	*C-reactive protein gene*	Sudanese	[137]
Hemoglobin variants and *Plasmodium falciparum* malaria in children under 5 years of age living in a high and seasonal malaria transmission area of Burkina Faso	Cross-sectional	452	Lower risk of clinical malaria	*Hb AC* and *AS genotype*	Burkinabé	[138]
Common genotypic polymorphisms in glutathione S-transferases in mild and severe falciparum malaria in Tanzanian children.	Case–control	157	Protect against severe falciparum malaria in children	*Glutathione S-transferases—GSTP1* and *GSTM1*	Tanzanians	[139]
Plasmodium falciparum malaria in south-west Nigerian children: Is the polymorphism of *ICAM-1* and *E-selectin* genes contributing to the clinical severity of malaria?	Cross-sectional	223	Increases risk of severe malaria	*G allele at the ICAM-1 exon 6*	Nigerians	[140]
Malaria severity: Possible influence of the *E670G PCSK9* polymorphism: A preliminary case–control study in Malian children	Case–control	752	Susceptibility to severe malaria	*E670G Proprotein Convertase Subtilisin/Kexin Type 9 (PCSK9)*	Malians	[141]

(Continued)

Table 11.3 (Continued)

Main Description/Outcome of the Study	*Study Design*	*Number of Individuals Analyzed*	*Infectious Condition and Associated Phenotype*	*Gene(s) Variants Implicated*	*Population*	*Reference*
High plasma levels of *HLA-G* are associated with low birth weight and with an increased risk of malaria in infancy.	Cross-sectional	165	Susceptibility to malaria in infants	*HLA-G*	Beninois	[142]
Novel genetic polymorphisms associated with severe malaria and under selective pressure in north-eastern Tanzania.	Case–control and family based	914	Protective effect against severe malaria	*HbS, interleukin receptors (IL-23R, IL-12RBR2)*, and *kelch-like protein KLHL3*	Tanzanians	[143]
Toll-like receptor (*TLR*) polymorphisms in African children: Common *TLR-4* variants predispose to severe malaria.	Case–control	870	Predisposes to severe malaria in children	*TLR-4*	Ghanaians	[144]
Association of candidate gene polymorphisms and TGF-β/IL-10 levels with malaria in three regions of Cameroon: A case–control study.	Case–control	1,862	Protective effect of against severe malaria	*HbS* and nitrogen oxide synthase 2 (*NOS2*)	Cameroonians	[145]
Transforming growth factor β 2 and *heme oxygenase 1* genes are risk factors for the cerebral malaria syndrome in Angolan children.	Cross-sectional	430	Risk factors for the cerebral malaria syndrome	Transforming growth factor beta 2 and heme oxygenase 1	Angolan	[146]
Cyclooxygenase-2 haplotypes influence the longitudinal risk of malaria and severe malarial anemia in Kenyan children from a holoendemic transmission region.	Longitudinal	1,081	Susceptibility to malaria and severe malarial anemia	Cyclooxygenase-2 (*COX-2*) promoter	Kenyans	[147]

Table 11.3 (Continued)

Main Description/Outcome of the Study	*Study Design*	*Number of Individuals Analyzed*	*Infectious Condition and Associated Phenotype*	*Gene(s) Variants Implicated*	*Population*	*Reference*
Mutations of complement lectin pathway genes *MBL2* and *MASP2* associated with placental malaria.	Case–control	269	Susceptibility to placental malaria	Mutations of complement lectin pathway genes *MBL2* and *MASP2*	Ghanaians	[148]
High frequency of the Duffy-negative genotype and absence of *Plasmodium vivax* infections in Ghana.	Retrospective	952	Protection against *Plasmodium vivax*	Duffy-negative	Ghanaians	[149]
CD209 and not CD28 or STAT6 polymorphism mediates clinical malaria and parasitemia among children from Nigeria.	Cross-sectional	531	Susceptibility to malaria infections	*CD209* polymorphisms	Nigerians	[150]
Genetic variants of tumor necrosis factor-α −308G/A (rs1800629) but not Toll-interacting proteins or vitamin D receptor genes enhances susceptibility and severity of malaria infection.	Case–control	297	Susceptibility and severity of malaria infections	Tumor necrosis factor-α	Nigerians	[151]
Pyruvate kinase and Fcγ receptor gene copy numbers associated with malaria phenotypes.	Case–control	373	Associated with malaria severity	Pyruvate Kinase and Fcγ Receptor Gene Copy Numbers	Gabonese	[152]
Association of *HLA* alleles with *Plasmodium falciparum* severity in Malian children.	Case–control	435	Susceptibility to cerebral malaria	*HLA* alleles	Malians	[153]

(Continued)

Table 11.3 (Continued)

Main Description/Outcome of the Study	*Study Design*	*Number of Individuals Analyzed*	*Infectious Condition and Associated Phenotype*	*Gene(s) Variants Implicated*	*Population*	*Reference*
Association of *HLA-G* 3′ UTR polymorphisms with response to malaria infection: A first insight.	Longitudinal	504	Associated with asymptomatic infection	*HLA-G* 3'UTR polymorphisms	Senegalese	[154]
Human genetic polymorphisms and asymptomatic *Plasmodium falciparum* malaria in Gabonese schoolchildren.	Cross-sectional	158	Protection against clinical malaria.	Glucose-6-phosphate dehydrogenase (*G6PD*) & *TNFalpha*	Gabonese	[155]

responses to these public health challenges [156]. Genomic technologies have revolutionized infectious disease control by enabling real-time surveillance, informing targeted treatments, mapping transmission networks, and accelerating vaccine development—as exemplified by the rapid sequencing of SARS-CoV-2 and the unprecedented speed of COVID-19 vaccine design and rollout [157]. Genomic surveillance involves sequencing the genomes of pathogens from clinical or environmental samples to monitor changes in their genetic makeup [158].

In Africa, this approach has become increasingly valuable for real-time epidemic tracking and understanding pathogen evolution. The growing adoption of NGS platforms across the continent has greatly enhanced this capability. As illustrated in Figure 11.2, these platforms were crucial during the COVID-19 pandemic and now serve as a backbone for ongoing and future surveillance. Notably, genomic tools were rapidly deployed in response to Marburg virus outbreaks in Rwanda (September 2024), Tanzania (March 2023), Equatorial Guinea (February 2023), and Ghana (July 2022); Mpox in the Democratic Republic of the Congo (January 2022); and Ebola virus disease in Uganda (January 2025) [159].

Genomic surveillance of malaria, HIV, and TB provides critical insights that directly inform drug and vaccine development. For malaria, sequencing *Plasmodium falciparum* has enabled the detection of mutations in genes such as *kelch13*, *pfcrt*, and *pfmdr1*, which are associated with resistance to artemisinin and partner drugs. This information guides the development of next generation antimalarials and helps inform treatment policies to delay resistance. In HIV, genomic surveillance enables the identification of transmitted and acquired drug resistance mutations in key viral genes like *pol*. This informs antiretroviral therapy regimens and supports the development of new inhibitors targeting resistant strains. Additionally, it aids in identifying immune escape variants that influence vaccine efficacy, guiding iterative vaccine design efforts, including those involving broadly neutralizing antibodies. For TB, whole-genome sequencing of *M. tuberculosis* uncovers mutations conferring resistance to isoniazid, rifampicin, and second-line drugs, allowing for rapid diagnostics and personalized treatment. It also identifies lineage-specific variations that impact vaccine design, such as in BCG replacement strategies.

Overall, integrating genomics data enhances precision in tracking pathogen evolution, informs rational drug and vaccine design, and supports public health strategies tailored to local and global resistance patterns and pathogen diversity.

Figure 11.2 highlights the presence and diversity of NGS platforms currently operational across the African continent. Each colored circle represents a specific sequencing platform, with some locations hosting multiple technologies. The platforms include portable devices like MinION and iSeq 100, as well as high-throughput instruments such as NovaSeq 6000, PacBio Sequel II, and the more recent NovaSeq X.

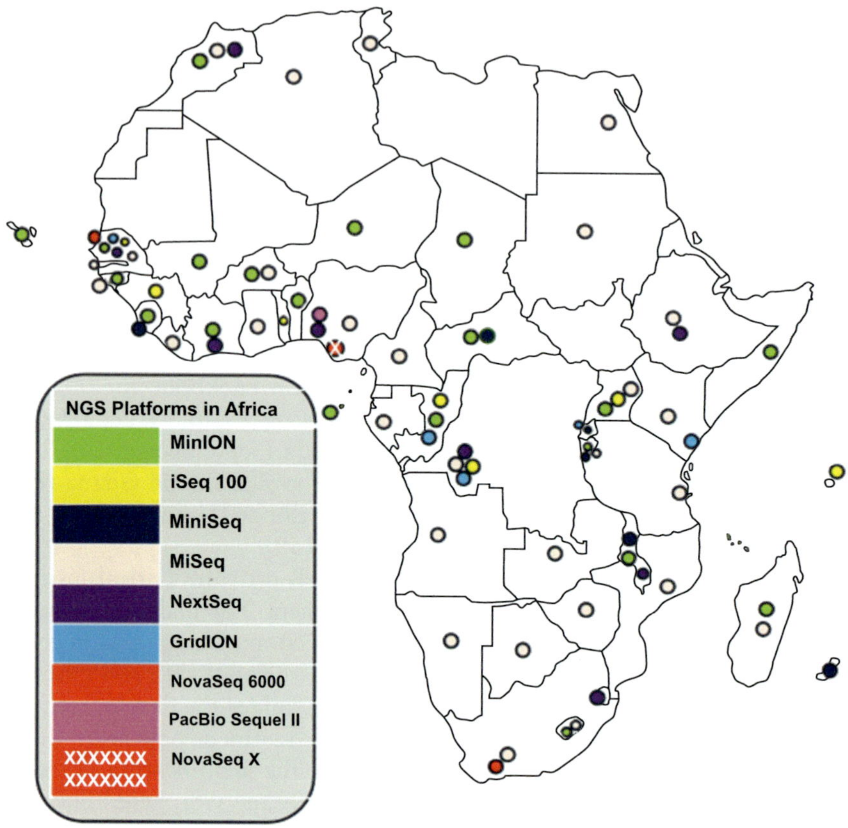

Figure 11.2 Geographic distribution of next-generation sequencing platforms across Africa.

These sequencing technologies have played a critical role in expanding Africa's genomic surveillance capacity—especially during the COVID-19 pandemic—and now underpin ongoing efforts in pathogen genomics, antimicrobial resistance monitoring, and outbreak response. The growing distribution of these platforms reflects the continent's increasing investment in precision public health and bioinformatics infrastructure.

11.5.1 HIV Genomic Surveillance

NGS has emerged as a critical tool in national public health laboratories across Africa for monitoring both pretreatment and acquired HIV drug resistance. Compared to traditional Sanger sequencing, NGS offers superior sensitivity by detecting low-frequency drug-resistant variants that might otherwise be missed. This heightened detection capability enables more accurate resistance profiling, which is essential for tailoring ART, optimizing treatment regimens, and ultimately improving clinical outcomes for people living with HIV [160]. A recent study conducted at Angola's national public health laboratory employed NGS to assess HIV-1 genetic diversity and pre-treatment drug resistance (PDR) among newly diagnosed individuals. The analysis revealed a high prevalence of PDR, especially against non-nucleoside reverse transcriptase inhibitors. These findings highlight the critical value of integrating NGS into routine national HIV surveillance programs to inform first-line treatment choices and strengthen the effectiveness of ART strategies [161]. Moreover, the implementation of NGS-based HIV drug resistance testing has been accelerated by the availability of user-friendly and scalable bioinformatics pipelines, such as HIV-DRIVES (i.e., HIV Drug Resistance Identification, Variant Evaluation, and Surveillance), which streamline data analysis and interpretation for clinical and public health laboratories [160]. These advancements contribute to more personalized treatment adjustments and improved public health outcomes by enabling timely identification of resistance patterns.

The global geographical subtype distribution of HIV-1 is evolving over time, and there has been a notable increase in newly emerging recombinants [162]. Africa harbors the greatest diversity of HIV-1 subtypes, especially subtypes A, C, and D, which vary in their transmission efficiency and response to ART. In Cameroon, genomic surveillance revealed region-specific patterns in the evolution of HIV-1, highlighting that each endemic area exhibits distinct viral dynamics. The findings also indicated a rising proportion of unique recombinant forms containing circulating recombinant form fragments, reflecting ongoing and significant shifts in the virus's genetic landscape [163]. In Uganda [164] and Kenya [165], genomic surveillance has revealed shifts in circulating HIV-1 subtypes due to migration and ART scale-up, with implications for diagnostic test performance and vaccine development. In South Africa, a study identified rapid disease progression among women infected with HIV-1 subtype C, underscoring the variability in disease outcomes and the influence of both viral and host genetic factors [166].

A series of studies in Kenya have used HIV-1 *pol* gene sequence data to elucidate transmission dynamics among key populations. One study analyzed 372 sequences from men who have sex with men (MSM) in Coastal Kenya, Nairobi, and Nyanza regions collected between 2006 and 2019. Phylogenetic and Bayesian phylogeographic analyses revealed 46 distinct transmission clusters, with approximately half exhibiting inter-regional spread—particularly from the Coast to Nairobi and Nyanza. Subtype A1 was most prevalent (72%), followed by subtype D and recombinant forms [165]. Another study, conducted during 2019–2020 among gay, bisexual, and other men who have sex with men (GBMSM) in Kisumu, Mombasa, and Kiambu counties, retrieved 36 sequences from dried blood spots, with

13 (36.1%) forming 8 phylogenetic clusters. Most clustered individuals (76.9%) reported finding partners online, and 84.6% tested for HIV less frequently than Kenyan guidelines recommend [167]. A third study in Coastal Kenya analyzed 658 sequences from MSM, injecting drug users (IDU), female sex workers (FSW), and heterosexual individuals (HET) collected between 2005 and 2019. It identified 61 transmission clusters, with 85% occurring within the same risk group. A large IDU-only cluster likely originated around 1987 [168], indicating a distinct sub-epidemic. Collectively, these findings underscore the need for geographically and population-targeted HIV prevention strategies that address distinct risk behaviors and networks.

A 2017 community-wide phylogenetic study in KwaZulu-Natal, South Africa, analyzed HIV transmission dynamics among 9,812 individuals aged 15–49. The study identified 90 transmission clusters, revealing that young women under 25 were often linked to older male partners aged 25–40, many of whom were unaware of their HIV-positive status and not on ART. These findings highlight the role of age-disparate relationships in driving HIV transmission and underscore the need for targeted interventions focusing on older men to reduce HIV incidence among young women [169]. The integration of NGS into HIV drug resistance monitoring in African countries has significantly improved the detection of both pretreatment and acquired resistance mutations. This supports the design of more effective ART regimens and strengthens efforts to control the HIV epidemic. Considering the evolving landscape of HIV subtypes in the region, continuous genomic surveillance remains essential for informing treatment strategies, guiding prevention efforts, and advancing vaccine development.

11.5.2 Malaria Genomic Surveillance

Plasmodium falciparum, the most virulent human malaria parasite, has demonstrated a formidable ability to develop resistance not only to antimalarial drugs but also, more recently, to molecular diagnostic assays [170]. Genomic surveillance tools—including whole-genome sequencing and targeted deep amplicon sequencing—are increasingly being used to detect resistance-associated mutations in key genes such as *pfcrt*, *pfmdr1*, *dhps*, and *kelch13*. These genomic insights are proving essential for guiding national malaria control strategies and informing timely updates to treatment policies. Notably, a pivotal study in Rwanda reported the emergence and local spread of the *kelch13* R561H mutation, a validated molecular marker associated with partial resistance to artemisinin [171]. This discovery has heightened concern within the global health community and prompted intensified surveillance as well as policy discussions on the continued suitability of artemisinin-based combination therapies (ACTs) as frontline treatments in affected regions.

Whole-genome sequencing of *Plasmodium falciparum* isolates from seven provinces in Mozambique revealed significant regional variation in drug resistance markers and parasite population structure, underscoring the importance of incorporating genomics data into national malaria control programs to inform treatment policies and optimize interventions [172]. Complementing these findings, a study of *P. falciparum* isolates from asymptomatic individuals in Ouélessébougou, Mali, demonstrated high genetic diversity and multiclonality, reflecting complex transmission dynamics. Recent samples (2019–2020) exhibited elevated markers of resistance to sulfadoxine-pyrimethamine and lumefantrine compared to older Malian and pan-African isolates. Furthermore, 21 genes under positive selection were identified, including *pfCelTOS* and *pfdblmsp2*, which are associated with parasite transmission and host cell invasion [173]. Together, these studies illustrate how genomic surveillance can uncover regional patterns of resistance and evolutionary adaptation, providing valuable insights for tailoring malaria control strategies in Africa.

Recent genomic surveillance studies across Africa have illuminated critical insights into *Plasmodium falciparum* diversity, drug resistance, and vaccine-relevant targets. In Ghana, researchers implemented a field-adaptable nanopore sequencing workflow to analyze 196 clinical samples, revealing continued susceptibility to chloroquine, persistent resistance to sulfadoxine-pyrimethamine, and no signs of artemisinin resistance. Notably, polymorphisms were detected in the circumsporozoite protein (CSP), a key vaccine antigen, highlighting the potential implications for vaccine efficacy [174]. Along the Kenya–Uganda border, whole-genome sequencing of 38 isolates revealed distinct subpopulations and complex transmission dynamics. A declining prevalence of the chloroquine resistance marker Pfcrt K76T and detection of the PfK13 V568I variant suggested evolving patterns of drug resistance [175].

In Thiès, Senegal, longitudinal surveillance over 4 years using molecular barcoding and whole-genome sequencing identified clustered transmission within households and decreasing genetic similarity with increasing distance—pinpointing potential hotspots for targeted interventions [176]. On Mount Cameroon, a genomic analysis of 220 isolates across altitudinal gradients demonstrated high genetic diversity, low population structure, and substantial gene flow. A large proportion of mixed-genotype infections indicated high outcrossing potential. Furthermore, 94 antigenic genes—many of which are vaccine targets—were under balancing selection, with 17 showing signs of recent directional selection [177]. Collectively, these studies underscore the value of integrating genomic surveillance with epidemiology to guide malaria control, treatment policy, and vaccine design in diverse African contexts.

At a continental level, expanding genomic surveillance across Africa is essential to monitor emerging and re-emerging resistance patterns, guide evidence-based policy decisions, and support the development of regionally tailored treatment and elimination strategies in the face of evolving malaria threats.

11.5.3 Tuberculosis Genomic Surveillance

M. tuberculosis exhibits distinct phylogeographic patterns across Africa, with various lineages and sub-lineages showing region-specific distributions. These patterns reflect the long-standing co-evolution of *M. tuberculosis* with human populations and are shaped by historical and contemporary human migration within and beyond the continent. Whole-genome sequencing has emerged as a powerful tool for elucidating these evolutionary dynamics, enabling the tracking of lineage-specific transmission patterns and the early detection of MDR-TB cases. The study curated and analyzed a large-scale dataset of 13,753 *M. tuberculosis* genomes from 26 African countries, providing a comprehensive resource for understanding TB in the region. Most isolates belonged to Lineage 4, though significant representation from Lineages 1, 2, and 3 reflected the continent's phylogenetic diversity. Researchers identified 157 known drug resistance-associated mutations across 12 genes, along with novel variants potentially linked to resistance [178]. Standardized phylogenetic classifications and annotations were applied, making the dataset suitable for surveillance, comparative genomics, and resistance monitoring. The findings offer critical insights to inform TB control programs and guide future genomics research across Africa. Through whole-genome sequencing, public health programs can better understand transmission networks, monitor drug resistance mutations, and tailor interventions to local epidemiological contexts.

Recent genomic studies in Africa have significantly advanced our understanding of *M. tuberculosis* transmission dynamics, drug resistance, and lineage diversity, with important implications for public health. In South Africa, large-scale genomic surveillance has been instrumental in mapping the spread of the LAM4/KZN MDR-TB strain, particularly in KwaZulu-Natal province [179]. One study characterized the evolutionary trajectory of the extensively drug-resistant (XDR) F15/LAM4/KZN lineage, while another used whole-genome sequencing to reconstruct its emergence and decades-long dissemination [180]. These findings have informed targeted public health responses to MDR-TB outbreaks in the region.

In Tanzania, whole-genome sequencing analysis of 191 *M. tuberculosis* isolates revealed substantial strain diversity, with Lineage 3 (East African–Indian) being most prevalent, followed by Lineage 4 (Euro-American). Drug-resistance mutations were more frequently observed in Lineage 4, and specific phylogenetic lineages were associated with HIV co-infection and prior TB treatment—highlighting the role of genomics in tailoring interventions to patient and lineage characteristics [181]. Complementing these efforts, a study in South Africa and Zambia demonstrated the feasibility of nanopore targeted sequencing (NTS) directly from sputum samples. NTS accurately predicted resistance to critical TB drugs such as rifampicin and isoniazid, with rapid turnaround times suitable for field deployment [182]. Together, these studies emphasize the transformative potential of genomics in improving TB diagnosis, tracking resistant strains, and guiding context-specific treatment strategies across Africa

A study conducted in Ghana analyzed 452 *M. tuberculosis* complex isolates collected between July 2012 and December 2015 using whole-genome sequencing and spatial analysis to investigate TB transmission dynamics. The researchers confirmed 314 clusters, with a median cluster size of 7.5 genomes, and estimated a recent transmission rate of 24.7%. Notably, a widespread Cameroon sub-lineage clone was predominantly found in the Ablekuma sub-district of Accra. Additionally, a recent transmission cluster associated with isoniazid resistance belonging to the Ghana sub-lineage of Lineage 4 was identified [183]. A study conducted in 2024 analyzed 68 rifampin-resistant *M. tuberculosis* isolates from 8 districts in Bangui and surrounding regions of the Central African Republic. The genomic analysis revealed that all isolates belonged to *M. tuberculosis* sensu stricto, encompassing 9 lineages, with L4.1.2.1 Haarlem and L4.6 Euro-American being the most prevalent. Antibiotic resistance profiling indicated that 86.7% of isolates exhibited at least one predicted resistance, while 13.2% remained susceptible [184].

11.6 PRECISION PUBLIC HEALTH GENOMICS

Precision public health genomics seeks to enhance disease prevention and control by integrating genomics data from both pathogens and human populations with near real-time epidemiological insights. This approach enables the design of targeted interventions that reflect the genetic diversity, transmission dynamics, and risk profiles specific to communities or regions. In Africa—where HIV, malaria, and TB continue to pose significant public health burdens—genomics surveillance is proving to be a powerful tool for developing more precise, effective, and equitable responses.

Genomics analysis of HIV-1 in African settings has revealed region-specific transmission networks and patterns of drug resistance. In South Africa, phylogenetic clustering has helped identify high-transmission hubs and key populations—such as men who have sex with men and adolescent girls that require targeted interventions. Sequencing of the *pol* gene has further allowed optimization of first-line ART regimens by identifying circulating resistance mutations, ensuring that treatment protocols are guided by local resistance profiles rather than generic global recommendations.

The use of genomic tools to track *Plasmodium falciparum* mutations—such as *kelch13* (linked to artemisinin resistance) and *pfcrt* (associated with chloroquine resistance)—has become integral to malaria control efforts. These data enable national malaria control programs to adapt treatment guidelines in response to emerging resistance. Moreover, host genomic factors, including variants in *G6PD* and the *Duffy antigen* gene, are informing decisions on drug safety and population susceptibility to malaria infection.

Whole-genome sequencing of *M. tuberculosis* has provided high-resolution insights into transmission dynamics and resistance mutations. In South Africa, genomic characterization of the LAM4/KZN strain of MDR-TB has informed strategic public health responses, including targeted contact tracing and deployment of novel treatment regimens.

Integrating both host and pathogen genomics data into public health surveillance offers African countries a pathway beyond one-size-fits-all interventions toward precision public health strategies. These approaches enable more efficient allocation of resources, informed therapeutic decision-making, and context-specific vaccination campaigns—each tailored to the genetic profiles of both the population and the circulating pathogens. Realizing the full potential of precision public health genomics in Africa, however, will require sustained investment in sequencing infrastructure, robust data governance frameworks, comprehensive bioinformatics training, and equitable access to genomic technologies.

11.7 FUTURE DIRECTIONS

As Africa continues to advance the application of genomic tools in public health, future frontiers will involve the integration of cutting-edge technologies that move beyond conventional sequencing. Among these, metagenomics has emerged as a powerful approach for characterizing the human microbiome and its role in shaping susceptibility to infectious diseases such as TB, malaria, and HIV. Recent studies underscore the gut microbiome's critical influence on both infection risk and treatment outcomes. In a 2025 longitudinal study conducted in Bamako, Mali, Diallo et al. investigated the effects of first-line TB therapy on gut microbial communities. The findings revealed profound and persistent disruptions in microbiota composition and function that extended up to 9 months post treatment. These changes were linked to significant metabolic and immunological alterations—including disturbances in tryptophan and tricarboxylic acid metabolism and shifts in cytokine profiles—suggesting that antibiotic-induced dysbiosis may impair immune function and increase the risk of TB recurrence [185].

Complementing these findings, a 2023 multi-omics study by Mandal et al. demonstrated that gut microbiota composition may also influence malaria severity. In experimental models, the introduction of *Bacteroides fragilis* into malaria-resistant mice induced hyperparasitemia. In parallel, analyses of human samples revealed that Ugandan children with severe malarial anemia had significantly higher gut *Bacteroides* abundance compared to those with asymptomatic infections. These results suggest a potential causal link between *Bacteroides* overrepresentation and severe malaria, identifying gut microbiome modulation as a promising therapeutic target [186]. Collectively, these studies underscore the emerging frontier of host–microbiome–pathogen interactions in determining disease susceptibility and clinical outcomes. Integrating microbiome insights into public health genomics may unlock innovative strategies for prevention, treatment, and long-term recovery, particularly in high-burden settings.

Single-cell genomics is emerging as a transformative tool for unraveling host–pathogen interactions with unprecedented resolution. In the context of HIV, recent studies have applied single-cell multi-omics to simultaneously profile transcription factor accessibility, transcriptomes, surface proteins, and both HIV-1 DNA and RNA within memory $CD4^+$ T cells from individuals during viremia and following suppressive antiretroviral therapy. This approach revealed distinct epigenetic, transcriptional, and surface protein signatures distinguishing latent from transcriptionally active HIV-infected cells, thereby illuminating cellular programs that contribute to HIV-1 persistence [185]. In malaria research, single-cell RNA sequencing has provided insight into how specific immune cell subsets respond to *Plasmodium* infection, advancing our understanding of naturally acquired immunity and informing vaccine design [186]. Similarly, in TB, single-cell technologies have facilitated the dissection of granuloma heterogeneity and host immune responses, offering a detailed view of the localized immune dynamics during *M. tuberculosis* infection [187].

Adapting these technologies to African contexts will require sustained investments in capacity building, the development of local infrastructure, and the establishment of equitable data-sharing frameworks. These efforts are essential to fully harnessing their potential in advancing precision public health strategies on the continent.

The pathogen-specific innate immune response constitutes a critical first line of defense and is strongly influenced by host genetic diversity [188, 189]. Shaped by evolutionary pressures, the human immune system displays substantial inter-individual and population-level variation. However, universal public health interventions—such as vaccination and antimicrobial therapy—often overlook this underlying genetic heterogeneity [190]. Consequently, vaccine efficacy can differ significantly across geographic regions. For instance, attenuated malaria vaccines have shown nearly 100% protection in high-income settings but only 20–50% efficacy in certain African populations [191]. Genetic studies indicate that thousands of weak-effect genomic loci may collectively explain up to half of the observed variation in immune responses. This immunogenetic diversity is largely driven by polymorphic genes and modulated by environmental exposures that shape immune system configurations [192].

Collaborative efforts to eliminate Africa's three major infectious diseases—malaria (parasitic), HIV (viral), and TB (bacterial)—through vaccines remain a public health priority. The development of an effective malaria vaccine has been a global health priority for nearly 6 decades. In 2021, the World Health Organization approved RTS,S/AS01, the first malaria vaccine to reach large-scale implementation. Since its introduction, over two million children in Ghana, Kenya, and Malawi have received the vaccine. Several other countries, including Uganda, Ethiopia, Guinea, Cameroon, and Mali, are preparing for rollout. RTS,S/AS01 has demonstrated a favorable safety profile and has been associated with reductions in severe malaria cases and childhood mortality in pilot implementation settings. Despite these benefits, the vaccine's overall efficacy remains moderate. Eradication efforts are further challenged by the high genetic diversity of both *Plasmodium* parasites and African human populations [198], which may limit the vaccine's effectiveness across diverse settings. Notably, RTS,S/AS01 remains the only licensed vaccine targeting a parasitic disease [193].

Similarly, HIV vaccine development has encountered repeated setbacks over the past 40 years, despite early optimism. Nevertheless, these challenges have yielded valuable insights that have informed the development of vaccines for other emerging infections, including COVID-19. Encouragingly, a recent early-phase clinical trial of the eOD-GT8 60-mer HIV nanoparticle vaccine demonstrated that it was safe and induced robust immune responses in nearly all 36 healthy adult participants in the United States [199, 200]. Ongoing studies are now evaluating its efficacy in African populations, including trials in Rwanda and South Africa [201]. In parallel, the multisite phase I HIV-CORE 006 trial is being conducted in Kenya, Uganda, and Zambia to assess the safety and immunogenicity of HIVconsvX, a mosaic vaccine candidate designed to elicit broad immune responses against diverse HIV-1 strains [202]. At the same time, combination prevention strategies—such as integrating vaccines with pre-exposure prophylaxis—are gaining traction. Building on the success of COVID-19 mRNA platforms, several trials are also exploring mRNA-based HIV vaccines [203], although their immunological efficacy across diverse populations remains to be fully determined [204].

TB remains the world's leading cause of death from a bacterial infectious disease, claiming 1.6 million lives and affecting over 10.6 million people in 2021. This persistent global burden is driven, in part, by the continued lack of highly effective diagnostics and vaccines. A promising candidate, M72/AS01E—a recombinant protein-based vaccine—is currently in late-phase clinical evaluation. If proven effective, it would represent the first new TB vaccine in more than a century [205, 206]. The ongoing trial involves approximately 26,000 participants across more than 50 sites in Africa and Southeast Asia. However, vaccination alone will not be sufficient to curb the epidemic. There remains an urgent need for improved diagnostic tools and shorter, more effective treatment regimens. Notably, more than a dozen additional TB vaccine candidates are currently at various stages of development and evaluation [206].

In the global effort to eradicate infectious diseases, both pathogen and human genomics remain central to unraveling host–pathogen interactions across diverse populations and ecological settings. Africa has made notable progress in this domain through initiatives such as H3Africa, which has significantly advanced genomic infrastructure, workforce training, and large-scale studies on infectious diseases [102, 198, 207]. Building on this foundation, the establishment of the African Genomic Centres of Excellence represents a promising step toward further strengthening genomic capacity, guiding precision public health interventions, and training the next generation of African genomic scientists [101]. Finally, the Wellcome Trust is set to lead the largest population-based genomics research initiative in Africa through the African Population Cohort Consortium. Currently in its co-creation phase, the African Population Cohort Consortium brings together African scientists, policymakers, and community leaders to harness and advance genomic expertise across the continent [208].

REFERENCES

[1] SDG Target 3.3 Communicable diseases, https://www.who.int/data/gho/data/themes/topics/sdg-target-3_3-communicable-diseases (accessed 16 September 2023).

[2] Century I of M (US) C on G and the PH in the 21st, Hernandez LM. Priorities. *In: Implications of Genomics for Public Health: Workshop Summary*. National Academies Press (US), https://www.ncbi.nlm.nih.gov/books/NBK83756/ (2005, accessed 3 September 2023).

[3] Hedrick PW. Population genetics of malaria resistance in humans. *Heredity* 2011; 107: 283–304.

[4] Kwiatkowski DP. How malaria has affected the human genome and what human genetics can teach us about Malaria. *Am J Hum Genet* 2005; 77: 171–192.

[5] Driss A, Hibbert JM, Wilson NO, et al. Genetic polymorphisms linked to susceptibility to malaria. *Malaria Journal* 2011; 10: 271.

[6] Hill AV. The genomics and genetics of human infectious disease susceptibility. *Annu Rev Genomics Hum Genet* 2001; 2: 373–400.

[7] Bellamy RJ, Hill AV. Host genetic susceptibility to human tuberculosis. *Novartis Found Symp* 1998, 217: 3–13; discussion 13–23.

[8] Kachuri L, Chatterjee N, Hirbo J, et al. Principles and methods for transferring polygenic risk scores across global populations. *Nat Rev Genet* 2023; 1–18.

[9] Kidenya BR, Mboowa G. Inclusiveness of the all of us research program improves polygenic risk scores and fosters genomic medicine for all. *Commun Med* 2024; 4: 1–3.

[10] Zhang H, Zhan J, Jin J, et al. A new method for multiancestry polygenic prediction improves performance across diverse populations. *Nat Genet* 2023; 1–12.

[11] Karlsson EK, Kwiatkowski DP, Sabeti PC. Natural selection and infectious disease in human populations. *Nat Rev Genet* 2014; 15: 379–393.

[12] Campbell MC, Tishkoff SA. African genetic diversity: Implications for human demographic history, modern human origins, and complex disease mapping. *Annual Review of Genomics and Human Genetics* 2008; 9: 403–433.

[13] Anderson RM, May RM. Coevolution of hosts and parasites. *Parasitology* 1982; 85: 411–426.

[14] Cox FE. History of the discovery of the malaria parasites and their vectors. *Parasites & Vectors* 2010; 3: 5.

[15] Daniel TM. The history of tuberculosis. *Respiratory Medicine* 2006; 100: 1862–1870.

[16] Beyrer C. A pandemic anniversary: 40 years of HIV/AIDS. *The Lancet* 2021; 397: 2142–2143.

[17] He W, Neil S, Kulkarni H, et al. Duffy antigen receptor for chemokines mediates trans-infection of HIV-1 from red blood cells to target cells and affects HIV-AIDS susceptibility. *Cell Host Microbe* 2008; 4: 52–62.

[18] Lipsitch M, Moxon ER. Virulence and transmissibility of pathogens: What is the relationship? *Trends in Microbiology* 1997; 5: 31–37.

[19] Leung NHL. Transmissibility and transmission of respiratory viruses. *Nat Rev Microbiol* 2021; 19: 528–545.

[20] Windels EM, Wampande EM, Joloba ML, et al. HIV co-infection is associated with reduced Mycobacterium tuberculosis transmissibility in sub-Saharan Africa. 2023; 2023.09.12.557301.

[21] Samson M, Libert F, Doranz BJ, et al. Resistance to HIV-1 infection in Caucasian individuals bearing mutant alleles of the CCR-5 chemokine receptor gene. *Nature* 1996; 382: 722–725.

[22] Solloch UV, Lang K, Lange V, et al. Frequencies of gene variant CCR5-Δ32 in 87 countries based on next-generation sequencing of 1.3 million individuals sampled from 3 national DKMS donor centers. *Human Immunology* 2017; 78: 710–717.

[23] Kim MB, Giesler KE, Tahirovic YA, et al. CCR5 receptor antagonists in preclinical to phase II clinical development for treatment of HIV. *Expert Opin Investig Drugs* 2016; 25: 1377–1392.

[24] Jensen B-EO, Knops E, Cords L, et al. In-depth virological and immunological characterization of HIV-1 cure after CCR5Δ32/Δ32 allogeneic hematopoietic stem cell transplantation. *Nat Med* 2023; 29: 583–587.

[25] McGuire AL, Gabriel S, Tishkoff SA, et al. The road ahead in genetics and genomics. *Nat Rev Genet* 2020; 21: 581–596.

[26] A $100 genome? New DNA sequencers could be a 'game changer' for biology, medicine, https://www.science.org/content/article/100-genome-new-dna-sequencers-could-be-game-changer-biology-medicine (accessed 1 October 2023).

[27] Tuberculosis (TB), https://www.who.int/news-room/fact-sheets/detail/tuberculosis (accessed 3 September 2023).

[28] TB is once again the deadliest disease in Africa—what went wrong | Gavi, the Vaccine Alliance, https://www.gavi.org/vaccineswork/tb-once-again-deadliest-disease-africa-what-went-wrong (accessed 3 September 2023).

[29] Aravindan PP. Host genetics and tuberculosis: Theory of genetic polymorphism and tuberculosis. *Lung India* 2019; 36: 244.

[30] Li X, Yang Y, Zhou F, et al. SLC11A1 (NRAMP1) polymorphisms and tuberculosis susceptibility: Updated systematic review and meta-analysis. *PLoS One* 2011; 6: e15831.

[31] Thye T, Vannberg FO, Wong SH, et al. Genome-wide association analyses identifies a susceptibility locus for tuberculosis on chromosome 18q11.2. *Nat Genet* 2010; 42: 739–741.

[32] Thye T, Nejentsev S, Intemann CD, et al. MCP-1 promoter variant –362C associated with protection from pulmonary tuberculosis in Ghana, West Africa. *Hum Mol Genet* 2009; 18: 381–388.

[33] Curtis J, Luo Y, Zenner HL, et al. Susceptibility to tuberculosis is associated with variants in the ASAP1 gene encoding a regulator of dendritic cell migration. *Nature Genetics* 2015; 47: 523.

[34] Bellamy R, Ruwende C, Corrah T, et al. Variations in the NRAMP1 gene and susceptibility to tuberculosis in West Africans. *N Engl J Med* 1998; 338: 640–644.

[35] Hall NB, Igo RP, Malone LL, et al. Polymorphisms in TICAM2 and IL1B are associated with TB. *Genes Immun* 2015; 16: 127–133.

[36] Daya M, van der Merwe L, van Helden PD, et al. Investigating the role of gene-gene interactions in TB susceptibility. *PLoS One* 2014; 10: e0123970.

[37] Asante-Poku A, Morgan P, Osei-Wusu S, et al. Genetic analysis of TB susceptibility variants in Ghana reveals candidate protective loci in SORBS2 and SCL11A1 genes. *Front Genet* 2021; 12: 729737.

[38] Ben-Selma W, Harizi H, Bougmiza I, et al. Polymorphisms in the RANTES gene increase susceptibility to active tuberculosis in Tunisia. *DNA Cell Biol* 2011; 30: 789–800.

[39] Zaki HY, Leung KH, Yiu WC, et al. Common polymorphisms in TLR4 gene associated with susceptibility to pulmonary tuberculosis in the Sudanese. *Int J Tuberc Lung Dis* 2012; 16: 934–940.

[40] Möller M, Kwiatkowski R, Nebel A, et al. Allelic variation in BTNL2 and susceptibility to tuberculosis in a South African population. *Microbes Infect* 2007; 9: 522–528.

[41] Bezerra OC de L, Alvarado-Arnez LE, Mabunda N, et al. Putative pathogen-selected polymorphisms in the PKLR gene are associated with mycobacterial susceptibility in Brazilian and African populations. *PLOS Negl Trop Dis* 2021; 15: e0009434.

[42] Qrafli M, Amar Y, Bourkadi J, et al. The CYP7A1 gene rs3808607 variant is associated with susceptibility of tuberculosis in Moroccan population. *Pan Afr Med J*; 18. Epub ahead of print 5 January 2014 https://doi.org/10.11604/pamj.2014.18.1.3397.

[43] Bruiners N, Schurz H, Daya M, et al. A regulatory variant in the C1Q gene cluster is associated with tuberculosis susceptibility and C1qA plasma levels in a South African population. *Immunogenetics* 2020; 72: 305–314.

[44] Möller M, Nebel A, Valentonyte R, et al. Investigation of chromosome 17 candidate genes in susceptibility to TB in a South African population. *Tuberculosis (Edinb)* 2009; 89: 189–194.
[45] Bornman L, Campbell SJ, Fielding K, et al. Vitamin D receptor polymorphisms and susceptibility to tuberculosis in West Africa: A case-control and family study. *J Infect Dis* 2004; 190: 1631–1641.
[46] Abate E, Blomgran R, Verma D, et al. Polymorphisms in CARD8 and NLRP3 are associated with extrapulmonary TB and poor clinical outcome in active TB in Ethiopia. *Sci Rep* 2019; 9: 3126.
[47] Ben-Selma W, Harizi H, Letaief M, et al. Age- and gender-specific effects on NRAMP1 gene polymorphisms and risk of the development of active tuberculosis in Tunisian populations. *Int J Infect Dis* 2012; 16: e543–e550.
[48] Velez DR, Wejse C, Stryjewski ME, et al. Variants in toll-like receptors 2 and 9 influence susceptibility to pulmonary tuberculosis in Caucasians, African-Americans, and West Africans. *Hum Genet* 2010; 127: 65–73.
[49] Schurz H, Glanzmann B, Bowker N, et al. Deciphering Genetic Susceptibility to Tuberculous Meningitis. *Frontiers in Neurology* 2022; 13, https://www.frontiersin.org/articles/10.3389/fneur.2022.820168 (accessed 25 June 2023).
[50] Lombard Z, Dalton D-L, Venter PA, et al. Association of HLA-DR, -DQ, and vitamin D receptor alleles and haplotypes with tuberculosis in the Venda of South Africa. *Hum Immunol* 2006; 67: 643–654.
[51] Hassuna NA, El Feky M, Hussein AARM, et al. Interleukin-18 and interferon-γ single nucleotide polymorphisms in Egyptian patients with tuberculosis. *PLoS One* 2021; 16: e0244949.
[52] Arji N, Busson M, Iraqi G, et al. Genetic diversity of TLR2, TLR4, and VDR loci and pulmonary tuberculosis in Moroccan patients. *J Infect Dev Ctries* 2014; 8: 430–440.
[53] Qrafli M, Asekkaj I, Bourkadi JE, et al. New variant identified in major susceptibility locus to tuberculosis on chromosomal region 8q12-q13 in Moroccan population: A case control study. *BMC Infect Dis* 2017; 17: 712.
[54] Olvany JM, Sausville LN, White MJ, et al. CLEC4E (Mincle) genetic variation associates with pulmonary tuberculosis in Guinea-Bissau (West Africa). *Infec Genet Evol* 2020; 85: 104560.
[55] Meyer CG, Reiling N, Ehmen C, et al. TLR1 variant H305L associated with protection from pulmonary tuberculosis. *PLoS One* 2016; 11: e0156046.
[56] Baker AR, Zalwango S, Malone LL, et al. Genetic susceptibility to tuberculosis associated with cathepsin Z haplotype in a Ugandan household contact study. *Human Immunology* 2011; 72: 426–430.
[57] Gupta A, Montepiedra G, Gupte A, et al. Low vitamin-D Levels combined with PKP3-SIGIRR-TMEM16J host variants is associated with tuberculosis and death in HIV-infected and -exposed infants. *PLoS One* 2016; 11: e0148649.
[58] Ali AH, Omer AA, Saeed NS, et al. Influence of interferon-gamma Receptor 1 gene polymorphisms on the susceptibility to pulmonary tuberculosis among Sudanese population. *Int J Mycobacteriol* 2018; 7: 26–31.
[59] Lamsyah H, Rueda B, Baassi L, et al. Association of PTPN22 gene functional variants with development of pulmonary tuberculosis in Moroccan population. *Tissue Antigens* 2009; 74: 228–232.

[60] Grant AV, El Baghdadi J, Sabri A, et al. Age-dependent association between pulmonary tuberculosis and common TOX variants in the 8q12–13 linkage region. *Am J Hum Genet* 2013; 92: 407–414.
[61] Salie M, Daya M, Lucas LA, et al. Association of toll-like receptors with susceptibility to tuberculosis suggests sex-specific effects of TLR8 polymorphisms. *Infect Genet Evol* 2015; 34: 221–229.
[62] Sobota RS, Stein CM, Kodaman N, et al. A locus at 5q33.3 confers resistance to tuberculosis in highly susceptible individuals. *Am J Hum Genet* 2016; 98: 514–524.
[63] Ghermi M, Reguieg S, Attab K, et al. Interferon-γ (+874 T/A) and interleukin-10 (–1082 G/A) genes polymorphisms are associated with active tuberculosis in the Algerian population of Oran's city. *Indian J Tuberc* 2021; 68: 221–229.
[64] Mhmoud N, Fahal A, Wendy van de Sande WJ. Association of IL-10 and CCL5 single nucleotide polymorphisms with tuberculosis in the Sudanese population. *Trop Med Int Health* 2013; 18: 1119–1127.
[65] Herb F, Thye T, Niemann S, et al. ALOX5 variants associated with susceptibility to human pulmonary tuberculosis. *Human Molecular Genetics* 2008; 17: 1052–1060.
[66] Wamala D, Buteme HK, Kirimunda S, et al. Association between human leukocyte antigen class II and pulmonary tuberculosis due to mycobacterium tuberculosis in Uganda. *BMC Infect Dis* 2016; 16: 23.
[67] Möller M, Flachsbart F, Till A, et al. A functional haplotype in the 3'untranslated region of TNFRSF1B is associated with tuberculosis in two African populations. *Am J Respir Crit Care Med* 2010; 181: 388–393.
[68] Arji N, Busson M, Iraqi G, et al. The MCP-1 (CCL2) -2518 GG genotype is associated with protection against pulmonary tuberculosis in Moroccan patients. *J Infect Dev Ctries* 2012; 6: 73–78.
[69] Sabri A, Grant AV, Cosker K, et al. Association study of genes controlling IL-12-dependent IFN-γ immunity: STAT4 alleles increase risk of pulmonary tuberculosis in Morocco. *J Infect Dis* 2014; 210: 611–618.
[70] Ben-Selma W, Ben-Abderrahmen Y, Boukadida J, et al. IL-10R1 S138G loss-of-function polymorphism is associated with extrapulmonary tuberculosis risk development in Tunisia. *Mol Biol Rep* 2012; 39: 51–56.
[71] Mboowa G, Mwesigwa S, Katagirya E, et al. The Collaborative African Genomics Network (CAfGEN): Applying genomic technologies to probe host factors important to the progression of HIV and HIV-tuberculosis infection in sub-Saharan Africa. *AAS Open Res* 2018; 1: 3.
[72] Schliekelman P, Garner C, Slatkin M. Natural selection and resistance to HIV. *Nature* 2001; 411: 545–546.
[73] Ford N, Calmy A, Mills EJ. The first decade of antiretroviral therapy in Africa. *Global Health* 2011; 7: 33.
[74] Towards universal access: scaling up priority HIV/AIDS interventions in the health sector: Progress report 2009, https://apps.who.int/iris/handle/10665/44217 (accessed 17 September 2023).
[75] Nicol E, Mehlomakulu V, Jama NA, et al. Healthcare provider perceptions on the implementation of the universal test-and-treat policy in South Africa: A qualitative inquiry. *BMC Health Serv Res* 2023; 23: 293.

[76] Global AIDS update 2019—Communities at the centre, https://www.unaids.org/en/resources/documents/2019/2019-global-AIDS-update (accessed 17 September 2023).
[77] Gomez F, Hirbo J, Tishkoff SA. Genetic variation and adaptation in Africa: Implications for human evolution and disease. *Cold Spring Harb Perspect Biol* 2014; 6: a008524.
[78] 40 years after the discovery of HIV. *Institut Pasteur*, https://www.pasteur.fr/en/press-area/press-documents/40-years-after-discovery-hiv (2023, accessed 17 September 2023).
[79] Variation in African genomes linked to control of HIV. *Nature*. Epub ahead of print 2 August 2023. https://doi.org/10.1038/d41586-023-02277-2.
[80] Williamson C, Loubser SA, Brice B, et al. Allelic frequencies of host genetic variants influencing susceptibility to HIV-1 infection and disease in South African populations. *AIDS* 2000; 14: 449.
[81] 'Immune' Nairobi women also have 'killer' T-cells. *AIDS Anal Afr* 1995; 5: 3.
[82] 'Killer T-cells' give Gambian sex-workers the cutting-edge on HIV. *AIDS Anal Afr* 1995; 5: 3.
[83] Fowke KR, Nagelkerke NJ, Kimani J, et al. Resistance to HIV-1 infection among persistently seronegative prostitutes in Nairobi, Kenya. *Lancet* 1996; 348: 1347–1351.
[84] Slogrove AL, Goetghebuer T, Cotton MF, et al. Pattern of infectious morbidity in HIV-exposed uninfected infants and children. *Frontiers in Immunology* 2016; 7, https://www.frontiersin.org/articles/10.3389/fimmu.2016.00164 (accessed 6 July 2023).
[85] Slogrove AL, Powis KM, Johnson LF, et al. Estimates of the global population of children who are HIV-exposed and uninfected, 2000–18: A modelling study. *Lancet Glob Health* 2020; 8: e67–e75.
[86] Ramaley PA, French N, Kaleebu P, et al. Chemokine-receptor genes and AIDS risk. *Nature* 2002; 417: 140.
[87] Nyiro B, Amanya SB, Bayiyana A, et al. Reduced CCR5 expression among Uganda HIV controllers. *Retrovirology* 2023; 20: 8.
[88] Kløverpris HN, Leslie A, Goulder P. Role of HLA Adaptation in HIV Evolution. *Frontiers in Immunology* 2016; 6, https://www.frontiersin.org/articles/10.3389/fimmu.2015.00665 (accessed 17 September 2023).
[89] McLaren PJ, Porreca I, Iaconis G, et al. Africa-specific human genetic variation near CHD1L associates with HIV-1 load. *Nature* 2023; 620: 1025–1030.
[90] Real LM, Sáez ME, Corma-Gómez A, et al. A metagenome-wide association study of HIV disease progression in HIV controllers. *iScience* 2023; 26: 107214.
[91] Luo M, Czarnecki C, Nebroski M, et al. KIR3DL1 alleles and their epistatic interactions with human leukocyte antigen class I influence resistance and susceptibility to HIV-1 acquisition in the Pumwani sex worker cohort. *AIDS* 2018; 32: 841.
[92] Teixeira SLM, de Sá NBR, Campos DP, et al. Association of the HLA-B*52 allele with non-progression to AIDS in Brazilian HIV-1-infected individuals. *Genes Immun* 2014; 15: 256–262.
[93] Passaes CPB, Cardoso CC, Caetano DG, et al. Association of single nucleotide polymorphisms in the Lens Epithelium-Derived Growth Factor (LEDGF/p75)

with HIV-1 infection outcomes in Brazilian HIV-1+ individuals. *PLoS One* 2014; 9: e101780.

[94] Liu R, Paxton WA, Choe S, et al. Homozygous defect in HIV-1 coreceptor accounts for resistance of some multiply-exposed individuals to HIV-1 infection. *Cell* 1996; 86: 367–377.

[95] Horton RE, McLaren PJ, Fowke K, et al. Cohorts for the study of HIV-1—Exposed but uninfected individuals: Benefits and limitations. *J Infect Dis* 2010; 202: S377–S381.

[96] Zass L, Johnston K, Benkahla A, et al. Developing clinical phenotype data collection standards for research in Africa. *Global Health, Epidemiology and Genomics* 2023; 2023: e6693323.

[97] McLaren PJ, Fellay J. HIV-1 and human genetic variation. *Nat Rev Genet* 2021; 22: 645–657.

[98] Mann CJ. Observational research methods. Research design II: cohort, cross sectional, and case-control studies. *Emerg Med J* 2003; 20: 54–60.

[99] Bekker L-G, Beyrer C, Mgodi N, et al. HIV infection. *Nat Rev Dis Primers* 2023; 9: 1–21.

[100] CE-APCC, https://ce-apcc.org/home (accessed 22 September 2023).

[101] Caelers D. Plan for network of genomics centres of excellence across Africa. *Nat Africa*. Epub ahead of print 26 February 2023. https://doi.org/10.1038/d44148-023-00052-z.

[102] Lombard Z, Landouré G. Could Africa be the future for genomics research? *Nature* 2023; 614: 30–33.

[103] Nthobatsang K, Ncenga TL, Mwangi-Woto C, et al. Lack of RH2 gene expression may have influenced the HIV pandemic in sub-Saharan Africa. *AIDS* 2021; 35: 769–776.

[104] Wachira D, Lihana R, Okoth V, et al. Chemokine coreceptor-2 gene polymorphisms among HIV-1 infected individuals in Kenya. *Dis Markers* 2015; 2015: e952067.

[105] XieWen, AgnielDenis, ShevchenkoAndrey, et al. Genome-wide analyses reveal gene influence on HIV disease progression and HIV-1C acquisition in Southern Africa. *AIDS Res Hum Retroviruses*. Epub ahead of print 1 June 2017. https://doi.org/10.1089/aid.2016.0017.

[106] Madlala P, Gijsbers R, Christ F, et al. Association of polymorphisms in the LEDGF/p75 gene (PSIP1) with susceptibility to HIV-1 infection and disease progression. *AIDS* 2011, 25: 1711–1719.

[107] Madlala P, Singh R, An P, et al. Association of polymorphisms in the regulatory region of the cyclophilin a gene (PPIA) with gene expression and HIV/AIDS disease progression. *J Acquir Immune Defic Syndr* 2016; 72: 465–473.

[108] Dias BDC, Paximadis M, Martinson N, et al. The impact of bone marrow stromal antigen-2 (BST2) gene variants on HIV-1 control in black South African individuals. *Infect Genet Evol* 2020; 80: 104216.

[109] Lacap PA, Huntington JD, Luo M, et al. Associations of human leukocyte antigen DRB with resistance or susceptibility to HIV-1 infection in the Pumwani Sex Worker Cohort. *AIDS* 2008; 22: 1029.

[110] Kuhn L, Schramm DB, Donninger S, et al. African infants' CCL3 gene copies influence perinatal HIV transmission in the absence of maternal nevirapine. *AIDS* 2007; 21: 1753.

[111] Hennig BJ, Velez-Edwards DR, Schim van der Loeff MF, et al. CD4 intragenic SNPs associate with HIV-2 plasma viral load and CD4 count in a community-based study from Guinea-Bissau, West Africa. *J Acquir Immune Defic Syndr* 2011; 56: 1–8.
[112] Ball TB, Ji H, Kimani J, et al. Polymorphisms in IRF-1 associated with resistance to HIV-1 infection in highly exposed uninfected Kenyan sex workers. *AIDS* 2007; 21: 1091.
[113] Nkenfou CN, Mekue LCM, Nana CT, et al. Distribution of CCR5-Delta32, CCR5 promoter 59029 A/G, CCR2–64I and SDF1–3'A genetic polymorphisms in HIV-1 infected and uninfected patients in the west region of Cameroon. *BMC Res Notes* 2013; 6: 288.
[114] Thomsen D, Erikstrup C, Jespersen S, et al. The influence of human leukocyte antigen-types on disease progression among HIV-2 infected patients in Guinea-Bissau. *AIDS* 2018; 32: 721.
[115] Price H, Lacap P, Tuff J, et al. A TRIM5α exon 2 polymorphism is associated with protection from HIV-1 infection in the Pumwani sex worker cohort. *AIDS* 2010; 24: 1813.
[116] Petersen DC, Glashoff RH, Shrestha S, et al. Risk for HIV-1 infection associated with a common CXCL12 (SDF1) polymorphism and CXCR4 variation in an African population. *JAIDS J Acquired Immune Defic Syndr* 2005; 40: 521.
[117] Walley AJ, Aucan C, Kwiatkowski D, et al. Interleukin-1 gene cluster polymorphisms and susceptibility to clinical malaria in a Gambian case–control study. *Eur J Hum Genet* 2004; 12: 132–138.
[118] Fact sheet about malaria, https://www.who.int/news-room/fact-sheets/detail/malaria (accessed 22 September 2023).
[119] de Mendonça VRR, Goncalves MS, Barral-Netto M. The host genetic diversity in malaria infection. *J Trop Med* 2012; 2012: e940616.
[120] Greenwood B, Marsh K, Snow R. Why do some African children develop severe malaria? *Parasitol Today* 1991; 7: 277–281.
[121] Gong L, Parikh S, Rosenthal PJ, et al. Biochemical and immunological mechanisms by which sickle cell trait protects against malaria. *Malar J* 2013; 12: 317.
[122] Esoh K, Wonkam A. Evolutionary history of sickle-cell mutation: Implications for global genetic medicine. *Hum Mol Genet* 2021; 30: R119–R128.
[123] Piel FB, Hay SI, Gupta S, et al. Global burden of sickle cell anaemia in children under five, 2010–2050: Modelling based on demographics, excess mortality, and interventions. *PLOS Medicine* 2013; 10: e1001484.
[124] Timmann C, Thye T, Vens M, et al. Genome-wide association study indicates two novel resistance loci for severe malaria. *Nature* 2012; 489: 443–446.
[125] Jallow M, Teo YY, Small KS, et al. Genome-wide and fine-resolution association analysis of malaria in West Africa. *Nat Genet* 2009; 41: 657–665.
[126] Band G, Rockett KA, Spencer CCA, et al. A novel locus of resistance to severe malaria in a region of ancient balancing selection. *Nature* 2015; 526: 253–257.
[127] Kariuki SN, Williams TN. Human genetics and malaria resistance. *Hum Genet* 2020; 139: 801–811.

[128] Natama HM, Rovira-Vallbona E, Krit M, et al. Genetic variation in the immune system and malaria susceptibility in infants: a nested case–control study in Nanoro, Burkina Faso. *Malar J* 2021; 20: 94.

[129] Manjurano A, Sepúlveda N, Nadjm B, et al. USP38, FREM3, SDC1, DDC, and LOC727982 gene polymorphisms and differential susceptibility to severe malaria in Tanzania. *J Infect Dis* 2015; 212: 1129–1139.

[130] Ammar AI, El-Hefnawy SM, Shehab-Eldeen S, et al. Plasmodium falciparum malaria susceptibility and severity: Influence of MyD88-adaptor-like gene (rs8177374) polymorphism. *Infect Drug Resist* 2022; 15: 6815–6827.

[131] Mackinnon MJ, Mwangi TW, Snow RW, et al. Heritability of malaria in Africa. *PLoS Med* 2005; 2: e340.

[132] Sikora M, Ferrer-Admetlla A, Laayouni H, et al. A variant in the gene FUT9 is associated with susceptibility to placental malaria infection. *Hum Mol Genet* 2009; 18: 3136–3144.

[133] Feintuch CM, Tare A, Cusumano LR, et al. Type I interferon receptor variants in gene regulatory regions are associated with susceptibility to cerebral malaria in Malawi. *Am J Trop Med Hyg* 2018; 98: 1692–1698.

[134] Thiam F, Diop G, Coulonges C, et al. G6PD and HBB polymorphisms in the Senegalese population: Prevalence, correlation with clinical malaria. *PeerJ* 2022; 10: e13487.

[135] Munde EO, Raballah E, Okeyo WA, et al. Haplotype of non-synonymous mutations within IL-23R is associated with susceptibility to severe malaria anemia in a P. falciparum holoendemic transmission area of Kenya. *BMC Infect Dis* 2017; 17: 291.

[136] Fry AE, Griffiths MJ, Auburn S, et al. Common variation in the ABO glycosyltransferase is associated with susceptibility to severe Plasmodium falciparum malaria. *Hum Mol Genet* 2008; 17: 567–576.

[137] Giha HA, Nasr A, Ekström M, et al. Association of a single nucleotide polymorphism in the C-reactive protein gene (–286) with susceptibility to plasmodium falciparum malaria. *Mol Med* 2010; 16: 27–33.

[138] Bougouma EC, Tiono AB, Ouédraogo A, et al. Haemoglobin variants and Plasmodium falciparum malaria in children under five years of age living in a high and seasonal malaria transmission area of Burkina Faso. *Malar J* 2012; 11: 154.

[139] Kavishe RA, Bousema T, Shekalaghe SA, et al. Common genotypic polymorphisms in glutathione S transferases in mild and severe falciparum malaria in Tanzanian children. *Am J Trop Med Hyg* 2009; 81: 363–365.

[140] Amodu OK, Gbadegesin RA, Ralph SA, et al. Plasmodium falciparum malaria in south-west Nigerian children: is the polymorphism of ICAM-1 and E-selectin genes contributing to the clinical severity of malaria? *Acta Trop* 2005; 95: 248–255.

[141] Arama C, Diarra I, Kouriba B, et al. Malaria severity: Possible influence of the E670G PCSK9 polymorphism: A preliminary case-control study in Malian children. *PLoS One* 2018; 13: e0192850.

[142] Sadissou I, d'Almeida T, Cottrell G, et al. High plasma levels of HLA-G are associated with low birth weight and with an increased risk of malaria in infancy. *Malar J* 2014; 13: 312.

[143] Ravenhall M, Campino S, Sepúlveda N, et al. Novel genetic polymorphisms associated with severe malaria and under selective pressure in North-eastern Tanzania. *PLoS Gen* 2018; 14: e1007172.
[144] Mockenhaupt FP, Cramer JP, Hamann L, et al. Toll-Like Receptor (TLR) polymorphisms in African children: Common TLR-4 variants predispose to severe malaria. *Proc Natl Acad Sci U S A* 2006; 103: 177–182.
[145] Apinjoh TO, Anchang-Kimbi JK, Njua-Yafi C, et al. Association of candidate gene polymorphisms and TGF-beta/IL-10 levels with malaria in three regions of Cameroon: A case-control study. *Malar J* 2014; 13: 236.
[146] Sambo MR, Trovoada MJ, Benchimol C, et al. Transforming growth factor beta 2 and heme oxygenase 1 genes are risk factors for the cerebral malaria syndrome in Angolan children. *PLoS One* 2010; 5: e11141.
[147] Anyona SB, Hengartner NW, Raballah E, et al. Cyclooxygenase-2 haplotypes influence the longitudinal risk of malaria and severe malarial anemia in Kenyan children from a holoendemic transmission region. *J Hum Genet* 2020; 65: 99–113.
[148] Holmberg V, Onkamo P, Lahtela E, et al. Mutations of complement lectin pathway genes MBL2 and MASP2 associated with placental malaria. *Malar J* 2012; 11: 61.
[149] Brown CA, Pappoe-Ashong PJ, Duah N, et al. High frequency of the Duffy-negative genotype and absence of Plasmodium vivax infections in Ghana. *Malar J* 2021; 20: 99.
[150] Morenikeji OB, Metelski JL, Hawkes ME, et al. CD209 and Not CD28 or STAT6 polymorphism mediates clinical malaria and parasitemia among children from Nigeria. *Microorganisms* 2020; 8: 158.
[151] Ojurongbe O, Funwei RI, Snyder TJ, et al. Genetic variants of tumor necrosis factor-α -308G/A (rs1800629) but not Toll-interacting proteins or vitamin D receptor genes enhances susceptibility and severity of malaria infection. *Immunogenetics* 2018; 70: 135–140.
[152] Faik I, van Tong H, Lell B, et al. Pyruvate kinase and Fcγ receptor gene copy numbers associated with malaria phenotypes. *J Infect Dis* 2017; 216: 276–282.
[153] Lyke KE, Fernández-Viña MA, Cao K, et al. Association of HLA alleles with Plasmodium falciparum severity in Malian children. *Tissue Antigens* 2011; 77: 562–571.
[154] Garcia A, Milet J, Courtin D, et al. Association of HLA-G 3'UTR polymorphisms with response to malaria infection: a first insight. *Infect Genet Evol* 2013; 16: 263–269.
[155] Mombo L-E, Ntoumi F, Bisseye C, et al. Human genetic polymorphisms and asymptomatic Plasmodium falciparum malaria in Gabonese schoolchildren. *Am J Trop Med Hyg* 2003; 68: 186–190.
[156] Gardy JL, Loman NJ. Towards a genomics-informed, real-time, global pathogen surveillance system. *Nat Rev Genet* 2018; 19: 9–20.
[157] Krammer F. SARS-CoV-2 vaccines in development. *Nature* 2020; 586: 516–527.
[158] Kanyerezi S, Guerfali FZ, Anzaku AA, et al. Wastewater metagenomics in Africa: Opportunities and challenges. *PLOS Glob Public Health* 2024; 4: e0004044.

[159] Mboowa G, Kakooza F, Egesa M, et al. The rise of pathogen genomics in Africa. *F1000Res* 2024; 13: 468.

[160] Kanyerezi S, Sserwadda I, Ssemaganda A, et al. HIV-DRIVES: HIV drug resistance identification, variant evaluation, and surveillance pipeline. *Access Microbiol* 2024; 6: 000815.v3.

[161] Sebastião CS, Abecasis AB, Jandondo D, et al. HIV-1 diversity and pre-treatment drug resistance in the era of integrase inhibitor among newly diagnosed ART-naïve adult patients in Luanda, Angola. *Sci Rep* 2024; 14: 15893.

[162] Williams A, Menon S, Crowe M, et al. Geographic and population distributions of Human Immunodeficiency Virus (HIV)–1 and HIV-2 circulating subtypes: A systematic literature review and meta-analysis (2010–2021). *J Infect Dis* 2023; 228: 1583–1591.

[163] Powell R, Barengolts D, Mayr L, et al. The evolution of HIV-1 diversity in rural cameroon and its implications in vaccine design and trials. *Viruses* 2010; 2: 639–654.

[164] Kim S, Kigozi G, Martin MA, et al. Increasing intra- and inter-subtype HIV diversity despite declining HIV incidence in Uganda. *medRxiv* 2024; 2024.03.14.24303990.

[165] Nduva GM, Otieno F, Kimani J, et al. Phylogeographic assessment reveals geographic sources of HIV-1 dissemination among men who have sex with men in Kenya. *Front Microbiol*; 13. Epub ahead of print 9 March 2022. https://doi.org/10.3389/fmicb.2022.843330.

[166] Mlisana K, Werner L, Garrett NJ, et al. Rapid disease progression in HIV-1 subtype C–infected South African women. *Clin Infect Dis* 2014; 59: 1322–1331.

[167] Cholette F, Lazarus L, Macharia P, et al. HIV phylogenetic clusters point to unmet hiv prevention, testing and treatment needs among men who have sex with men in Kenya. *BMC Infect Dis* 2024; 24: 1323.

[168] Nduva GM, Hassan AS, Nazziwa J, et al. HIV-1 transmission patterns within and between risk groups in coastal Kenya. *Sci Rep* 2020; 10: 6775.

[169] de Oliveira T, Kharsany ABM, Gräf T, et al. Transmission networks and risk of HIV infection in KwaZulu-Natal, South Africa: A community-wide phylogenetic study. *Lancet HIV* 2017; 4: e41–e50.

[170] Fola AA, Feleke SM, Mohammed H, et al. Plasmodium falciparum resistant to artemisinin and diagnostics have emerged in Ethiopia. *Nat Microbiol* 2023; 8: 1911–1919.

[171] Uwimana A, Legrand E, Stokes BH, et al. Emergence and clonal expansion of in vitro artemisinin-resistant Plasmodium falciparum kelch13 R561H mutant parasites in Rwanda. *Nat Med* 2020; 26: 1602–1608.

[172] da Silva C, Boene S, Datta D, et al. Targeted and whole-genome sequencing reveal a north-south divide in P. falciparum drug resistance markers and genetic structure in Mozambique. *Commun Biol* 2023; 6: 1–11.

[173] Vanheer LN, Mahamar A, Manko E, et al. Genome-wide genetic variation and molecular surveillance of drug resistance in Plasmodium falciparum isolates from asymptomatic individuals in Ouélessébougou, Mali. *Sci Rep* 2023; 13: 9522.

[174] Girgis ST, Adika E, Nenyewodey FE, et al. Drug resistance and vaccine target surveillance of Plasmodium falciparum using nanopore sequencing in Ghana. *Nat Microbiol* 2023; 8: 2365–2377.

[175] Osborne A, Mańko E, Waweru H, et al. Plasmodium falciparum population dynamics in East Africa and genomic surveillance along the Kenya-Uganda border. *Sci Rep* 2024; 14: 18051.

[176] Sy M, Deme AB, Warren JL, et al. Plasmodium falciparum genomic surveillance reveals spatial and temporal trends, association of genetic and physical distance, and household clustering. *Sci Rep* 2022; 12: 938.

[177] Apinjoh TO, Ajonina MU, Abera D, et al. Genomic analysis of Plasmodium falciparum isolates across different altitudinal zones along the slope of Mount Cameroon. *Front Malar*; 1. Epub ahead of print 31 July 2023. https://doi.org/10.3389/fmala.2023.1075755.

[178] Laamarti M, El Fathi Lalaoui Y, Elfermi R, et al. Afro-TB dataset as a large scale genomic data of Mycobacterium tuberuclosis in Africa. *Sci Data* 2023; 10: 212.

[179] Ioerger TR, Koo S, No E-G, et al. Genome analysis of multi- and extensively-drug-resistant tuberculosis from KwaZulu-Natal, South Africa. *PLoS One* 2009; 4: e7778.

[180] Cohen KA, Abeel T, McGuire AM, et al. Evolution of extensively drug-resistant tuberculosis over four decades: Whole genome sequencing and dating analysis of mycobacterium tuberculosis Isolates from KwaZulu-Natal. *PLoS Med* 2015; 12: e1001880.

[181] Mutayoba BK, Michael Hoelscher null, Heinrich N, et al. Phylogenetic lineages of tuberculosis isolates and their association with patient demographics in Tanzania. *BMC Genomics* 2022; 23: 561.

[182] Schwab TC, Joseph L, Moono A, et al. Field evaluation of nanopore targeted next-generation sequencing to predict drug-resistant tuberculosis from native sputum in South Africa and Zambia. *J Clin Microbiol*; 63: e01390–24.

[183] Asare P, Otchere ID, Bedeley E, et al. Whole genome sequencing and spatial analysis identifies recent tuberculosis transmission hotspots in Ghana. *Front Med (Lausanne)* 2020; 7: 161.

[184] Jolly B, Saad J, Farra A, et al. A snapshot of genomic diversity and transmission clusters of rifampin-resistant Mycobacterium tuberculosis complex in the Central African Republic. *Tuberculosis (Edinb)* 2025; 152: 102627.

[185] Wei Y, Davenport TC, Collora JA, et al. Single-cell epigenetic, transcriptional, and protein profiling of latent and active HIV-1 reservoir revealed that IKZF3 promotes HIV-1 persistence. *Immunity* 2023; 56: 2584–2601.e7.

[186] Dooley NL, Chabikwa TG, Pava Z, et al. Single cell transcriptomics shows that malaria promotes unique regulatory responses across multiple immune cell subsets. *Nat Commun* 2023; 14: 7387.

[187] Gideon HP, Hughes TK, Tzouanas CN, et al. Multimodal profiling of lung granulomas in macaques reveals cellular correlates of tuberculosis control. *Immunity* 2022; 55: 827–846.e10.

[188] Iwasaki A, Medzhitov R. Control of adaptive immunity by the innate immune system. *Nat Immunol* 2015; 16: 343–353. doi:10.1038/ni.3123.

[189] Netea MG, Domínguez-Andrés J, Barreiro LB, Chavakis T, Divangahi M, Fuchs E, et al. Defining trained immunity and its role in health and disease. *Nat Rev Immunol* 2020; 20: 375–388. doi:10.1038/s41577-020-0285-6.

[190] Agnandji ST, Lell B, Fernandes JF, Abossolo BP, Methogo BGNO, Kabwende AL, et al. A phase 3 trial of RTS,S/AS01 malaria vaccine in African infants. *N Engl J Med* 2012; 367: 2284–2295. doi:10.1056/NEJMoa1208394.
[191] Mordmüller B, Surat G, Lagler H, et al. Safety, immunogenicity, and protective efficacy against controlled human malaria infection of Plasmodium falciparum sporozoite vaccine in Tanzanian adults. *Am J Trop Med Hyg.* 2018;99(2):338–349. doi:10.4269/ajtmh.17-1014. PMCID: PMC6090339
[192] Tan J, Sack BK, Oyen D, Zenklusen I, Piccoli L, Barbieri S, et al. A public antibody lineage that potently inhibits malaria infection through dual binding to the circumsporozoite protein. *Nat Med* 2018; 24: 401–407. doi:10.1038/nm.4513.
[198] Takala SL, Plowe CV. Genetic diversity and malaria vaccine design, testing and efficacy: preventing and overcoming "vaccine resistant malaria." *Parasite Immunol* 2009; 31: 560–573. doi:10.1111/j.1365-3024.2009.01138.x.
[199] International AIDS Vaccine Initiative. First-in-human clinical trial confirms novel HIV vaccine approach developed by IAVI and Scripps Research. *IAVI*; 2021. Available from: https://www.iavi.org/press-release/first-in-human-clinical-trial-confirms-novel-hiv-vaccine-approach-developed-by-iavi-and-scripps-research/
[200] Leggat DJ, Cohen KW, Willis JR, Fulp WJ, deCamp AC, Kalyuzhniy O, et al. Vaccination induces HIV broadly neutralizing antibody precursors in humans. *Science* 2022; 378: eadd6502. doi:10.1126/science.add6502.
[201] Chinyenze K, Nduati E, Muturi-Kioi V. Accelerating HIV vaccine development through meaningful engagement of local scientists and communities. *Curr Opin HIV AIDS*. 2023;18(6):284–289. doi:10.1097/COH.0000000000000815
[202] ClinicalTrials.gov. HIV-CORE 006: A study of HIVconsvX vaccines in HIV-negative adults. ClinicalTrials.gov identifier: NCT04553016. Available from: https://clinicaltrials.gov/study/NCT04553016
[203] World Health Organization. Tuberculosis vaccines. Geneva: World Health Organization. Available from: https://www.who.int/teams/global-programme-on-tuberculosis-and-lung-health/research-innovation/vaccines
[204] Parks KR, Moodie Z, Allen MA, Yen C, Furch BD, MacPhee KJ, et al. Vaccination with mRNA-encoded membrane-anchored HIV envelope trimers elicited tier 2 neutralizing antibodies in a phase 1 clinical trial. *Sci Transl Med* 2025; 17: eady6831. doi:10.1126/scitranslmed.ady6831.
[205] Tait DR, Hatherill M, Van Der Meeren O, Ginsberg AM, Van Brakel E, Salaun B, et al. Final analysis of a trial of M72/AS01E vaccine to prevent tuberculosis. *N Engl J Med* 2019; 381: 2429–2439. doi:10.1056/NEJMoa1909953.
[206] Gates Medical Research Institute. Gates MRI initiates phase 3 clinical trial of tuberculosis vaccine candidate. *Gates MRI*. Available from: https://www.gatesmri.org/mri-initiates-phase-3-clinical-trial-tuberculosis-vaccine-candidate/
[207] Mulder N, Abimiku A, Adebamowo SN, de Vries J, Matimba A, Olowoyo P, et al. H3Africa: current perspectives. *Pharmacogenomics Pers Med* 2018; 11: 59–66. doi:10.2147/PGPM.S141546.
[208] Ramsay M, Sankoh O, members of the African Population Cohorts Consortium. A new era for health research: the birth of the African Population Cohorts Consortium. *Lancet Glob Health* 2022; 10: e1054–e1055. Available from: https://www.ahri.org/a-new-era-for-health-research-the-birth-of-the-african-population-cohorts-consortium/

Chapter 12

Genomics in Africa—What Can We Expect in the Next Ten Years?

Nicola Mulder[1] *and Michèle Ramsay*[2*]

[1]Computational Biology Division, Department of Integrative Biomedical Sciences, Faculty of Health Sciences, Cape Town, Institute of Infectious Disease and Molecular Medicine, University of Cape Town, Cape Town, South Africa
[2]Sydney Brenner Institute for Molecular Bioscience, Faculty of Health Sciences, University of the Witwatersrand, Johannesburg, South Africa

12.1 INTRODUCTION

African human genomes hold exceptional promise for the discovery of novel solutions to global health challenges. Realizing their potential requires deep exploration of the vast genetic diversity on the continent and purposeful interventions supported by investments from education and research institutions, government departments, and industry. A focus on innovative, affordable, and context-specific approaches to generating big data on the phenotype, environment, and genetic makeup of hundreds of thousands of individuals from many populations across the vast continent is essential. We need to develop feasible and implementable interventions and solutions for public precision health that will benefit Africa and the world.

Africa's current population of over 1.5 billion is anticipated to rise to ~2.5 billion by 2050 (https://www.worldometers.info/ and https://www.statista.com/statistics/, accessed April 27, 2025). The emerging workforce across the continent will build on the infrastructure we put in place and foundational knowledge we unlock over the next decade.

Political instability, power struggles, natural climate disasters, and general poverty drive the burden of disease in Africa, yet its populations have demonstrated extraordinary resilience partly born out of hundreds of thousands of years of adaptation. Not least is the genetic adaptation to change, be it infectious diseases, climate, or conflict. The SARS-CoV-2 pandemic is a case in point, where the forecast was dire, yet on a global scale it affected African populations less than anticipated (Bamgboye et al. 2021). This does not negate the devastation of deaths and their impact on families and communities, but begs the question—why? The potential contributors include a younger population with fewer comorbidities and less obesity, a sparser

*Corresponding author:** Michele.Ramsay@wits.ac.za (Michèle Ramsay)

 DOI: 10.1201/9781003436584-12

population distribution with a predominantly rural and outdoor lifestyle, previous exposure to coronaviruses, a higher exposure to multiple pathogens possibly resulting in a stronger immune system, and universal vaccine coverage. It is most likely that genetic diversity and adaptation to pathogens over long periods played an important role. Unfortunately, we do not have representative data or robust scientific evidence from African countries to explore all these hypotheses. Investigating large, exposed cohorts with genetic data and COVID-19 outcomes at scale was an opportunity largely missed in Africa, despite some modest efforts (May et al. 2022; Müller et al. 2025).

Delineating diversity across Africa is complex. Genetic diversity in African populations has been well studied over several decades, providing insights into population migration and admixture (Campbell et al. 2014; Choudhury et al. 2020; Fan et al. 2023; Ragsdale et al. 2023). However, diversity in Africa also includes the political landscape (54 countries, each with their own health infrastructure and legal systems), communities bound by language and culture (over 2,000 different ethnolinguistic groupings often straddling political boundaries (https://www.ethnologue.com/)), geographic ecosystems defined by climate and climate change (ranging from deserts to tropical forests), and socioeconomic strata (with a range of Gini equality index coefficients, ranging from 37.4 in 2021 in Burkina Faso to 63.0 in 2014 in South Africa: https://data.worldbank.org/indicator/, accessed April 27, 2025). All these factors come into play when addressing the question, *What can we expect (in terms of genomics in Africa) in the next ten years?*

12.2 A VISION FOR GENOMICS IN AFRICA OVER THE NEXT TEN YEARS

A vision is bold but holds little sway if not grounded on an understanding of the current landscape. This is true for genomics in Africa. What we have learned and built over the past decade will help us understand how we can leverage that progress for the next ten years towards making a positive and accelerated impact on the health and wellbeing of populations in Africa.

Before outlining a vision, what do we know about the genomics of African populations? Importantly, we can only know what we have studied. The data we have do not cover much of Africa. Swathes of the continent in all five regions (north, west, central, east, and south) have not been explored from a genetic perspective and where we have started, we have not covered the complexity of the populations we know to be present. The international literature and purportedly global health commissions and guidelines consider that representing Africa through a small number of modestly sized studies adequately provides insights into *Africa*. This is misleading. There is ample data to show that populations in different parts of Africa are highly differentiated by population structure based on genetic evidence, health burden, and access to healthcare. Scholars, including geneticists, anthropologists, and public health experts have debated strategies for adequately covering continental Africa with reasonable regional representation. They all fall short of achieving this goal, and we do not have an answer. More is needed—more countries included, more ethnolinguistic groups covered, more data generated, more funding and infrastructure—to develop more knowledge and health-relevant insights.

As a start, by aligning with global recommendations, guidelines, and standards, African countries have an opportunity to become trusted partners. However, African investigators need to be involved in developing such documents and need to ensure that they are appropriate and compatible, as well as feasible, considering local resources and cultural contexts. In 2021 the World Health Organization (WHO) established the Science Council, which chose genomics as its first project, resulting in a call to action. The report titled "Accelerating Access to Genomics for Global Health" (published in 2022; https://www.who.int/publications/i/item/9789240052857) made recommendations under four broad themes: Advocacy, collaboration, implementation, and attention to ethical, legal, and social issues raised by genomics. The WHO technical advisory group for genomics (TAG-G) was convened and sketched a vision and road map for global genomics (Ambrosino et al. 2024) and developed principles for genomics data generation, access, use, and sharing following a consultative process that also involved African participants (https://www.who.int/publications/i/item/9789240102149, accessed April 27, 2025).

With the deficit of representative African data in international genomics databases, we diminish the probability that Africa populations will benefit from precision medicine initiatives. Precision medicine will continue to be targeted on non-African populations, and there is good evidence that the approaches and models will have limited transferability to African populations and that the "transferability" will vary across different African regions.

In 2003, Collins et al. (2003) identified six important crosscutting elements that are relevant to the use of genomics: Resources; technology development; computational biology; training; ethical, legal, and social implications; and education. A survey in 2021 by the EU-Africa PerMed study (https://www.euafrica-permed.eu/, accessed April 27, 2025) demonstrated that there is limited capacity for genomics and precision (or personalised) medicine research across Africa when considering the following metrics: Health research governance, finance, resources, publications, and collaborations. In addition to the more developed capacity in South Africa and Egypt, several countries are beginning to build resources and knowledge, including Algeria, Cameroon, Ghana, Kenya, Morocco, Nigeria, Tanzania, Uganda, Tunisia, and Zimbabwe. The continent is peppered with pockets of excellence in genomics in many countries, not just those mentioned here. However, the research needs to translate into clinical practice, and clinical practice needs to be supported by a national health infrastructure and adequate investment and funding.

Kamp et al. investigated the level of maturity for the implementation of genetic services across Africa (Kamp et al. 2021). Such services are extremely rare, are sometimes only available for a limited number of diseases, are often provided by research laboratories that rely on international partners, are often funded by international research grants for consortia, lack adequately trained genetic counsellors, have poor laboratory infrastructure and equipment, and have limited local data to adapt genetic tests to the mutation profiles of local populations (Kamp et al. 2022). Over the next decade and beyond, each one of these areas needs to be addressed and strengthened to meet the need for genetics-informed healthcare in African countries.

A step in the right direction is the efforts towards national genomics projects in Africa. One of their aims is to understand the genetic diversity in their countries. They are all aware of the extreme population diversity, and many of the champions of these initiatives are aiming for representative sampling based on geography and ethnic groups. The genetic similarity across populations and countries will impact the transferability of genetic tests with potential clinical impact. African genomics has provided insights that would not have been possible from the study of non-African populations in isolation.

To develop global standards for genomics data annotation, storage, and sharing, the Global Alliance for Genomics and Health (GA4GH), initiated in 2013, has become a trusted international community building on and adapting to knowledge from existing frameworks and communities of practice (https://www.ga4gh.org/). It is important for Africa to contribute to this effort and to adopt these recommendations to ensure that our populations can benefit and that our data can benefit others.

Over the past two decades we have witnessed the success of projects such as HapMap, 1000 Genomes Project, and the Simons Genome Project, which have all made their data openly available. National biobank initiatives including the UK Biobank and the US All of Us have made their data accessible through managed-access approaches, and various levels of access are provided by the Estonian Biobank, FinnGen, DeCode, BioBank Japan, Chinese Kadoorie Biobank, Qatar Genome Programme, and others. In Africa, South Africa started a fledgling national genome programme in 2011 and published a pilot study highlighting the genetic diversity and genetic population structure in the country (Choudhury et al. 2017), and two countries in North Africa have published their national biobanking marker papers: Egypt (Elmonem et al. 2024) and Tunisia (Hamdi et al. 2024). Several other countries, including Senegal, Tunisia, DRC, and Kenya, have initiated smaller projects, and South Africa is planning its 110K Human Genomes Project. It is vital for sustainability that national programmes are funded, at least in part, by their governments. African countries need to expand their efforts to develop capacity for ethically sound, responsible, and legally compliant processes for generating African omics data and managing it in a manner that makes it findable, accessible, interoperable, and reusable– FAIR.

Success over the next decade would be characterized by leadership on different levels: Building a stronger and more sustainable African Society of Human Genetics, implementing the plans for Genomics Centres of Excellence (see subsequent discussion), developing the African Bioinformatics Institute (ABI), integrating genetics and genomics into healthcare systems, and obtaining national commitments to supporting genomics. We need to understand how we move from pockets of excellence to service delivery rooted in national health services and infrastructures. Progress will be different across the continent, and some efforts would be country specific while others would leverage regional and even continental structures like Africa CDC and the African Union. Considerable effort is needed in community engagement at all levels, education and advocacy, and intent to action.

12.3 REALIZING THE POTENTIAL OF GENOMICS CENTRES OF EXCELLENCE IN AFRICA

National genome projects are essential for increasing the availability of reference genomes within a country, and most also plan to collect data for clinical use. However, the reality is that currently, most genomics data for African populations are being generated through research projects, and the results often fail to make their way into clinical applications. The Genomics Centres of Excellence in Africa (GenCoE) is an initiative being developed to address this gap and will adhere to the recommendations published by the WHO TAG-G, while building on the capacity developed by initiatives such as the H3Africa consortium (Mulder et al. 2018). It aims to develop centres of excellence across the continent consisting of collaborations between academic institutions and public health institutes or other governmental healthcare centres to close the circle between genomics research and clinical translation. Full centres should take on a capacity development role to bring developing centres up to full capacity over time and ensure equitable access to genomics for health. The GenCoEs will focus on three key areas: Genomics medicine for relevant diseases, pandemic preparedness and response for infectious diseases, and genomics workforce development.

The added value that GenCoE aims to bring is the coordination of the centres in data collection and sharing and promoting cooperation in solving health challenges across country borders. Seed funding to establish the GenCoE Coordinating Centre has been secured and a pilot project has been established through a partnership with industry to sequence 25,000 African genomes to supplement the existing African reference panel and for collaborative research for novel discoveries. This GenCoE-PAGRA African whole genome sequence (WGS) Moonshot project (PAGRA, or Partnership for Accelerating Genomics Research in Africa, an alliance of pharmaceutical companies) is building the required infrastructure to ensure that all 25,000 genomes are sequenced, stored, and analysed on the continent. The partners are working with sample providers to develop a framework for the generation and responsible sharing of genomics data with industry, underpinned by clear ethical and legal principles and accompanied by capacity development. Through this and future projects, GenCoE aims to use learnings from H3Africa, and work with new initiatives, including the African Population Cohorts Consortium (https://ce-apcc.org/) and the ABI to build a workforce, infrastructure, and valuable datasets for translational genomics research in Africa.

12.4 STRENGTHENING THE GENOMICS DATA INFRASTRUCTURE IN AFRICA

As large-scale genomics accelerates on the continent, there are some key questions and challenges related to the data generation and infrastructure that need to be considered. Some of these are discussed in the following sections.

12.4.1 Technologies

Advances in technologies for genomics data generation have accelerated, providing more choices than ever before, at lower costs. Genotyping arrays are still the most cost effective, but expanding the dataset of SNPs requires comprehensive reference panels. With the cost of WGS decreasing, including long read sequencing, it will soon be feasible for WGS to be the technology of choice for both research with large sample sizes and individual patient diagnosis. The drawback, however, is the larger cost of data storage and analysis for WGS and the associated need for, and expense of, bioinformatics expertise (https://medium.com/truwl/what-is-the-cost-of-bioinformatics-a-look-at-bioinformatics-pricing-and-costs-1e4c1c3bcb4f). Therefore, genotyping arrays are still valuable as a cost-effective technology for larger cohorts and can be developed into affordable custom arrays for population-specific diagnoses and screening for more common pathogenic variants.

Accurate processing and variant calling of WGS or whole exome sequencing relies on an appropriate reference genome to align reads to. To date, this has been the linear reference genome (currently GRCh38), but it is well known that this reference and its annotation is not representative, particularly of the diversity found in African genomes. To address this, as well as gaps in the current reference, the Human Pangenome Reference Consortium (HPRC; https://humanpangenome.org/) is creating telomere-to-telomere (T2T) genomes and pangenome reference graphs, which capture haplotype diversity across multiple populations (Wang et al. 2022). The HPRC is partnering with groups across the globe to expand this diversity to include underrepresented populations. In the AfriGen-D project (https://www.afrigen-d.org/) that spun out of H3ABioNet, African researchers are using long read sequence data to generate an African pangenome reference graph and African T2T genomes. These have been tested and were shown to identify novel variants in patient short read WGS data. This suggests that future WGS data analysis should embrace population-specific reference graphs for alignment and variant calling. However, a drawback is that the graphs cannot currently be shared publicly due to the lack of appropriate participant consents. The project is, therefore, working with the GA4GH on developing technical solutions for federated variant calling from distributed reference graphs. As these technologies and the graph-based variant calling and annotation tools advance, population pangenome reference graphs will become the reference of choice (Taylor et al. 2024). However, we must work towards community engagement to discuss the benefits of increasing the availability of datasets used to generate the reference graphs and their accessibility.

12.4.2 Data Infrastructure

Although H3Africa and H3ABioNet substantially improved the research infrastructure in many institutions across Africa, the data infrastructure is struggling to keep pace with the generation of genomic and associated data. International large-scale genomics projects are increasingly moving to commercial cloud solutions for data storage and processing, but with Internet-related challenges and limited commercial cloud infrastructure physically located on the continent, its adoption in Africa is low. Therefore, researchers rely on local computing facilities, which are generally inadequate. Public health institutes and particularly national healthcare facilities often have even more limited computing facilities and seldom have dedicated systems administrators who can manage genomics-related software. Another challenge is the ability (both technical and legal) to share genomic and health data across institutions or countries, an imperative during health emergencies.

If the future is one where Africa closes the gap in genomics data with the global North, then the expansion of data infrastructure on the continent must keep pace. This needs to include increased network speeds, larger compute facilities for data storage and processing, and, where necessary, platforms for sharing and reuse of data. One solution is to purchase one of the many complete software solutions for data storage, processing, and sharing, but they usually rely on high-speed Internet, use commercial clouds, and have expensive annual license fees. Nevertheless, Africa should not duplicate efforts but leverage existing solutions and customize them to their needs. The GA4GH standards and tools facilitate this, as they cover the breadth of data access control, data discovery, pipelines, and deployment of workflows on different computing environments. These have been successfully adopted by many national and large-scale genomics programs in other continents (Rehm et al. 2021). In Africa, the eLwazi Open Data

Science Platform (https://elwazi.org/) of the DS-I Africa consortium (https://dsi-africa.org/) is using GA4GH standards to provide a data analysis platform, working towards facilitating federated data analysis across datasets stored in different countries. The eLwazi catalogue (https://catalog.elwazi.org/#/datasets) is helping to make African datasets discoverable, complementing the H3Africa catalogue (https://catalog.h3africa.org/), which facilitates access requests for H3Africa data.

With the implementation of new data privacy laws in African countries, sharing human data across borders may become even more challenging, requiring a shift towards data visiting, where analysis tools are sent to the data, avoiding the need to download and move datasets. This requires the establishment of trusted/secure research environments (TREs), which host data and allow authorized users to interact with the data only within the TRE. In an ideal scenario, interoperable TREs could be established in different countries to host their own genomics data, but in reality, not all African countries have the expertise or infrastructure to host a TRE, and centralized, merged datasets could be extremely valuable. Additionally, raw genomics datasets are often not required for secondary use, but rather curated data, which are harmonized and/or aggregated. Though this requires resources for curation and data preparation, integrating genomics datasets from across the continent has the potential to create an invaluable data asset with numerous applications in research and translation. One example is the African Genome Variation Database (https://agvd.h3abionet.org), which provides regional-level minor allele frequency data for African populations by aggregating WGS data from multiple sources. AGVD is working towards joining the federated gnoMAD network of allele frequency databases, which are based on individual level data that cannot be widely shared. In the future, the ABI will be uniquely positioned to support and coordinate data infrastructure and African data resources across the continent.

12.4.3 Genomics Applications

Genomics data are valuable for unravelling the genetic diversity across populations; however, they are of infinitely more value when accompanied by demographic and standardized deep phenotyping and other clinical data. There are many resources available for supporting standard data collection, including the GA4GH standards and numerous ontologies and controlled vocabularies related to genomics experiments, phenotypes, and diseases. While genomics research studies often develop in isolation, adoption of these standards will significantly increase the FAIR compliance and utility of the data generated for downstream use.

National genomics projects should have a mandate to adopt international standards to ensure the data are interoperable. Standardized phenotype and other data associated with participants (including environment, lifestyle, imaging data) increase the potential for use of the data for novel discoveries in isolation as well as when merged with other genomics datasets. With the current scarcity in genomics data for African populations and generally small sample sizes, there is usually a need to merge datasets that do exist. This can only be done successfully if the datasets are FAIR compliant.

Genomics data collections with associated phenotype data provide opportunities for research work to increase our knowledge of the genetic basis for diseases in our populations as well as for clinical utility. A genotype-to-phenotype database for African populations would be invaluable for clinical geneticists trying to solve undiagnosed cases. Unfortunately, the reluctance of researchers to share genomic and phenotypic data has thus far limited our ability to create such a resource. For translation of research outputs into clinical practice, it is important that researchers forge partnerships with industry to increase the potential impact of genomics data on healthcare and personalized medicine.

12.5 BUILDING HUMAN GENETICS SERVICES ACROSS AFRICA

Genetic services are built on knowledge about the impact of genetic variation on diseases in individuals, families, and communities. Broadly, it is referred to as genomic medicine (specifically delineating the impact of genetic variation on the phenotype), precision medicine (including genomics and all other information that informs diagnosis, treatment, and prevention), and personalized medicine (often used synonymously with precision medicine, but where some make the distinction that this applies to an individual, whereas the former could also apply to public health scenarios).

Genomic medicine is premised on understanding genetic contributions to diseases and includes accurate diagnosis, genetic counselling, enabling reproductive options, and guiding treatment approaches. Different scenarios require different approaches. Each is outlined here with a brief outline of the status in Africa and what the future may hold.

12.5.1 Rare Genetic Disorders (Monogenic or Mendelian Disorders)

There are many examples highlighting both allelic (different mutations in the same gene) and locus (mutations in different genes) heterogeneity for monogenic disorders in African populations (Krause et al. 2018). Examples include the genetic aetiology of deafness (Adadey et al. 2022), Huntington's disease (Anderson et al. 2025), cystic fibrosis (Padoa et al. 1999), spinal muscular atrophy (Labrum et al. 2007; Vorster et al. 2020), neurological conditions (Bamba et al. 2024; Cissé et al. 2024), and many others. The implications are that diagnostic mutation screening panels and gene panels developed from data derived predominantly from European populations could miss causal mutations in many African families and lead to false conclusions about a genetic diagnosis. For this reason, research on African communities is essential, and a potential focus on the more common monogenic disorders, while providing more accurate prevalence, could make a difference through preventative strategies, aligned with local cultural norms. About 80% of sickle cell anaemia cases occur in sub-Saharan Africa (Adigwe et al. 2023). Most affected families have little access to even the most basic treatment (hydroxyurea) to alleviate the symptoms, and access to the gene editing cure is unaffordable and inaccessible to African patients (estimated at over USD1 million per case and requiring access to transplant facilities) (Makani et al. 2025). Research studies have investigated possible genetic contributions to treatment with hydroxyurea (Nkya et al. 2025) and genetic modifiers of disease severity (Wonkam et al. 2025). In some countries, prevention through premarital screening for carrier status to inform partner choice is available and in others prenatal testing. The uptake is influenced by cultural beliefs, availability, and affordability.

Variant interpretation, based largely on knowledge from European and Asian studies, hampers the identification of pathogenic variants in African populations with many variants of unknown (or unexplainable) significance. African populations have provided important insights where variants previously annotated as pathogenic are shown to be common in unaffected African individuals, leading to reclassification as benign. There are several rare disease initiatives across Africa, and their expansion will inform future genetic services.

12.5.2 Common Multifactorial Complex Disorders

The development of polygenic risk scores (PRS) for complex diseases with genetic susceptibility and environmental triggers and contributors are being implemented for prevention strategies in some developed countries. PRS is essentially a tool to stratify populations, placing groups of individuals on a spectrum from low to high genetic risk. Public health interventions would target those at high risk (e.g., for earlier surveillance for breast cancer). Their current utility is debated, but commercial entities are enthusiastically promoting these tests and African countries have not been spared, even though there is limited evidence of accurate transferability across populations. Genome-wide association studies (GWAS), which inform the development of PRS, are based on large samples (currently many hundreds of thousands, and in some cases several million, participants). As pointed out in dozens of publications, continental African studies are few and sample sizes very small (in the thousands of participants only), thereby limiting their impact in shaping PRS appropriate for African communities. Although there is now a trend towards the development of PRS based on multi-ethnic discovery populations, which perform better in a diverse range of target populations, African genomes remain underrepresented and their genetic risk alleles undiscovered (Kamiza et al. 2022; Chikowore et al. 2024; Kullo et al. 2024). It has also been shown that absolute risk needs to include other risk factors (e.g., integrated risk scores; Kamp et al. 2024). Two populations with the same genetic background but living in different parts of the world, where access to healthcare and general environments differ (van der Linden et al. 2019), will have different risk profiles and genetic associations, putting them on a spectrum of risk categories.

While the largest single African cohorts with genomics data at just over 10,000 participants and the CHAIR (Owolabi et al. 2019) and CARDINAL (Adebamowo et al. 2022) combined cohorts at over 50,000, the thin spread of populations across Africa is woefully inadequate. Even though smaller sample sizes from Africa have been shown to add value above expectation, their limited power means that many associated loci with moderate to small contributions to a trait remain undiscovered. A future focus on larger cohorts with genomics data is critical and efforts are underway, for example, through the APCC to build larger continental African cohorts with rich phenotype, behaviour, and environmental data (https://ce-apcc.org/; Ramsay et al. 2024).

12.5.3 Cancer Genomics

There are four areas of cancer research and implementation that are pertinent: (1) Monogenic or familial cancer predisposition with high-impact cancer mutations; (2) complex multifactorial and usually sporadic and later onset cancers; (3) therapeutic approaches to somatic mutations in cancer and how these inform treatment decisions; and (4) pharmacogenomic variants that influence the effects and outcomes of specific classes of medication depending on genetic variation in pharmacogenes (this is briefly highlighted in the subsequent section).

The first two areas require data from African populations and the third and fourth can successfully leverage international literature and guidelines, as somatic mutation risk profiles can be universally applied to guide treatment, with the caveat that clinicians need to be mindful of potential pharmacogenetic variations that are more common in African populations.

There is limited literature on familial cancers in African populations, with an emphasis on breast and colorectal cancers, and Lynch syndrome. Allelic heterogeneity has been demonstrated and mutation screening methods that are developed in European populations would lead to false conclusions about the involvement of a specific gene and could lead to inappropriate treatment approaches.

Recent publications on the complex genetic aetiology of sporadic cancers in African populations have highlighted different sentinel SNPs at known loci and novel cancer-associated loci with the following examples: oesophageal cancer (Chen et al. 2023), prostate cancer (Janivara et al. 2024), and breast cancer (Hayat et al. 2025). However, the sample sizes remain modest and the variability across different African regions remains to be explored further.

12.5.4 Pharmacogenomics

Understanding the effects and frequency of genetic variants in pharmacogenes across African populations provide an opportunity for public precision health (Dandara et al. 2019; Twesigomwe et al. 2025). Affordable and accessible screening at a population level has the potential for fiscal savings by recommending and implementing the use of appropriate first-line medications and avoiding costly adverse drug reactions. Excellent examples include the use of Efavirenz for HIV treatment (Masimirembwa et al. 2016) and increasing the dose of Tamoxifen in women with breast cancer who have the CYP2D6*17 variant (Kanji et al. 2023). These approaches would impact mortality and morbidity to support better outcomes. While several groups are tackling this, including the African Pharmacogenomics Consortium, basic research and clinical trials are not commensurate with the disease burden in Africa.

12.6 STRENGTHENING THE HUMAN GENOMICS COMMUNITY IN AFRICA

The vision for the future of genomics in Africa must include the availability of genomic and phenotypic data representative of all African populations and a catalogue of validated population-specific genotype-to-phenotype relationships to enable the equitable use of genomics data for health and wellbeing. Since populations are not confined by existing country borders, it is essential that the genomics community collaborates across African countries. The H3Africa initiative demonstrated the value in coordination, sharing experiences, and developing unified resources (Lombard and Landouré, 2023). With resources generally being limited in low- to middle-income countries, countries should work together to build the infrastructure required for implementing genomics for discovery and healthcare. This requires education and training to promote the power of genomics among clinicians and communities and the development of a workforce for the broad range of genomics competencies required. Genomics should also be incorporated into more clinical trials to increase the data and our understanding of the role of genetics in drug response.

The African Society of Human Genetics, through its annual conferences, brings together the African genomics community. The Society should continue to play a vital role in providing an avenue for sharing of genomics research and best practices across the continent. Africa can also learn from existing initiatives in the global North. For example, Europe has recognized the value in sharing genomics and health data, evidenced by the introduction of the European Health Data Space regulation (https://health.ec.europa.eu/ehealth-digital-health-and-care/european-health-data-space-regulation-ehds_en) and the establishment of the 1+ Million Genomes Initiative and the Genomics Data Infrastructure (https://gdi.onemilliongenomes.eu/). These initiatives aims to enable the sharing of genomic and health data across borders in Europe. African countries can learn from such cross-border initiatives, and this should be supported at the level of the African Union.

12.7 CONCLUSION

Genomics research is on the rise across Africa. Africa is a continent with multiple disease burdens including high rates of infection by HIV, TB, and malaria and an increase in non-communicable diseases like diabetes, hypertension, heart disease, and cancers. The projected steep population growth across sub-Saharan Africa highlights the imperative for precision medicine approaches, including genomics medicine, to maximize health benefit from knowledge gained through genomics research. Developing and expanding Genomic Centres of Excellence across Africa that work closely with the healthcare workforce and ministries of health in multiple countries has the potential to generate knowledge with clinical utility. Reaching those most in need will require stakeholder engagement, sensitization to the power of genomics, education and skill development in clinical and laboratory genetics, a critical mass of well-trained bioinformaticists, clinical geneticists and counsellors, dedicated personnel, and laboratory and clinical infrastructure, aligned with national health policies in individual countries to support these activities.

Actualizing a vision for health-focused genomics in Africa is far more challenging than the often-repeated broad statements such as "African genomics promises global insights into the biology of health and disease," "African genomics provides an opportunity for novel drug discovery," and "African discoveries will benefit not only those living in Africa but all people across the world." It requires an intentional approach and political will from individual nations to integrate genomics into existing healthcare infrastructures and to develop sustainable funding mechanisms. What is clear is that the deep evolutionary roots of our species are represented in populations across Africa, and we need to tap into their genetic diversity as that diversity has enabled our survival and will continue to do so as we adapt to the changing world around us.

REFERENCES

Adadey SM, Wonkam-Tingang E, Aboagye ET, Quaye O, Awandare GA, Wonkam A. Hearing loss in Africa: Current genetic profile. Hum Genet. 2022 Apr;141(3–4):505–517. https://doi.org/10.1007/s00439-021-02376-y. Epub 2021 Oct 5. PMID: 34609590; PMCID: PMC9034983.

Adebamowo CA, Adeyemo A, Ashaye A, Akpa OM, Chikowore T, Choudhury A, Fakim YJ, Fatumo S, Hanchard N, Hauser M, Mitchell B, Mulder N, Ofori-Acquah SF, Owolabi M, Ramsay M, Tayo B, VasanthKumar AB, Zhang Y, Adebamowo SN. Polygenic risk scores for CARDINAL study. Nat Genet. 2022 May;54(5):527–530. https://doi.org/10.1038/s41588-022-01074-3. PMID: 35513726; PMCID: PMC9907721.

Adigwe OP, Onoja SO, Onavbavba G. A critical review of sickle cell disease burden and challenges in Sub-Saharan Africa. J Blood Med. 2023 May 31;14:367–376. https://doi.org/10.2147/JBM.S406196. PMID: 37284610; PMCID: PMC10239624.

Ambrosino E, Abou Tayoun AN, Abramowicz M, Zilfalil BA, Boughtwood T, Hamdi Y, Hubbard T, Kato K, Lopes-Cendes I, Majumder PP, Mascalzoni D, Ndiaye R, Ramsay M, Repetto GM, Shotelersuk V, Taylor S, Reeder JC, Ross AL. The WHO genomics program of work for equitable implementation of human genomics for global health. Nat Med. 2024 Oct;30(10):2711–2713. https://doi.org/10.1038/s41591-024-03225-x. PMID: 39227441.

Anderson DG, Krause A, Margolis RL. Huntington Disease-Like 2. 2004 Jan 30 [updated 2025 Apr 10]. In: Adam MP, Feldman J, Mirzaa GM, Pagon RA, Wallace SE, Amemiya A, editors. GeneReviews® [Internet]. Seattle (WA): University of Washington, Seattle; 1993–2025. PMID: 20301701.

Bamba S, Sidibé L, Diallo SH, Cissé L, Dembélé K, Yalcouyé A, Ji W, Dembélé ME, Diarra S, Maiga ADB, Traoré O, Diallo S, Mefoung SE, Touré A, Koné A, Jeffries L, Guinto CO, Mis EK, Fischbeck KH, Khokha MK, Lakhani SA, Landouré G. Case report: Novel variants cause developmental and epileptic encephalopathy in three unrelated families from Mali. Front Genet. 2024 Nov 18;15:1412442. https://doi.org/10.3389/fgene.2024.1412442. PMID: 39624497; PMCID: PMC11609193.

Bamgboye EL, Omiye JA, Afolaranmi OJ, Davids MR, Tannor EK, Wadee S, Niang A, Were A, Naicker S. COVID-19 pandemic: Is Africa different? J Natl Med Assoc. 2021 Jun;113(3):324–335. https://doi.org/10.1016/j.jnma.2020.10.001. Epub 2020 Nov 3. PMID: 33153755; PMCID: PMC7607238.

Campbell MC, Hirbo JB, Townsend JP, Tishkoff SA. The peopling of the African continent and the diaspora into the new world. Curr Opin Genet Dev. 2014 Dec;29:120–132. https://doi.org/10.1016/j.gde.2014.09.003. PMID: 25461616; PMCID: PMC4308437.

Chen WC, Brandenburg JT, Choudhury A, Hayat M, Sengupta D, Swiel Y, Babb de Villiers C, Ferndale L, Aldous C, Soo CC, Lee S, Curtis C, Newton R, Waterboer T, Sitas F, Bradshaw D, Abnet CC, Ramsay M, Parker MI, Singh E, Lewis CM, Mathew CG. Genome-wide association study of esophageal squamous cell cancer identifies shared and distinct risk variants in African and Chinese populations. Am J Hum Genet. 2023 Oct 5;110(10):1690–1703. https://doi.org/10.1016/j.ajhg.2023.08.007. Epub 2023 Sep 5. PMID: 37673066; PMCID: PMC10577073.

Chikowore T, Läll K, Micklesfield LK, Lombard Z, Goedecke JH, Fatumo S, Norris SA, Magi R, Ramsay M, Franks PW, Pare G, Morris AP. Variability of polygenic prediction for body mass index in Africa. Genome Med. 2024 May 30;16(1):74. https://doi.org/10.1186/s13073-024-01348-x. PMID: 38816834; PMCID: PMC11140909.

Choudhury A, Aron S, Botigué LR, Sengupta D, Botha G, Bensellak T, Wells G, Kumuthini J, Shriner D, Fakim YJ, Ghoorah AW, Dareng E, Odia T, Falola O, Adebiyi E, Hazelhurst S, Mazandu G, Nyangiri OA, Mbiyavanga M, Benkahla A, Kassim SK, Mulder N, Adebamowo SN, Chimusa ER, Muzny D, Metcalf G, Gibbs RA, TrypanoGEN Research Group, Rotimi C, Ramsay M, H3Africa Consortium, Adeyemo AA, Lombard Z, Hanchard NA. High-depth African genomes inform human migration and health. Nature. 2020 Oct;586(7831):741–748. https://doi.org/10.1038/s41586-020-2859-7. Epub 2020 Oct 28. Erratum in: Nature. 2021 Apr;592(7856):E26. https://doi.org/10.1038/s41586-021-03286-9. PMID: 33116287; PMCID: PMC7759466.

Choudhury A, Ramsay M, Hazelhurst S, Aron S, Bardien S, Botha G, Chimusa ER, Christoffels A, Gamieldien J, Sefid-Dashti MJ, Joubert F, Meintjes A, Mulder N, Ramesar R, Rees J, Scholtz K, Sengupta D, Soodyall H, Venter P, Warnich L, Pepper MS. Whole-genome sequencing for an enhanced understanding of genetic variation among South Africans. Nat Commun. 2017 Dec 12;8(1):2062. https://doi.org/10.1038/s41467-017-00663-9. PMID: 29233967; PMCID: PMC5727231.

Cissé L, Bamba S, Diallo SH, Ji W, Dembélé ME, Yalcouyé A, Coulibaly T, Traoré I, Jeffries L, Diarra S, Maiga ADB, Diallo S, Nimaga K, Touré A, Traoré O, Kotioumbé M, Mis EK, Cissé CAK, Guinto CO, Fischbeck KH, Khokha MK, Lakhani SA, Landouré G. Genetic profile of progressive myoclonic epilepsy in Mali reveals novel findings. Front Neurol. 2024 Sep 25;15:1455467. https://doi.org/10.3389/fneur.2024.1455467. PMID: 39385815; PMCID: PMC11461190.

Collins FS, Green ED, Guttmacher AE, Guyer MS. US National human genome research institute. A vision for the future of genomics research. Nature. 2003 Apr 24;422(6934):835–847. https://doi.org/10.1038/nature01626. Epub 2003 Apr 14. PMID: 12695777.

Dandara C, Masimirembwa C, Haffani YZ, Ogutu B, Mabuka J, Aklillu E, Bolaji O, H3Africa. African pharmacogenomics consortium: Consolidating pharmacogenomics knowledge, capacity development and translation in Africa: Consolidating pharmacogenomics knowledge, capacity development and translation in Africa. AAS Open Res. 2019 Jun 4;2:19. https://doi.org/10.12688/aasopenres.12965.1. PMID: 32382701; PMCID: PMC7194139.

Elmonem MA, Soliman NA, Moustafa A, Gad YZ, Hassan WA, Taha T, El-Feky G, Sakr M, Amer K. The Egypt genome project. Nat Genet. 2024 Jun;56(6):1035–1037. https://doi.org/10.1038/s41588-024-01739-1. PMID: 38684896.

Fan S, Spence JP, Feng Y, Hansen MEB, Terhorst J, Beltrame MH, Ranciaro A, Hirbo J, Beggs W, Thomas N, Nyambo T, Mpoloka SW, Mokone GG, Njamnshi A, Folkunang C, Meskel DW, Belay G, Song YS, Tishkoff SA. Whole-genome sequencing reveals a complex African population demographic history and signatures of local adaptation. Cell. 2023 Mar 2;186(5):923–939.e14. https://doi.org/10.1016/j.cell.2023.01.042. PMID: 36868214; PMCID: PMC10568978.

Hamdi Y, Trabelsi M, Ghedira K, Boujemaa M, Ben Ayed I, Charfeddine C, Souissi A, Rejeb I, Kammoun Rebai W, Hkimi C, Neifar F, Jandoubi N, Mkaouar R, Chaouch M, Bennour A, Kamoun S, Chaker Masmoudi H, Abid N, Mezghani Khemakhem M, GTCA Consortium, Masmoudi S, Saad A, BenJemaa L, Ben-Kahla A, Boubaker S, Mrad R, Kamoun H, Abdelhak S, Gribaa M, Belguith N, Kharrat N, Hmida D, Rebai A. Genome Tunisia project: Paving the way for precision medicine in North Africa. Genome Med. 2024 Aug 27;16(1):104. https://doi.org/10.1186/s13073-024-01365-w. PMID: 39187811; PMCID: PMC11348534.

Hayat M, Chen WC, Babb de Villiers C, Hyuck Lee S, Curtis C, Newton R, Waterboer T, Sitas F, Bradshaw D, Muchengeti M, Singh E, Lewis CM, Ramsay M, Mathew CG, Brandenburg JT. Genome-wide association study identifies common variants associated with breast cancer in South African Black women. Nat Commun. 2025 Apr 14;16(1):3542. https://doi.org/10.1038/s41467-025-58789-0. PMID: 40229280; PMCID: PMC11997036.

Janivara R, Chen WC, Hazra U, Baichoo S, Agalliu I, Kachambwa P, Simonti CN, Brown LM, Tambe SP, Kim MS, Harlemon M, Jalloh M, Muzondiwa D, Naidoo D, Ajayi OO, Snyper NY, Niang L, Diop H, Ndoye M, Mensah JE, Abrahams AOD, Biritwum R, Adjei AA, Adebiyi AO, Shittu O, Ogunbiyi O, Adebayo S, Nwegbu MM, Ajibola HO, Oluwole OP, Jamda MA, Pentz A, Haiman CA, Spies PV, van der Merwe A, Cook MB, Chanock SJ, Berndt SI, Watya S, Lubwama A, Muchengeti M, Doherty S, Smyth N, Lounsbury D, Fortier B, Rohan TE, Jacobson JS, Neugut AI, Hsing AW, Gusev A, Aisuodionoe-Shadrach OI, Joffe M, Adusei B, Gueye SM, Fernandez PW, McBride J, Andrews C, Petersen LN, Lachance J, Rebbeck TR. Heterogeneous genetic architectures of prostate cancer susceptibility in sub-Saharan Africa. Nat Genet. 2024 Oct;56(10):2093–2103. https://doi.org/10.1038/s41588-024-01931-3. Epub 2024 Oct 2. PMID: 39358599.

Kamiza AB, Toure SM, Vujkovic M, Machipisa T, Soremekun OS, Kintu C, Corpas M, Pirie F, Young E, Gill D, Sandhu MS, Kaleebu P, Nyirenda M, Motala AA, Chikowore T, Fatumo S. Transferability of genetic risk scores in African populations. Nat Med. 2022 Jun;28(6):1163–1166. https://doi.org/10.1038/s41591-022-01835-x. Epub 2022 Jun 2. PMID: 35654908; PMCID: PMC9205766.

Kamp M, Krause A, Ramsay M. Has translational genomics come of age in Africa? Hum Mol Genet. 2021 Oct 1;30(20):R164–R173. https://doi.org/10.1093/hmg/ddab180. PMID: 34240178.

Kamp M, Pain O, Lewis CM, Ramsay M. Ancestry-aligned polygenic scores combined with conventional risk factors improve prediction of cardiometabolic outcomes in African populations. Genome Med. 2024 Aug 26;16(1):106. https://doi.org/10.1186/s13073-024-01377-6. PMID: 39187845; PMCID: PMC11346299.

Kamp M, Pain O, May A, Lewis CM, Ramsay M. Clinicians' perceptions towards precision medicine tools for cardiovascular disease risk stratification in South Africa. J Pers Med. 2022 Aug 24;12(9):1360. https://doi.org/10.3390/jpm12091360. PMID: 36143145; PMCID: PMC9505828

Kanji CR, Nyabadza G, Nhachi C, Masimirembwa C. Pharmacokinetics of tamoxifen and its major metabolites and the effect of the African ancestry specific CYP2D6*17 variant on the formation of the active metabolite, endoxifen.

J Pers Med. 2023 Jan 31;13(2):272. https://doi.org/10.3390/jpm13020272. PMID: 36836506; PMCID: PMC9961245.

Krause A, Seymour H, Ramsay M. Common and founder mutations for monogenic traits in Sub-Saharan African populations. Annu Rev Genomics Hum Genet. 2018 Aug 31;19:149–175. https://doi.org/10.1146/annurev-genom-083117-021256. PMID: 30169122.

Kullo IJ, Conomos MP, Nelson SC, Adebamowo SN, Choudhury A, Conti D, Fullerton SM, Gogarten SM, Heavner B, Hornsby WE, Kenny EE, Khan A, Khera AV, Li Y, Martin I, Mercader JM, Ng M, Raffield LM, Reiner A, Rowley R, Schaid D, Stilp A, Wiley K, Wilson R, Witte JS, Natarajan P, Polygenic Risk Methods in Diverse Populations (PRIMED) Consortium. The PRIMED consortium: Reducing disparities in polygenic risk assessment. Am J Hum Genet. 2024 Dec 5;111(12):2594–2606. https://doi.org/10.1016/j.ajhg.2024.10.010. Epub 2024 Nov 18. PMID: 39561770; PMCID: PMC11639095.

Labrum R, Rodda J, Krause A. The molecular basis of Spinal Muscular Atrophy (SMA) in South African black patients. Neuromuscul Disord. 2007 Oct;17(9–10):684–692. https://doi.org/10.1016/j.nmd.2007.05.005. Epub 2007 Jul 12. PMID: 17627822.

Lombard Z, Landouré G. Could Africa be the future for genomics research? Nature. 2023 Feb;614(7946):30–33. https://doi.org/10.1038/d41586-023-00222-x. PMID: 36720945.

Makani J, Bukini D, Malangahe S, Arnold SD, Patel S, Rifai A, Moshi G. Access to gene therapy for sickle cell disease in Africa. EBioMedicine. 2025 Apr 16;115:105713. https://doi.org/10.1016/j.ebiom.2025.105713. Epub ahead of print. PMID: 40245493.

Masimirembwa C, Dandara C, Leutscher PD. Rolling out efavirenz for HIV precision medicine in Africa: Are we ready for pharmacovigilance and tackling neuropsychiatric adverse effects? OMICS. 2016 Oct;20(10):575–580. https://doi.org/10.1089/omi.2016.0120. Epub 2016 Sep 14. PMID: 27627692.

May AK, Seymour H, Etheredge H, Maher H, Nunes MC, Madhi SA, Sokhela SM, Venter WDF, Martinson N, Nabeemeeah F, Cohen C, Moyes J, Walaza S, Tempia S, Kleynhans J, von Gottberg A, Nel J, Dawood H, Variava E, Tollman S, Kahn K, Herbst K, Wong EB, Tiemessen CT, van Blydenstein A, Murray L, Venter M, Fabian J, Ramsay M. Coronavirus host genomics study: South Africa (COVIGen-SA). Glob Health Epidemiol Genom. 2022 Oct 6;2022:7405349. https://doi.org/10.1155/2022/7405349. PMID: 36263375; PMCID: PMC9560830.

Mulder N, Abimiku A, Adebamowo SN, de Vries J, Matimba A, Olowoyo P, Ramsay M, Skelton M, Stein DJ. H3Africa: Current perspectives. Pharmgenomics Pers Med. 2018 Apr 10;11:59–66. https://doi.org/10.2147/PGPM.S141546. PMID: 29692621; PMCID: PMC5903476.

Müller T, Dzanibe S, Day C, Mpangase PT, Chimbetete T, Pedretti S, Schwager S, Gray CM, Sturrock E, Peter J. Integrated renin angiotensin system dysregulation and immune profiles predict COVID-19 disease severity in a South African cohort. Sci Rep. 2025 Apr 14;15(1):12799. https://doi.org/10.1038/s41598-025-96161-w. PMID: 40229302; PMCID: PMC11997227.

Nkya S, Nzunda C, Saukiwa E, Kaywanga F, Buchard E, Solomon D, Christopher H, Ngowi D, Johansen J, Urio F, Mgaya J, Kindole C, Yonazi M, Karim S, Alimohamed MZ, Sangeda RZ, Chamba C, Dandara C, Novelli E, Chimusa

ER, Makani J. Exploring pharmacogenetic factors influencing hydroxyurea response in tanzanian sickle cell disease patients: A genomic medicine approach. Pharmacogenomics J. 2025 Apr 23;25(3):11. https://doi.org/10.1038/s41397-025-00372-3. PMID: 40268903.

Owolabi MO, Akpa OM, Made F, Adebamowo SN, Ojo A, Adu D, Motala AA, Mayosi BM, Ovbiagele B, Adebamowo C, Tayo B, Rotimi C, Akinyemi R, Gebregziabher M, Sarfo F, Wahab KW, Parekh RS, Engel ME, Chisala C, Peprah E, Mensah G, Wiley K, Troyer J, Ramsay M; as members of the CVD Working Group of the H3Africa Consortium. Data resource profile: Cardiovascular H3Africa innovation resource (CHAIR). Int J Epidemiol. 2019 Apr 1;48(2):366–367g. https://doi.org/10.1093/ije/dyy261. PMID: 30535409; PMCID: PMC6469307.

Padoa C, Goldman A, Jenkins T, Ramsay M. Cystic fibrosis carrier frequencies in populations of African origin. J Med Genet. 1999 Jan;36(1):41–44. PMID: 9950364; PMCID: PMC1762947.

Ragsdale AP, Weaver TD, Atkinson EG, Hoal EG, Möller M, Henn BM, Gravel S. Publisher correction: A weakly structured stem for human origins in Africa. Nature. 2023 Aug;620(7972):E11. https://doi.org/10.1038/s41586-023-06433-6. Erratum for: Nature. 2023 May;617(7962):755–763. https://doi.org/10.1038/s41586-023-06055-y. PMID: 37460744; PMCID: PMC10512153.

Ramsay M, Crampin AC, Bawah AA, Gitau E, Herbst K. The value proposition of coordinated population cohorts across Africa. Annu Rev Biomed Data Sci. 2024 Aug;7(1):277–294. https://doi.org/10.1146/annurev-biodatasci-020722-015026. PMID: 39178423.

Rehm HL, Page AJH, Smith L, Adams JB, Alterovitz G, Babb LJ, Barkley MP, Baudis M, Beauvais MJS, Beck T, Beckmann JS, Beltran S, Bernick D, Bernier A, Bonfield JK, Boughtwood TF, Bourque G, Bowers SR, Brookes AJ, Brudno M, Brush MH, Bujold D, Burdett T, Buske OJ, Cabili MN, Cameron DL, Carroll RJ, Casas-Silva E, Chakravarty D, Chaudhari BP, Chen SH, Cherry JM, Chung J, Cline M, Clissold HL, Cook-Deegan RM, Courtot M, Cunningham F, Cupak M, Davies RM, Denisko D, Doerr MJ, Dolman LI, Dove ES, Dursi LJ, Dyke SOM, Eddy JA, Eilbeck K, Ellrott KP, Fairley S, Fakhro KA, Firth HV, Fitzsimons MS, Fiume M, Flicek P, Fore IM, Freeberg MA, Freimuth RR, Fromont LA, Fuerth J, Gaff CL, Gan W, Ghanaim EM, Glazer D, Green RC, Griffith M, Griffith OL, Grossman RL, Groza T, Guidry Auvil JM, Guigó R, Gupta D, Haendel MA, Hamosh A, Hansen DP, Hart RK, Hartley DM, Haussler D, Hendricks-Sturrup RM, Ho CWL, Hobb AE, Hoffman MM, Hofmann OM, Holub P, Hsu JS, Hubaux JP, Hunt SE, Husami A, Jacobsen JO, Jamuar SS, Janes EL, Jeanson F, Jené A, Johns AL, Joly Y, Jones SJM, Kanitz A, Kato K, Keane TM, Kekesi-Lafrance K, Kelleher J, Kerry G, Khor SS, Knoppers BM, Konopko MA, Kosaki K, Kuba M, Lawson J, Leinonen R, Li S, Lin MF, Linden M, Liu X, Udara Liyanage I, Lopez J, Lucassen AM, Lukowski M, Mann AL, Marshall J, Mattioni M, Metke-Jimenez A, Middleton A, Milne RJ, Molnár-Gábor F, Mulder N, Munoz-Torres MC, Nag R, Nakagawa H, Nasir J, Navarro A, Nelson TH, Niewielska A, Nisselle A, Niu J, Nyrönen TH, O'Connor BD, Oesterle S, Ogishima S, Wang VO, Paglione LAD, Palumbo E, Parkinson HE, Philippakis AA, Pizarro AD, Prlic A, Rambla J, Rendon A, Rider RA, Robinson PN, Rodarmer KW, Rodriguez LL, Rubin AF, Rueda M, Rushton GA, Ryan RS,

Saunders GI, Schuilenburg H, Schwede T, Scollen S, Senf A, Sheffield NC, Skantharajah N, Smith AV, Sofia HJ, Spalding D, Spurdle AB, Stark Z, Stein LD, Suematsu M, Tan P, Tedds JA, Thomson AA, Thorogood A, Tickle TL, Tokunaga K, Törnroos J, Torrents D, Upchurch S, Valencia A, Guimera RV, Vamathevan J, Varma S, Vears DF, Viner C, Voisin C, Wagner AH, Wallace SE, Walsh BP, Williams MS, Winkler EC, Wold BJ, Wood GM, Woolley JP, Yamasaki C, Yates AD, Yung CK, Zass LJ, Zaytseva K, Zhang J, Goodhand P, North K, Birney E. GA4GH: International policies and standards for data sharing across genomic research and healthcare. Cell Genom. 2021 Nov 10;1(2):100029. https://doi.org/10.1016/j.xgen.2021.100029. PMID: 35072136; PMCID: PMC8774288.

Taylor DJ, Eizenga JM, Li Q, Das A, Jenike KM, Kenny EE, Miga KH, Monlong J, McCoy RC, Paten B, Schatz MC. Beyond the human genome project: The age of complete human genome sequences and pangenome references. Annu Rev Genomics Hum Genet. 2024 Aug;25(1):77–104. https://doi.org/10.1146/annurev-genom-021623-081639. Epub 2024 Aug 6. PMID: 38663087; PMCID: PMC11451085.

Twesigomwe D, Mazhindu TA, Nagy M, Agesa G, Scholefield J, Masimirembwa C. Pharmacogenomics in Africa: A potential catalyst for precision medicine in genetically diverse populations. Annu Rev Genomics Hum Genet. 2025 Mar 27. https://doi.org/10.1146/annurev-genom-121323-104008. Epub ahead of print. PMID: 40146997.

van der Linden EL, Meeks K, Beune E, de-Graft Aikins A, Addo J, Owusu-Dabo E, Mockenhaupt FP, Bahendeka S, Danquah I, Schulze MB, Spranger J, Klipstein-Grobusch K, Tetteh Appiah L, Smeeth L, Stronks K, Agyemang C. The prevalence of metabolic syndrome among Ghanaian migrants and their homeland counterparts: The Research on Obesity and type 2 Diabetes among African Migrants (RODAM) study. Eur J Public Health. 2019 Oct 1;29(5):906–913. https://doi.org/10.1093/eurpub/ckz051. PMID: 31220248; PMCID: PMC6761842.

Vorster E, Essop FB, Rodda JL, Krause A. Spinal muscular atrophy in the Black South African population: A matter of rearrangement? Front Genet. 2020 Feb 13;11:54. https://doi.org/10.3389/fgene.2020.00054. PMID: 32117462; PMCID: PMC7033609.

Wang T, Antonacci-Fulton L, Howe K, Lawson HA, Lucas JK, Phillippy AM, Popejoy AB, Asri M, Carson C, Chaisson MJP, Chang X, Cook-Deegan R, Felsenfeld AL, Fulton RS, Garrison EP, Garrison NA, Graves-Lindsay TA, Ji H, Kenny EE, Koenig BA, Li D, Marschall T, McMichael JF, Novak AM, Purushotham D, Schneider VA, Schultz BI, Smith MW, Sofia HJ, Weissman T, Flicek P, Li H, Miga KH, Paten B, Jarvis ED, Hall IM, Eichler EE, Haussler D, Human Pangenome Reference Consortium. The human pangenome project: A global resource to map genomic diversity. Nature. 2022 Apr;604(7906):437–446. https://doi.org/10.1038/s41586-022-04601-8. Epub 2022 Apr 20. PMID: 35444317; PMCID: PMC9402379.

Wonkam A, Esoh K, Levine RM, Ngo Bitoungui VJ, Mnika K, Nimmagadda N, Dempsey EAD, Nkya S, Sangeda RZ, Nembaware V, Morrice J, Osman F, Beer MA, Makani J, Mulder N, Lettre G, Steinberg MH, Latanich R, Casella JF, Drehmer D, Arking DE, Chimusa ER, Yen JS, Newby GA, Antonarakis SE. FLT1 and other candidate fetal haemoglobin modifying loci in sickle cell disease in African ancestries. Nat Commun. 2025 Mar 1;16(1):2092. https://doi.org/10.1038/s41467-025-57413-5. PMID: 40025045; PMCID: PMC11873275.

Index

Note: Page numbers in *italics* indicate a figure and page numbers in **bold** indicate a table on the corresponding page.

For Product Safety Concerns and Information please contact our EU representative GPSR@taylorandfrancis.com Taylor & Francis Verlag GmbH, Kaufingerstraße 24, 80331 München, Germany